Fundamentals of
Beam Physics

J.B. ROSENZWEIG

Department of Physics and Astronomy
University of California, Los Angeles

OXFORD
UNIVERSITY PRESS

OXFORD
UNIVERSITY PRESS

Great Clarendon Street, Oxford OX2 6DP

Oxford University Press is a department of the University of Oxford.
It furthers the University's objective of excellence in research, scholarship,
and education by publishing worldwide in

Oxford New York

Auckland Bangkok Buenos Aires Cape Town Chennai
Dar es Salaam Delhi Hong Kong Istanbul Karachi Kolkata
Kuala Lumpur Madrid Melbourne Mexico City Mumbai Nairobi
São Paulo Shanghai Taipei Tokyo Toronto

Oxford is a registered trade mark of Oxford University Press
in the UK and in certain other countries

Published in the United States
by Oxford University Press Inc., New York

A catalogue record for this title is available from the British Library

Library of Congress Cataloging in Publication Data
(Data available)

ISBN 0 19 852554 0

10 9 8 7 6 5 4 3 2 1

Typeset by Newgen Imaging Systems (P) Ltd., Chennai, India
Printed in Great Britain
on acid-free paper by The Bath Press, Avon

Preface

This book was developed as part of my efforts to build and enhance the accelerator and beam physics program in the UCLA Department of Physics and Astronomy. This program includes an active research program in advanced accelerators (particle accelerators based on lasers and/or plasmas) and free-electron lasers (conversely, lasers based on accelerators and particle beams) and is essentially a student-oriented enterprise. Its focus is primarily on graduate education and training, but the laboratory activities of the group also include a strong showing of undergraduate students. The desire to introduce both incoming graduate students and interested undergraduates to the physics of beams led me to develop a course, Physics 150, to formally provide the background needed to enter the field.

In teaching this senior-level course the first few times, I found that I was relying on a combination of excerpts from a variety of accelerator and laser physics texts and my own notes. Because the hybrid nature of the UCLA research program in beams is reflected in the course, it was simply not possible to use a single text for the course's source material. As one might imagine, in the mixing of written references, the notation, level, and assumed background varied widely from reading to reading. In addition, many existing texts and references in this area are geared towards a practitioner of accelerator physics working at a major accelerator facility. Thus this type of text has an emphasis that is heavy on the physics, engineering methods, and technical jargon specific to large accelerators at the high-energy frontier. The needs of this professional reader are inherently a bit different than the senior level university student, however. As such, previous texts typically have given less orientation to basic physics concepts than the university student needs in order to be properly introduced to the subject. My desire to clarify the written introduction of beam physics as it is practiced at UCLA to undergraduate students led directly to the production of this book.

The contents of this book were also flavored by my desire to create the compact introduction to beam physics that I wish I could have had—hopefully the reader will benefit from the resulting weight put on the points I have found to be least clear or intuitive in my journeys in the field. The present book is therefore written with the student constantly in mind, and has been structured to give a unified discussion of a variety of subjects that may seem to be, on the surface, disparate. The intent of the book is to provide a coherent introduction to the ideas and concepts behind the physics of particle beams. As such, the book begins, after some introductory historical and conceptual comments, with a review of relativity and mechanics. This discussion is intended to build up our sets of physics tools, by placing a few standard approaches to modern dynamics, such as Hamiltonian and phase space-based analyses, in the context of relativistic motion. We then give a presentation of charged particle dynamics in various combinations of simple magnetostatic and electrostatic field configurations, providing another unifying set of basic tools for understanding more complex scenarios in beam physics. Also, at a higher conceptual level, we examine the physics of circular accelerators using simple extensions of the principles developed first in the context linear accelerators, and then use similar approaches to analyzing both transverse optics and acceleration (longitudinal) dynamics. The adopted emphasis on fundamental, unified tools is motivated by the challenges of modern accelerators and their applications as encountered in the laboratory. In present, state-of-the-art beam physics labs, the experimental systems display an increasing wealth of physical phenomena, that require a physicist's insight to understand.

One of the more unique aspects of this text is that its unified approach is extended to include a discussion of the connection between the methods of charged particle beam optics and descriptions of the physics of paraxial light beams such as lasers. This unification of concepts, between wave-based light beams and classical charged particle beams, is also motivated by experimental challenges arising from two complementary sources. The first is that lasers are increasingly used as critical components in accelerators—for example, they are used to produce intense, picosecond electron pulses in devices termed photoinjectors, two of which are found at UCLA. The second was already mentioned above. In advanced accelerators and free-electron lasers, the concepts of accelerators, particle beams, and lasers are, in fact, merged. These cutting edge subjects in beam physics are what provide the intellectual impetus behind my research program at UCLA. It should not be surprising that such subjects find their way into this introductory text in a number of different ways.

This book is also structured so that an abbreviated course consisting of the first four chapters may give an introduction to single particle dynamics in accelerators. Additionally, Chapter 5 introduces the physics of beam distributions (collections of many particles), a subject that is quite necessary if one wishes to apply this text in practice. Further, the notions developed in Chapters 1–5 can then be used to give the basis for the material on photon beams in Chapter 8. Chapters 6 and 7, which discuss the technical subjects of magnets as well as waveguides and accelerator cavities, can stand virtually by themselves. They do, however, complete the set of basic material offered here, and it is hoped that they prove useful in practice as an introductory guide to the design and use of accelerator laboratory components.

To aid in streamlining the approach to learning from this text, sections that contain general "review" material are marked with the ⋆ symbol next to their title. For an advanced student, these sections (the reviews of relativity and mechanics fall into this category) might be omitted on first reading. Other sections are marked by an asterix (*), and contain material which can be considered in some way tangential to the main exposition of topics in beam physics. Such sections, while forming important components of the book as a whole, and may be referenced elsewhere in the text, may contain material that is too lengthy or deep for a fast initial reading. In an alternative approach, much of the material in both these special sections would be included in appendices. Here they are included in the main body of text, both to improve the logical flow of the book, and to allow illustrative exercises for the student to be included.

The exercises in this book, included at the end of each chapter, are meant to be an integral part of the exposition, as some important topics are actually covered in the exercises. In order to provide a guide to approaching the exercises, and to help emphasize their importance, worked solutions to roughly one-third of the problems are included in Appendix A.

The subjects introduced in this book are related in both obvious and subtle ways. To aid in tying threads of the text together for the reader, a short summary is included at each chapter end.

Numerous acknowledgments are in order. One must first find your interest sparked by a field of inquiry; my present colleague David Cline provided the spark for me when I was yet a student. My initial training in accelerator theory was most heavily influenced by Fred Mills (then at Fermilab), and his style can be seen in many of the approaches taken to analyzing beam dynamics in this text. Other friends, mentors and collaborators of particular note, who have given me stimulus to go deeper into the subjects presented here are: Pisin Chen, Richard Cooper, Luca Serafini, and Jim Simpson. A special debt of gratitude is due to my close colleague, Claudio Pellegrini, who has, with our students and post-docs, built the UCLA beam physics program with me.

Perhaps the highest level of thanks must go, however to those students at UCLA who have provided components of the background, motivation, and critical feedback to finish this text. A no-doubt incomplete listing of the graduate students (many of whom are now professional colleagues) follows: Ron Agusstson, Scott Anderson, Gerard Andonian, Kip Bishofberger, Salime Boucher, Nick Barov, Eric Colby, Xiadong Ding, Joel England, Spencer Hartman, Mark Hogan, Pietro Musumeci, Sven Reiche, Soren Telfer, Andrei Terebilo, Matt Thompson, Gil Travish and Aaron Tremaine. Special thanks are due to the undergraduate students who have aided in the editing of this text, Pauline Lay and Maria Perrelli.

As a university professor with a large, active research program, my textbook writing has generally been performed after the "day job" is done. Therefore, I must also thank my family (my wife, Judy, and children, Max, Julia and Ian) for their patience, and occasional cheers—mainly for attempts at artistic graphics—as this project began to unfold at home.

James Rosenzweig
Los Angeles, 2002

Contents

Introduction to beam physics

This textbook, the first of a planned two-set volume on the physics of accelerators and beams, is intended to provide a comprehensive introduction to the physical principles underlying the theory and application of particle beam, accelerator, and photon beam physics, within the context of a one semester (or two quarter) undergraduate course. Its emphasis lies in providing the basic conceptual and analytical tools underpinning further study in the field. The course of study is presented from a unified viewpoint, with connections drawn between what may at first seem to be disparate topics made wherever possible.

The book begins, in this chapter, with an overview of the basic concepts needed to start the discussion of particle beams—collections of charged particles all traveling in nearly the same direction with the nearly the same (possibly relativistic) speed—and accelerators. These concepts include Lagrangian and Hamiltonian approaches to mechanics, and how these methods are applied naturally in the context of relativistic charged particle motion. After this formal re-introduction to some powerful analysis tools, we then proceed to examine the motion of charged particles in static electric and magnetic fields, with the purpose of acquiring a basic understanding of the relevant categories of motion. From these building blocks, we then take up a series of topics in particle beam physics: linear transverse oscillations, acceleration and longitudinal motion in linear and circular devices, and envelope descriptions of beams. In this initial volume, these topics are discussed from the viewpoint of collections of nearly non-interacting particles, where the forces generated by the particles' collective electromagnetic fields are too small to be of interest. On the other hand, much of modern accelerator physics is concerned with intense beams that have very strong self-forces, and display characteristics of plasmas (ionized gases); the physics of such systems is beyond the scope of this text, but will be addressed in the following volume. A description of the topics covered in this second volume is given in Appendix B.

After the introductory survey of particle and beam dynamics in the first five chapters of this volume, we subsequently examine some aspects of relevant technologies. In particular, we concentrate on the features of physics and engineering methods used in accelerator magnet and electromagnetic accelerating systems most directly related to the material presented on charged particle motion. Our investigation is then extended to include the comparisons between single particle and collective descriptions of charged particle beam optics on the one hand, and ray and wave optics in coherent electromagnetic (light) beams, such as lasers, on the other. With the introduction of electromagnetic radiation in the text, the discussion progresses to encompass aspects of charged particle radiation processes and their effect on charged particle motion.

1.1   History and uses of particle accelerators

The history of particle accelerators is one of physics and technology at the cutting edge, as the desire to use increasingly higher-energy particles for basic research in physics has led to vigorous innovation and experimentation. These efforts have caused explosive growth in the field of particle accelerators, which at the dawn of the new millennium has established itself as a fundamental area of research in its own right, with its own research journals and societies. It is, however, a cross-disciplinary field, having many connections to other sub-disciplines within the world of physics. Areas of inquiry in which particle beam physicists have made a significant impact, or have borrowed techniques from, over the years include high-energy physics, nuclear physics, nonlinear dynamics, medical physics, plasma physics, and coherent radiation and X-ray sources. Particle beams are now an indispensable tool in these endeavors, with thousands of practitioners worldwide using them, and billions of dollars per year being spent by governments and industry to develop and improve them.

It was not always such! Let us review the history of particle accelerators in order to provide a context for our discussion and to introduce some concepts and terminology. This is not meant to be a self-contained discussion—some concepts are mentioned and can only be defined later—we only wish to sketch a description of the successive generations of accelerators so that, when we discuss them further in the text, the reader has an idea of the conceptual and chronological roles of these devices.

The history of the particle accelerator begins in the mid-nineteenth century with the development of the cathode ray tube, in which electrons are accelerated across a vacuum gap with an applied electrostatic potential. The need to create cathode ray tubes spurred the advance of many aspects of modern experimental physics, such as voltage sources and vacuum techniques. The cathode rays themselves were the subject of intense scrutiny, which led to the actual discovery of the electron and its properties. Notable experiments in this regard include the determination of the electron's charge-to-mass ratio by J.J. Thomson, and the discovery of the photoelectric effect by Lenard and Millikan.

The cathode ray tube evolved[1] over time and technological development to the electrostatic accelerator, which, instead of kV potentials, gave rise to MV potentials and the creation of electron beams with relativistic velocity. The technological innovations associated with electrostatic accelerators included several inventions such as the belt-charged Van der Graaf accelerator and the cascaded-voltage Cockcroft–Walton generator. These devices could be used to accelerate both electrons and heavier, ionized particles—allowing the birth of nuclear physics, and playing a key role in the quantum revolution of the 1920s and 1930s. During this time, radio-frequency linear accelerators were also studied, but did not become prominent tools in physics research until later.

The advancement of particle accelerators definitively hit its stride with the invention of the initial circular accelerators: the ion accelerator known as the cyclotron, and an electron accelerator termed the betatron. The betatron was proposed as early as 1924 by Wideroe and was made into experimental reality by Kerst and Serber in 1940. This device introduced acceleration based on electromagnetic induction and provided a demonstration of the principle of weak focusing (giving rise to simple transverse betatron oscillations, discussed in Sections 2.2 and 3.1). This transverse focusing effect was also developed in

[1] The cathode ray tube evolved most importantly from the general viewpoint into the television display and computer monitor. Much of this text was being written on a lap-top computer with a liquid crystal display, however, so this ancient accelerator technology may, after a century of use, be losing its dominance in this application.

the cyclotron, a machine with an expanding circular geometry (see Fig. 2.14 associated with Ex. 2.3). The cyclotron was also notably the first device to show acceleration based on **resonance** of particle motion with time-varying electromagnetic fields. Additionally, the cyclotron lends its name to the frequency of oscillation upon which this resonance is based, the well-known **cyclotron frequency**. This frequency is not constant (see Section 2.1) but turns out to vary noticeably when the particle becomes relativistic. For heavy particles, going above a few 100 MeV of kinetic energy required the invention of the synchrocyclotron, in which the frequency of the applied fields is varied in time. The cyclotron concept is still employed in many nuclear and medical accelerators, but for higher energies, such devices could not be used. This is due to iron-based magnets that must be used throughout the machine to bend the particles in circular or spiral orbits. At a certain point one cannot keep building larger magnets, due to the complication and expense involved.

After the Second World War (in which the cyclotron played a part in development of the atomic bomb[2]), radar technology pushed the invention of the radio-frequency linear accelerator, by allowing microwave powers high enough to directly accelerate particles in electromagnetic cavities. The **Alvarez drift-tube linear accelerator** (linac), a standing wave structure (see Figs 1.1 and 4.13), was the first of this category of accelerator, followed by the periodically loaded traveling wave (e.g. Fig. 4.1) structures typical of modern linacs. The **traveling wave linac** has allowed the construction of a 50-GeV electron accelerator at Stanford in which the quark structure of matter was first observed. Higher-energy linacs are now on the horizon, and may be the next frontier tool for discoveries in particle physics.

The circular accelerator also underwent a "revolution" in the post-war period due to the invention of the **synchrotron**. The synchrotron, in which the concept of phase focusing, or phase stability (characterized by longitudinal—in the direction of nominal beam motion—**synchrotron oscillations**), was fully developed, is a merging of the linac, in that it employs radio-frequency acceleration, with the circular accelerator and its associated bend magnets. In the synchrotron, unlike the cyclotron, particles always stay on approximately the same radius orbit. This is also true of the betatron, but, since (see Ex. 2.2) the acceleration in the betatron arises from electromagnetic induction, the entire interior area bounded by the particle orbit must have a time-varying magnetic field. The synchrotron, however, is free of the constraints on magnetic field of both the betatron and the cyclotron, so the bend magnets need only be placed near that orbit, not the entire device. This innovation, along with the implementation of **alternating gradient focusing** (also termed **strong focusing**, as opposed to the **weak focusing** of the betatron), has allowed very large energy synchrotrons to be built. One such device is the 0.9-TeV (1 TeV = 10^{12} eV) Tevatron at Fermi National Accelerator laboratory outside of Chicago with a radius of 1 km. One of course could not imagine the cost associated with using iron in the entire interior of this device! In fact the modern electromagnets employed in the Tevatron, which are shown in Fig. 1.2, are superconducting and as such do not rely on iron to achieve high fields.

The Tevatron is an example of a synchrotron that is operated as a **collider**, in which counter-propagating beams of equal energy particles and antiparticles are squeezed into sub-mm-sized collision regions located inside of the huge, sophisticated particle detectors used to analyze the debris produced in hard

[2]The principles of cyclotron motion were employed in radioactive isotope separation.

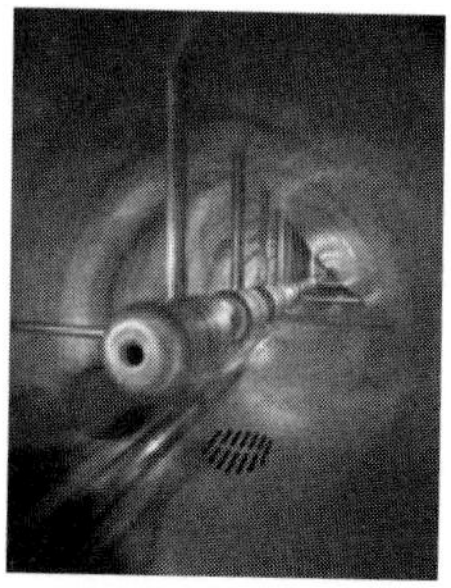

Fig. 1.1 The interior of an Alvarez drift-tube linear accelerator cavity, showing the drift tubes and their supports.

Fig. 1.2 Bend magnets associated with the Tevatron collider, presently the world's largest highest-energy synchroton.

Fig. 1.3 Aerial view of the Tevatron collider complex at Fermilab.

collisions. At the Tevatron, the top quark was recently discovered in such proton–antiproton ($p\bar{p}$) collider experiment; at the European laboratory CERN which built the first $p\bar{p}$ collider, the W and Z intermediate vector bosons were discovered some 15 years previously in a similar manner. An aerial view of the entire Tevatron complex at Fermilab is shown in Fig. 1.3. The Tevatron injection system includes a charged particle source, linear accelerator, and two smaller energy synchrotrons, as well as the collider ring itself. It also has beamlines through which high-energy protons extracted from the rings can be directed onto fixed targets, allowing experiments based on creation of secondary beams that consist of more exotic particles, such as muons and neutrinos.

The synchrotron also lends its name to the radiation produced by charged particles as they bend in magnetic fields—synchrotron radiation. This radiation is both a curse and a blessing. As an energy loss mechanism that has a strong dependence on the ratio of the particle energy to its rest energy (see Section 8.7), it practically limits the energy of electron synchrotrons to that currently achieved, around 100 GeV. On the other hand, synchrotron radiation derived from multi-GeV electron synchrotrons is the preferred source of hard x-rays for research purposes today, with over a dozen such major facilities (synchrotron light sources) world-wide. Synchrotron radiation also forms the physical basis of the free-electron laser; it can produce coherent radiation in both long- and short-wavelength regimes that are inaccessible to present laser sources based on quantum systems. Both the need to create collisions in high-energy colliders, and the desire to make a high-intensity free-electron laser imply that the beams involved must be not only energetic, but of very high quality. A measure of this quality is the phase space density of the beam, which is introduced in Section 1.5.

Today, particle accelerators, while a mature field, present considerable challenges to the physicist who must use and improve these tools. These challenges arise from the need in elementary particle experiments to move to ever increasing energies, a trend that is placed in doubt by the cost of future machines. As the present high-energy frontier machines cost well in excess of $\$10^9$, accelerator physicists are in the process of exploring much more compact and powerful accelerators based on new physical principles. These new acceleration techniques may include use of lasers, plasmas, or ultra-high-intensity charged particle beams themselves. Accelerators also promise to play a critical future role in short-wavelength radiation production, inertial fusion, advanced fission schemes, medical diagnosis, surgery and therapy, food sterilization, and transmutation of nuclear waste. These goals present new challenges worthy of the short, yet accomplished, history of the field.

Even with the present level of sophistication, the subject of accelerators can be initially approached in a straightforward way. The fundamental aspects of particle motion in accelerators can be appreciated from examination of simple configurations magnetostatic (or, less commonly, electrostatic) fields, which may be used to focus and guide the particles, and confined electromagnetic fields that allow acceleration. Moreover, analysis of charged particle dynamics in these physical systems has certain general characteristics, which are discussed in the remainder of the chapter. We begin this discussion by writing the basic equations governing the electromagnetic field, and then proceed to review aspects of methods in mechanics—Lagrangians and Hamiltonians, as well as special relativity. Based upon this discussion, we then introduce the description of beams as distributions in phase space. We finish the present chapter by examining the notion of the design trajectory and analysis of nearby "paraxial" trajectories.

We note that the following three sections are indicated (by the ★ symbol) as review, and therefore optional for a first reading of the text. In fact, most readers should benefit from the material presented, either as a review for those who are familiar with the methods discussed, or as a focused introduction to the uninitiated. In any case, the results contained in Sections 1.2–1.4 will be referenced often in the remainder of the text, and will thus need to be seriously examined sooner or later.

1.2   System of units, and the Maxwell equations★

In order to construct our analyses, we must begin by choosing a system of units. While classical electromagnetism in general, and particle beam physics in particular, are a bit more compactly written in cgs units, we use mks or SI units in this text. This is for two main reasons: (a) ease of translation of the results into laboratory situations, and (b) familiarity of undergraduate physics students, as well as engineers, with the mks system.

Several basic equations need to be introduced in units-specific context. These include the Maxwell equations:

$$\vec{\nabla} \cdot \vec{B} = 0, \tag{1.1}$$

$$\vec{\nabla} \cdot \vec{D} = \rho_e, \tag{1.2}$$

$$\vec{\nabla} \times \vec{H} = \frac{\partial \vec{D}}{\partial t} + \vec{J}_e, \tag{1.3}$$

and

$$\vec{\nabla} \times \vec{E} = -\frac{\partial \vec{B}}{\partial t}, \tag{1.4}$$

where ρ_e and $\vec{J}_e$ are the free electric charge density and current density, respectively, that are related by the equation of continuity

$$\vec{\nabla} \cdot \vec{J}_e + \frac{\partial \rho_e}{\partial t} = 0. \tag{1.5}$$

We will also make use of the following relations between the electromagnetic fields, the scalar potential ϕ_e and the vector potential $\vec{A}$,

$$\vec{E} = -\vec{\nabla}\phi_e - \frac{\partial \vec{A}}{\partial t}, \tag{1.6}$$

$$\vec{B} = \vec{\nabla} \times \vec{A}. \tag{1.7}$$

For completeness, we must also include the constitutive equations,

$$\vec{D} = \varepsilon(\vec{D})\vec{E} \quad \text{and} \quad \vec{B} = \mu(\vec{H})\vec{H}, \tag{1.8}$$

where $\varepsilon(\vec{D})$ and $\mu(\vec{H})$ are the electric permittivity and the magnetic permeability of a material, respectively. We will not encounter the first of Eq. (1.8) again in the text (except in the context of vacuum electrodynamics, where $\varepsilon(\vec{D}) = \varepsilon_0$, the permittivity of free space) until we discuss propagation of light in Chapter 8. The latter of Eq. (1.8) will be revisited when we discuss the design principles of electromagnets based on ferric materials. We note here that the constants $\varepsilon_0 = 8.85 \times 10^{-12} \mathrm{C}^2/\mathrm{N\,m}^2$ and $\mu_0 = 4\pi \times 10^{-7} \mathrm{N}/\mathrm{A}^2$ are related to the speed of light c by

$$c = (\varepsilon_0 \mu_0)^{-1/2} = 2.998 \times 10^8 \mathrm{m/s}. \tag{1.9}$$

In particle beam physics, one often wishes to use MeV (10^6 eV=1.6×10^{-13} J) as the unit of energy. In this case it is useful, when making calculations of applied acceleration due to an electric field, to quote the electric force (acceleration gradient) qE in terms of eV/m by simply absorbing the charge q, an integer multiple of e, into the units. This same position may be adopted in the context of applied magnetic forces if one notes that the force qvB also has units of eV/m when one absorbs the charge q and multiplies the magnetic field B in tesla (T) by the velocity v in m/s. Note that this implies that the commonly encountered level of 1 T static magnetic field is equivalent to a 299.8 MV/m static electric field in force for a relativistic ($v \approx c$) charged particle. This electric field exceeds typical breakdown limits on metallic surfaces by nearly two orders of magnitude, giving partial explanation to the predominance of magnetostatic devices over electrostatic devices for manipulation of charged particle beams.

When considering the self-forces of a collection of charged particles, the combination of constants $e^2/4\pi\varepsilon_0$ often arises. This quantity may be converted to our desired units by writing

$$\frac{e^2}{4\pi\varepsilon_0} = r_{\mathrm{c}} m_0 c^2, \tag{1.10}$$

where r_{c} is defined as the **classical radius** of the (assumed $|q| = e$) and m_0 is the rest mass of the particle. In the case of the electron, we have a rest energy $m_0 c^2$ in useful "high-energy physics" units of $m_e c^2 = 0.511$ MeV and a classical radius of $r_e = 2.82 \times 10^{-15}$ m. Thus, we may write $e^2/4\pi\varepsilon_0 = 1.44 \times 10^{-15}$ MeV m.

1.3   Variational methods and phase space★

The study of beam physics is based on the understanding of relativistic motion of charged particles under the influence of electromagnetic fields. Such fields are constrained by the relations shown in Eqs (1.1)–(1.7). Given the $\vec{E}$ and $\vec{B}$ fields, the analysis of charged particle dynamics can be performed, perhaps most naturally, using only differential equations derived from the Lorentz force equation,

$$\frac{\mathrm{d}\vec{p}}{\mathrm{d}t} = q(\vec{E} + \vec{v} \times \vec{B}). \tag{1.11}$$

While we will base many of our discussions of charged particle motion in this book on the Lorentz force equation, more powerful methods are also available that use variational principles, that is, Lagrangian and Hamiltonian analyses. These methods, which have traditionally been introduced at the graduate level, are now increasingly taught in undergraduate-level mechanics courses. The power of variational methods is found in their rigor, and in the clarity of the results obtained when such approaches are applied to problems naturally formulated in difficult coordinate systems, such as curvilinear or accelerating systems. Even in difficult cases, variational methods give a straightforward formalism that reliably yields the correct equations of motion.

We now give a short review of these methods, which will also prepare us, in a quite natural way, to discuss the roles of electromagnetic fields and special relativity in classical mechanics. This review is meant to clarify these subjects to the reader who is already conversant in variational methods and relativity. For one who has not studied these subjects before, the following discussion (the remainder of this chapter) may serve as an introduction, albeit a steep one, which may be supplemented by material recommended in the bibliography. It may be remarked that beam physics provides some of the most elegant and illustrative uses of advanced methods in dynamics, as well as the role of relativity in these dynamics, that are encountered in modern physics. Thus, even if this text serves as a first introduction to these subjects, it will be a physically relevant and, hopefully, rewarding discussion.

The discussion of variational methods nearly always is initiated by introduction of the **Lagrangian**, which in non-relativistic mechanics is given by

$$L(\vec{x}, \dot{\vec{x}}) = T - V. \tag{1.12}$$

Most commonly, the potential energy V is a function of the position (coordinates) $\vec{x}$ and the kinetic energy T is a function of the velocity $\dot{\vec{x}}$. Note that we use the notation $\vec{x}$ to indicate the set of M **generalized coordinates**[3] x_i ($i = 1, \ldots, M$), and the associated velocities are, thus, defined as $\dot{\vec{x}} \equiv d\vec{x}/dt$ (the compact notation $(\cdot) \equiv d/dt$ will be used in this text to indicate a total time derivative). The application of Lagrangian formalism, and the Hamiltonian formalism that is based upon it, to forces not derivable from a scalar potential V (such as magnetic forces) is discussed in the next section.

The equations of motion are derived from the Lagrangian by **Hamilton's principle**, or the principle of extreme action,

$$\delta \int_{t_1}^{t_2} L \, dt = 0. \tag{1.13}$$

The variation of coordinate and velocity components in the integral in Eq. (1.13), when at an extremum, yields a recipe that gives the **Lagrange–Euler** equations of motion,

$$\frac{d}{dt}\left(\frac{\partial L}{\partial \dot{x}_i}\right) - \frac{\partial L}{\partial x_i} = 0. \tag{1.14}$$

The power of these equations is first and foremost in that they rigorously generate forces of constraint and "fictitious forces" such as those arising from centripetal acceleration. This is a significant accomplishment, but one that is eventually overshadowed by the use of the Lagrangian to form the basis of constructing a **Hamiltonian** function,

$$H(\vec{x}, \vec{p}) \equiv \vec{p} \cdot \dot{\vec{x}} - L, \tag{1.15}$$

where the **canonical momenta** are defined through the Lagrangian by

$$p_i \equiv \frac{\partial L}{\partial \dot{x}_i}. \tag{1.16}$$

[3] A generalized coordinate is often a simple Cartesian distance (x, y, or z), but may also be an angle, as naturally found in cylindrical or spherical polar coordinate systems.

These momenta (momentum components) are new dependent variables in the formalism, replacing the role of the velocity components in the Lagrangian analysis. In the most familiar example, that of non-relativistic motion in Cartesian coordinates, the kinetic energy is $T = \frac{1}{2}m_0\dot{\vec{x}}^2 = \vec{p}^2/2m_0$, and the momenta are $p_i = m_0\dot{x}_i$, as expected.

In the Hamiltonian formalism, Hamilton's principle gives twice the number of equations of motion,

$$\dot{x}_i = \frac{\partial H}{\partial p_i}, \qquad \dot{p}_i = -\frac{\partial H}{\partial x_i}, \tag{1.17}$$

as the Lagrange–Euler equations. The first of Eq. (1.17) defines the velocity components in terms of the canonical momentum components; the second, governing the time evolution of the momentum, is a generalization of Eq. (1.11). The canonical momentum components and corresponding coordinates have a nearly symmetrical relationship with each other, and pairs of such variables are termed **canonically conjugate**. The space $(\vec{x}, \vec{p})$ of all such pairs is termed **phase space**. It should be noted that the canonical momentum is not necessarily identical to the more familiar **mechanical momentum** employed in Eq. (1.11). This point is returned to in Section 1.4.

The Hamiltonian formalism allows constants of the motion to be derived with little difficulty. From Eq. (1.17), it is apparent that if the Hamiltonian is independent of the coordinate, then the conjugate momentum component is a constant of the motion. Likewise, if the Hamiltonian is independent of the momentum component, then the conjugate coordinate is a constant of the motion. Further, the Hamiltonian obeys the relation

$$\dot{H} = \frac{\partial H}{\partial t}, \tag{1.18}$$

and therefore H is a constant of the motion if it is not explicitly dependent on the time t.

If the Hamiltonian is a constant of the motion, we can most often identify this constant as the total energy U of the system. The invariance of H allows one of the main tools of particle beam physics to be employed—the drawing of the so-called phase space maps. The phase space is the $2M$-dimensional space of all M pairs of coordinates and their canonically conjugate momenta. Phase space maps, an example of which is displayed in Fig. 1.4, are of a (x_i, p_i) trajectory drawn in two-dimensional projections of the full phase space. As such, these representations may more properly be called **phase plane** maps. Of course, in fully three-dimensional accelerators, we have $M = 3$, and there are three phase planes in which maps are drawn. In particle beam physics, in fact, one often deals with motion in which the variables in one phase plane are very nearly independent of any other phase plane variable. This state of affairs, in addition to the inherent ease of two-dimensional (as opposed to higher dimensional) visualization, makes phase plane descriptions popular as a tool for understanding particle beam dynamics.

The creation of phase plane maps such as Fig. 1.4 is accomplished analytically by using a time-independent Hamiltonian and plotting $H(x_i, p_i) = $ constant curves. In more complicated cases, phase plane maps are created by numerical solution of the equations of motion. The concept of phase space is central to

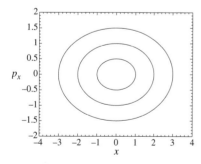

Fig. 1.4 Phase plane plot for simple harmonic oscillator orbits, corresponding to three different values of the Hamiltonian $H = 1/8, 1/2$, and 2, with $m = 1$ and, $\omega = 0.5$. The trajectories of the oscillator lie along ellipses described by these constant H curves.

the field of particle beam physics and certain results, such as the invariance of phase space density (see discussion Section 1.4) can only be clearly discussed in the context of Hamiltonian formalism.

Perhaps the most familiar example of this mapping technique is the one-dimensional non-relativistic simple harmonic oscillator. In this case (for simplicity indicating the coordinate $x_1 \Rightarrow x$) the one-dimensional Hamiltonian is of the form

$$H = \frac{1}{2m}[p_x^2 + m^2\omega^2 x^2], \tag{1.19}$$

where $\omega^2 = K/m$, and K is the oscillator strength or spatial gradient of the restoring force, $K = -F_x/x$. The phase plane maps associated with simple harmonic motion are thus ellipses in the two-dimensional (x, p_x) phase plane, as shown by the examples in Fig. 1.4. Note that the Hamiltonian alone does not indicate the direction in which the system traces out the ellipse, but examination of the force and velocity direction does—the direction of motion is clearly clockwise in phase space for this system.

The area of the phase plane ellipse is proportional to the value of the Hamiltonian associated with each trajectory, and is therefore also a constant of the motion. This area is given by

$$\oint p_x \, dx = \oint p_x \dot{x} \, dt = \oint (H + L) \, dt = U\tau. \tag{1.20}$$

where τ is the period of the oscillation. We shall see in Chapter 5 that the area associated with a closed trajectory in phase space forms a central place in the theory particle beam dynamics.

The phase plane map is of great use in visualizing the motion of charged particles beyond simple harmonic orbits (see Ex. 1.3) and is profitably employed even in cases when the Hamiltonian is not a constant of the motion. In the case of a time-varying Hamiltonian, one may not trivially generate plots like Fig. 1.4, but must often solve the equations of motion (Eq.(1.17)) first. Furthermore, if one solves these equations in such a case, it may not be illuminating, but rather confusing (e.g. Fig. 3.6), to use continuous lines in phase space to illustrate the motion as it advances continuously in time. For systems typical of circular accelerators, the Hamiltonian varies periodically in time t, however. A valuable strategy for phase plane plotting in this case is taking periodic "snap-shots" and plotting the instantaneous position in the phase plane once per Hamiltonian (not oscillation) period. This type of map is termed a **Poincare plot** (e.g. Fig. 3.7) and is discussed further in Chapter 3.

There are also manipulations of the phase space or phase plane variables that can be undertaken to create a description where the Hamiltonian is a constant of the motion in the new variables, where it was not constant in the old variables. To see the utility of this approach, consider an explicitly time-dependent Hamiltonian, in which the potential arises from a traveling wave so that the Hamiltonian can be written

$$H = \frac{1}{2m}[p_x^2 + G(x - v_\varphi t)], \tag{1.21}$$

where v_φ is the phase velocity of the wave in the x direction and G is an arbitrary function. The simplest way to make the Hamiltonian into a constant of the motion is to perform a mathematical transformation of the system description. Let us examine such a transformation of the coordinates, the Galilean transformation, where

$$\tilde{x} = x - v_\varphi t. \tag{1.22}$$

Now we must transform the Hamiltonian so that the coordinate's equation of motion remains correct. With the new canonical momentum set equal to the old, $\tilde{p}_x = p_x$,

$$\dot{\tilde{x}} = -\frac{\partial \tilde{H}}{\partial \tilde{p}_x} = \frac{\tilde{p}_x}{m} - v_\varphi = v - v_\varphi, \tag{1.23}$$

as expected for a Galilean transformation. To generate Eq. (1.23) as a correct canonical equation of motion (i.e. one derivable from Eq. (1.17)), the new Hamiltonian $\tilde{H}$ must transform from the old Hamiltonian H as

$$\tilde{H}(\tilde{x}, \tilde{p}_x) = H(\tilde{x}, \tilde{p}_x) - v_\phi \tilde{p}_x = \frac{1}{2m}[\tilde{p}_x^2 + G(\tilde{x})] - v_\phi \tilde{p}_x. \tag{1.24}$$

Now the new Hamiltonian $\tilde{H}(\tilde{x}, \tilde{p}_x)$ is explicitly independent of t and is thus a constant of the motion. The trajectory of the charged particle in this wave potential can, therefore, be visualized, as before, with a phase plane map created by the simple algebraic relationship between $\tilde{x}, \tilde{p}_x$, and $\tilde{H}$. For example, we may use Eq. (1.24) with a moving simple harmonic oscillator potential, $G(\tilde{x}) = \frac{1}{2}K\tilde{x}^2$. This leads to

$$\tilde{H}(\tilde{x}, \tilde{p}_x) = \frac{1}{2m}\left[\tilde{p}_x^2 + m^2\omega^2\tilde{x}^2 - 2p_\varphi \tilde{p}_x\right]$$

$$= \frac{1}{2m}\left[(\tilde{p}_x - p_\varphi)^2 + m^2\omega^2\tilde{x}^2\right] - T_\varphi, \tag{1.25}$$

where we have defined p_φ and T_φ as the (non-relativistic) momentum and kinetic energy associated with a particle of mass m traveling at the phase velocity v_φ. In this case, the constant $\tilde{H}$ curves associated with the motion shift upward in $\tilde{p}_x$ by p_φ, when compared those shown in Fig. 1.4, to as shown in Fig. 1.5.

Note that the phase plane plots for the moving simple harmonic oscillator potential can be made to look identical to the stationary potential plot by use of

Fig. 1.5 Phase plane plot for moving simple harmonic oscillator orbits corresponding to same limits in $\tilde{x}$ as those in x found in Fig. 1.4. Here $p_\varphi = 2$, with $m = 1$ and $\omega = 0.5$. The curves corresponding to the moving potential are represented by solid lines, and their counterparts from Fig. 1.4 are shown in dashed lines.

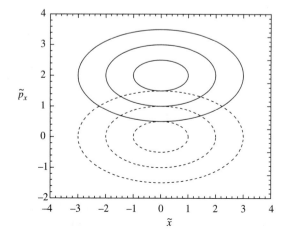

our Galilean transformation from x to $\tilde{x}$ and, further, by plotting $\delta p \equiv \tilde{p}_x - p_\varphi$. Each of the curves in the $\tilde{x}$ frame is associated with values (total energies) of the new Hamiltonian $\tilde{H} = H - T_\varphi$. The constant $-T_\varphi$ contains no information about the system's dynamics (cf. Eq. (1.17)), however, and may be ignored. At this point, we need to clarify that the transformation given by Eqs (1.22)–(1.24) is a purely mathematical change of variables, not a change of physical frame. We will make use of this type of **mathematical** transformation while discussing acceleration in traveling electromagnetic waves (Chapter 4). The **physical** change of frame is described, of course, not by a Galilean transformation but by a Lorentz transformation, as discussed in Section 1.4.

The type of variable transformation illustrated by Eqs (1.22)–(1.24) is termed a **canonical transformation** because it preserves the canonically conjugate relationship between the coordinate and the momentum. In general, one does not use such an ad hoc way of deriving the transformation but a more rigorous method based on **generating functions**, which are discussed in advanced mechanics textbooks. These functions come in a variety of types, depending on the variables to be transformed. The generating function always is dependent on a pair of variables per phase plane, some combination of the old and new canonically conjugate variables.

As an example, consider use of a generating function to transform the one-dimensional simple harmonic oscillator Hamiltonian, Eq. (1.19) to a particularly interesting new form. The well-known solutions for the motion in such a system are given by

$$x(t) = x_m \cos(\omega t + \theta_0), \quad \text{and} \quad p_x(t) = m\omega x_m \sin(\omega t + \theta_0). \tag{1.26}$$

We would like to transform the Hamiltonian to one that reflects the constant rate of advance in argument of the cosine function in Eq. (1.26), so we propose that the new coordinate be chosen as $\theta = \omega t + \theta_0$. In this case, we can use a generating function of the form $F(x, \theta)$ to transform the simple harmonic oscillator problem into a more useful form. According to Hamilton's principle, Eq. (1.27) must yield the following formal properties:

$$p_x = \frac{\partial}{\partial x} F(x, \theta), \quad J = \frac{\partial}{\partial \theta} F(x, \theta), \quad H' = H + \frac{\partial}{\partial t} F(x, \theta), \tag{1.27}$$

where J is the new momentum and H' is the new Hamiltonian. We can deduce from Eq. (1.27) that a proper generating function is given by $F(x, \theta) = \frac{1}{2} m\omega x^2 \cot(\theta)$. Using $F(x, \theta)$ to obtain the new momentum and Hamiltonian, we have

$$H' = J\omega, \tag{1.28}$$

which is a constant of the motion. Since ω is a constant, the momentum J, known as the **action**, is also a constant. The new canonically conjugate pair are termed **action-angle variables**. The action-angle description is important for analyzing perturbations to simple harmonic systems, a commonly encountered problem in particle beam physics. The action is, comparing Eqs (1.18) and (1.25), simply related to the area enclosed by the phase space trajectory,

$$J = \frac{1}{2\pi} \oint p_x \, dx. \tag{1.29}$$

The action is also generally known to be an **adiabatic invariant**, in that when the parameters of an oscillatory system are changed slowly, the action remains a constant. This can be illustrated by writing the differential equation for a slowly varying oscillator (with the mass factor set to $m = 1$) as

$$\ddot{x} + K(t)x = 0, \tag{1.30}$$

and substituting an assumed form of the solution

$$x(t) = C\sqrt{a(t)}\cos(\psi(t)). \tag{1.31}$$

After some manipulations, we obtain the relations

$$\dot{\psi} = \frac{1}{a} \quad \text{and} \quad \ddot{a} - \frac{\dot{a}^2 + 4}{2a} + 2Ka = 0. \tag{1.32}$$

The substitution given in Eq. (1.31) is commonly found in the theory of time-dependent oscillators, which is quite important in particle beam physics—it is explicitly used in an analysis in Chapter 5. The solution of the first equation in (1.32) is formally

$$\psi = \int \frac{dt}{a(t)}, \tag{1.33}$$

while the second of these equations' solution can be examined approximately. Assuming the first and second time derivatives of a are small ($\ddot{a}/a \ll K$ and $(\dot{a}/a)^2 \ll K)^4$, one has simply

$$a(t) \cong (K(t))^{-1/2} \quad \text{or} \quad x(t) = C(K(t))^{-1/4}\cos(\psi(t)). \tag{1.34}$$

The momentum corresponding to this approximate solution is

$$p_x(t) = \dot{x} \cong C(K(t))^{1/4}\sin(\psi(t)). \tag{1.35}$$

The area of the phase space ellipse, whose semimajor and semiminor axes are the maximum excursion in x and p_x, respectively, is again the value of the action at that time,

$$J = \tfrac{1}{2}p_{x,\max}x_{\max} = \tfrac{1}{2}C^2. \tag{1.36}$$

The action is independent of time and dependent only on the initial conditions (taken at $t = 0$) through the constant C^2. Thus, we have shown the **adiabatic invariance** of the action J for oscillators whose strength is slowly varying, a result pertinent to discussions in future chapters.

We emphasize at this juncture the primary role of the momenta in Hamiltonian methods, as opposed to the velocity components found in the Lagrangian formalism. This is inherent in both the structure of Hamiltonian and relativistic analyses, as is reviewed in Section 1.4.

[4] These conditions give a quantitave definition of the term "adiabatic".

1.4 Dynamics with special relativity and electromagnetism★

The Hamiltonian formulation of dynamics is naturally suited to analyzing relativistic systems. We shall see that this is because in the canonical approach to dynamics the roles of the coordinates, momenta, time, and energy have a rigorously defined relationship with each other. The relationships between these dynamical variables actually become clearer after one studies relativistic dynamics, where the ways in which all such variables transform from one inertial frame to another are emphasized.

The point of departure for the present discussion is precisely this transformation, which should be familiar to any reader of this text, the **Lorentz transformation**. This relation, which governs the transformation of coordinates and time from one inertial reference frame to another moving at constant speed $v_f = \beta_f c$ with respect to the first frame (along what we choose to be the z-axis), is written as

$$x' = x, \quad y' = y, \quad z' = \gamma_f(z - \beta_f ct), \quad ct' = \gamma_f(ct - \beta_f z), \qquad (1.37)$$

where the **Lorentz factor** $\gamma_f = (1 - \beta_f^2)^{-1/2}$. This Lorentz transformation acts upon the **space-time four-vector** $\vec{X} \equiv (x, y, z, ct)$, and preserves the length, or norm, of the four-vector. This norm, termed a **Lorentz invariant** because it is frame independent, is specified by the quantity

$$|\vec{X}|^2 = x^2 + y^2 + z^2 - (ct)^2. \qquad (1.38)$$

The invariance of the norm of the space-time four-vector is often the starting point of the derivation of the Lorentz transformation, as it indicates that the phase velocity c of spherical light waves in vacuum is independent of inertial reference frame.

The invariance property of the norm of $\vec{X}$ is therefore entirely equivalent to the property that the four-vector transforms between frames under the rules of a Lorentz transformation. Thus, a four-vector can be defined equivalently either as an object that obeys Lorentz transformations or one in which its norm, as defined by Eq. (1.38), is conserved during such a transformation. The invariance of four-vector norm is a key tool in performing analyses of relativistic dynamics.

The absolute value of $|\vec{X}|^2$, which refers to the "distance" in space-time between two events (or implicitly, one event and the origin), can be positive, negative, or zero. If it is positive, it is termed "space-like", as one may always transform to a frame where the events occur at the same time, but at a separated distance. If it is negative, it is termed "time-like", as one may always transform to a frame in which the events occur at the same point in space, but at separate times. Space-like pairs of events cannot be causally connected, because they are too far separated in space-time for light to propagate between them. If the norm of $\vec{X}$ is zero, the two events are exactly connected by a signal traveling at the speed of light. In this case, the events are said to be on each other's light cone.

Using Lorentz transformations of space and time, it can be trivially shown that properties of waves—the wave numbers (spatial frequencies) k_i and (temporal) frequency ω—form a four-vector. This is intuitively so, since the wave number simply measures spatial intervals while the frequency measures intervals in time. As an illustration of this derivation, consider a plane electromagnetic wave moving in the positive z-direction, with functional form $\cos[k_z z - \omega t]$. If one begins in a frame moving with velocity $\beta_f c$ in the z-direction, the inverse transformation

$$\cos\left[k_z \gamma_f(z' + \beta_f ct') - \omega \gamma_f\left(t' + \beta_f \frac{z'}{c}\right)\right]$$

$$= \cos\left[\gamma_f\left(k_z - \beta_f \frac{\omega}{c}\right)z' - \gamma_f(\omega - \beta_f k_z c)t'\right]$$

can be deduced from Eq. (1.37). We can thus see that k_z and ω (Lorentz) transform as z and t, respectively. Further, we know, therefore, that the norm of the four-vector $(\vec{k}c, \omega) = (k_x c, k_x c, k_x c, \omega)$ is

$$\sum_i k_i^2 c^2 - \omega^2 = \text{const.} \tag{1.39}$$

In the case where the constant on the right-hand side of Eq. (1.39) is zero, this can be recognized as the dispersion relation for vacuum electromagnetic waves. From a quantum-mechanical viewpoint, such waves correspond to massless photons (cf. Section 7.1). When the constant in Eq. (1.39) is not zero, a quantum-mechanical interpretation indicates that we are examining an object of non-vanishing rest mass, or rest energy.

The quantum mechanical indentifications of free particle momenta and energy in terms of wave properties, $p_i = \hbar k_i$ and $U = \hbar \omega$, lead us to the conclusion that momentum and energy must also form a four-vector, $\vec{P} \equiv (\vec{p}c, U) = \hbar(\vec{k}c, \omega)$, with the invariant

$$\vec{P}^2 \equiv \vec{p}^2 c^2 - U^2 = \sum_i p_i^2 c^2 - U^2 = \text{const.} \tag{1.40}$$

We must allow for the possibility of Lorentz transformation into the rest frame of the particle, in which case $\vec{p} = 0$ and the invariant can be identified as the rest energy of the particle. Thus,

$$\vec{p}^2 c^2 - U^2 = -(m_0 c^2)^2, \tag{1.41}$$

where m_0 is the rest mass of the particle. Note the norm of the momentum–energy four-vector is always negative (or "energy-like"), because the square of the rest energy is positive definite.

As an example of the utility of Lorentz transformations of the $(\vec{k}c, \omega)$ four-vector, we consider the process known as relativistic **Thomson backscattering** where a relativistic electron collides head-on with a photon (quantum of light, see Section 8.1), yielding a reversal of photon direction and an increase in the photon energy and momentum. This process is illustrated in Fig. 1.6, through a diagram of the initial and final momentum vectors of the electron and photon. The term **Thomson** is somewhat imprecisely[5] applied to this scattering process whenever the change in the electron momentum during collision is negligible. In the frame traveling with the electron, $\beta_f = v/c$ (v is the velocity of the electron), an oncoming photon of laboratory frequency ω has an observed frequency given by Lorentz transformation, $\omega' = \omega \gamma_f (1 + \beta_f)$. If we assume that this photon suffers a reversal of its momentum vector direction but no change in amplitude during collision, a second Lorentz transformation of the $(\vec{k}c, \omega)$ four-vector yields $\omega_s = \omega' \gamma_f (1 + \beta_f) = \omega \gamma_f^2 (1 + \beta_f)^2$. Thus, for a highly relativistic ($\beta_f \approx 1$) electron, the frequency (energy) of backscattered light is increased, $\omega_s \cong 4\gamma_f^2 \omega$. This scattering process is explored further in Exercise 1.6.

[5]It is more precise to term this process "inverse Compton scattering", as in the end we see that the photon energy, as observed in the laboratory frame, increases. This is the opposite of what happens in Compton scattering of photons off of electrons that are initially at rest. See Exercise 1.6 for further discussion of this point.

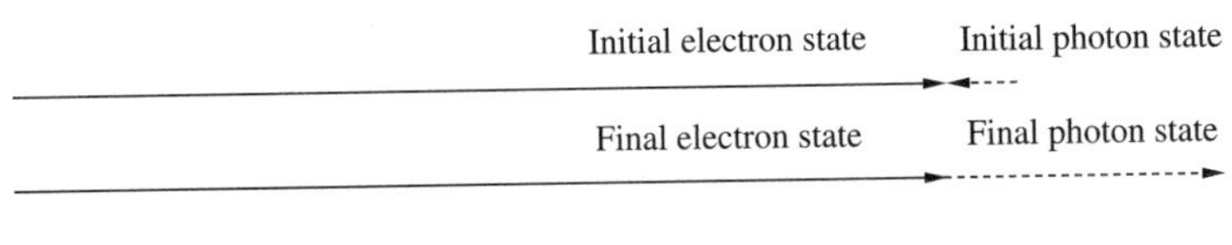

Fig. 1.6 Diagram of electron and photon momenta in initial and final states of the Thomson backscattering process. The violation of momentum conservation is exaggerated in this picture.

Initial electron state   Initial photon state

Final electron state   Final photon state

While we have now established the four-vector relationship between the energy and momentum of a physical system, we have not described what these quantities are in the familiar terms of mass and velocity. In doing so, we must obtain the well-known expressions of the non-relativistic limit, where the momentum $\vec{p} = m\vec{v}$ and energy $U = p^2/2m + \text{const}$. We must also preserve the most general relationship between the momentum and energy,

$$dU = \vec{v} \cdot d\vec{p}, \quad \text{or less specifically,} \quad \frac{dU}{dp} = v, \tag{1.42}$$

where $p = |\vec{p}|$ and $v = |\vec{v}|$. This result is derived by noting that the energy change on a particle is equal to the work performed on it, $dU = \vec{F} \cdot d\vec{l}$, where the force is assumed to still obey Newton's third law, $\vec{F} = d\vec{p}/dt$, and the differential length is $d\vec{l} = \vec{v} \cdot dt$. Differentiating Eq. (1.40) and combining with Eq. (1.42), we also obtain

$$\vec{v} = \frac{\vec{p}c^2}{U}. \tag{1.43}$$

Then, solving for the energy as a function of the velocity, we may write

$$\left(\left(\frac{v}{c}\right)^2 - 1 \right) U^2 = -(m_0 c^2)^4 \quad \text{or} \quad U = \gamma m_0 c^2, \tag{1.44}$$

where we have now defined the Lorentz factor associated with the particle motion as $\gamma \equiv (1 - (\vec{v}^2/c^2))^{-1/2}$. The Lorentz factor of a particle is, therefore, its total (mechanical) energy normalized to its rest energy, and the condition $\gamma \gg 1$ implies a particle that travels at nearly the speed of light. For electrons, having rest energy $m_e c^2 = 0.511\,\text{MeV}$, it is very easy to obtain a particle that travels nearly at the speed of light—megavolt-class electrostatic accelerators can accomplish this feat. However, for the other most commonly accelerated particle, the proton (with rest energy $m_p c^2 = 938\,\text{MeV}$), it is relatively difficult to impart enough energy (several GeV) to make the particle relativistic.

Using Eqs (1.43) and (1.44), we also now have an expression for the momentum vector,

$$\vec{p} = \gamma m_0 \vec{v} \equiv \vec{\beta} \gamma m_0 c, \tag{1.45}$$

which allows Eq. (1.40) to be written, after removing the common factor of $m_0 c^2$ in all quantities, as

$$\gamma^2 = \vec{\beta}^2 \gamma^2 + 1. \tag{1.46}$$

Equation (1.40) is valid not only for single particle systems, but also for general systems of many ($j = 1, \ldots, N$) objects, in which case we have

$$\left(\sum_j \vec{p}_j \right)^2 c^2 - \left(\sum_j U_j \right)^2 = \text{constant}. \tag{1.47}$$

For such systems it is still true that, if any of the objects have non-zero rest mass, one may transform to a coordinate system in which the total momentum of the system vanishes, $\sum_j p_j = 0$.

In this frame, the total energy of the system is obviously minimized. This fact allows straightforward calculation of the available energy for particle creation in high-energy physics experiments. Such a calculation serves as an illustrative example of the use of a Lorentz invariant norm.

In colliders, where charged particles and their antiparticles counter-circulate in rings (or collide after accelerating in opposing linacs), the colliding species, in general, have equal and opposite momentum. Therefore, the first term in Eq. (1.47) vanishes and all of the particle energy ($2U$) is available for creation of new particles. Thus, the Z^0 particle (with a rest energy of 91.8 GeV, 1 GeV = 10^9 eV) has been studied in detail using the LEP collider at CERN, by using electron and positron beams accelerated to 45.9 GeV and then collided. Before the era of the colliding beam machines, however, the frontier energies for exploring creation of new particles took place in fixed-target experiments where the beam particles struck stationary target particles. In the fixed target collision, one may calculate the Lorentz invariant on the right-hand side of Eq. (1.47) by evaluating the left-hand side in the lab frame,

$$p_b^2 c^2 - (U_b + m_t c^2)^2 = -m_t^2 c^4 - m_t^2 c^4 - 2\gamma_b m_b m_t c^4 = \text{constant}, \quad (1.48)$$

where the subscripts b and t indicate beam and target particles, respectively. In the center of momentum frame, however, the total momentum vanishes and the constant in Eq. (1.48) is seen to set a maximum on the total rest energy of particles created in the collision,

$$\sum_i m_{p,i} c^2 \leq \sqrt{2\gamma_b m_b m_t c^2}. \quad (1.49)$$

The maximum energy for particle creation in Eq. (1.49) occurs when the beam and target particles annihilate and the newly created particles are at rest in the center of momentum frame. Equation (1.49) clearly indicates why colliding beam machines are so important in exploring the energy frontier. At Fermilab, proton–antiproton collisions occur between counter-propagating beams of 900 GeV with up to 1.8 TeV available for creation of new particles.[6] If one, instead, substitutes a stationary proton as the target particle with a 900 GeV incident particle, the available energy for particle creation is only 41 GeV!

Equations (1.44) and (1.45) have introduced the relativistically correct momentum and energy. If one substitutes the relativistically correct form of the momenta into Eq. (1.9), it should be emphasized that this Lorentz force relation remains valid by construction. It is of interest to examine this vector equation of motion for the momentum in the case (discussed in detail in Section 2.1) where only a magnetic field is present. In such a scenario the energy of the particle is constant, and we may write

$$\gamma m_0 \frac{d\vec{v}}{dt} = q(\vec{v} \times \vec{B}). \quad (1.50)$$

Equation (1.45) displays, upon comparison with the non-relativistic version of Eq. (1.9), an effect known as the **transverse relativistic mass increase**, where the inertial mass under transverse (normal to the velocity) acceleration effectively behaves as though $m_0 \rightarrow \gamma m_0$. Note that this is also true for electric forces that are instantaneously transverse. For cases involving energy-changing acceleration (in which the acceleration is parallel to the velocity vector), the situation is different, as illustrated in Section 2.4.

[6]Because protons and antiprotons are hadrons, composed of substituent particles (quarks and gluons), the effective (likely to observe) energy available for particle creation is considerably smaller than the full beam particle energy, and the "physics reach" of a 1.8-TeV hadron collider may be only in the several hundreds of GeV.

With the results in this section thus far, obtained by emphasizing the invariance of the norms of a variety of four-vectors under Lorentz transformation, it is
straightforward to find other relations more typically derived directly by use of
Lorentz transformations. For example, let us examine the addition of velocities.
Using the notation $v_f = \beta_f c$ to designate the relative velocity of a new frame,
the velocity in the new frame of a particle whose velocity is v (parallel to v_f) in
the original frame is

$$
v' = \frac{p'c^2}{U'} = \frac{\gamma_f c(pc - \beta_f U)}{\gamma_f(U - \beta_f pc)} = \frac{\gamma_f \gamma m_0 c^3(\beta - \beta_f)}{\gamma_f \gamma m_0 c^2(1 - \beta_f \beta)} = \frac{\beta - \beta_f}{1 - \beta_f \beta}c. \quad (1.51)
$$

This derivation is, perhaps, more transparent than the more standard version
based on Lorentz transformation of the components of $\vec{X}$, because it begins
with more powerful concepts. It also points to an important facet of the theory
of special relativity—velocities do not play a central role as useful descriptions
of the motion since they do not form a part of a four-vector.

The momenta and energy, on the other hand, do form a four-vector. This is
an interesting state of affairs within the context of particle dynamics, because
these same quantities play a primary role in the Hamiltonian formulation of
mechanics. We now discuss how the two concepts, Hamiltonian and four-vector
dynamics, relate to each other in the context of charged particle dynamics in
electromagnetic fields.

We begin by noting that the Lorentz force (Eq. (1.9)) acting on a charged
particle is written, in terms of potentials, as

$$
\vec{F}_L = \frac{d\vec{p}}{dt} = q(\vec{E} + \vec{v} \times \vec{B})
$$

$$
= q\left[-\vec{\nabla}\phi_e - \frac{\partial\vec{A}}{\partial t} + \vec{v} \times (\vec{\nabla} \times \vec{A}) \right]
$$

$$
= q\left[-\vec{\nabla}\phi_e - \frac{\partial\vec{A}}{\partial t} - (\vec{v} \cdot \vec{\nabla})\vec{A} \right] = q\left[-\vec{\nabla}\phi_e - \frac{d\vec{A}}{dt} \right]. \quad (1.52)
$$

In the last line of this expression, we have used the definition of the total (partial
plus convective) time derivative.

The forces in electromagnetic fields are most generally derived not only from
a scalar (electrostatic) potential ϕ, but from an electromagnetic vector potential
as well; thus, we must generalize our approach to Hamiltonian analysis. In
particular, we note that Eq. (1.52) indicates that the equations of motion for
the momenta are not simply derivable from a conservative potential energy
function. If we define the canonical momenta to be $p_{c,i} = p_i + qA_i$, however,
the Hamilton equation of motion for this canonical momentum is correctly
obtained by the prescription

$$
\frac{dp_{c,i}}{dt} = -\frac{\partial H}{\partial x_i} = -q\frac{\partial\phi_e}{\partial x_i}, \quad (1.53)
$$

where the Hamiltonian contains only a conservative (electrostatic) potential
energy. At this point, what we have considered (á la Newton) to be the
momentum in the problem can now be seen to be a **mechanical** as opposed

to canonical momentum. The symmetry between the total "canonical" energy (value of the Hamiltonian) and the canonical momentum is clear—the total energy $H = U + V = U + q\phi_e$ (the numerical value of the Hamiltonian function) also has a portion arising from a potential (in this case, $q\phi_e$). Note that the remainder of the total energy (the "mechanical" component) also includes a rest energy component, $U = \gamma m_0 c^2 = T + m_0 c^2$, that is, the sum of rest and kinetic energies.

From this discussion, it should also be clear that the scalar and vector potentials, since they comprise components of the momentum–energy four-vector, also form a four-vector, $(\vec{A}c, \phi)$. Further, they are paired with the mechanical momenta and energy through the definitions of canonical momenta and energy $p_{c,i} = p_i + qA_i$ and $H = U + q\phi$. To complete our survey of electrodynamic four-vector quantities, we note that the equations governing the potentials can be written as

$$\left[\vec{\nabla}^2 - \frac{1}{c^2} \frac{\partial^2}{\partial t^2} \right] \left\{ \begin{matrix} \vec{A} \\ \phi_e \end{matrix} \right\} = - \left\{ \begin{matrix} \mu_0 \vec{J}_e \\ \frac{\rho_e}{\varepsilon_0} \end{matrix} \right\}. \tag{1.54}$$

Since the potentials form a four-vector, the sources $(\vec{J}_e, \rho_e c)$ form one as well. This result could also have been alternatively derived directly from charge conservation, and Lorentz transformation of lengths (and thus charge density and velocities.

As we have just discussed sources and potentials associated with electromagnetic fields, it is appropriate at this point to examine the transformation between inertial frames of these fields. Since the potentials form a four-vector $(\vec{A}c, \phi_e)$, an obvious starting point of the discussion is to discuss the Lorentz transformation of this four-vector. This relation, governing the transformation of potentials from one frame to another that moves at speed $\beta_f c$ along what we again choose to be the z-axis is written as

$$cA'_x = cA_x, \quad cA'_y = cA_y, \quad cA'_z = \gamma_f(cA_z - \beta_f \phi_e), \quad \phi'_e = \gamma_f(\phi_e - \beta_f cA_z) \tag{1.55}$$

The quantities A_x, A_y, A_z, and ϕ are all, in principle, functions of x, y, z, and t. In the new (primed) frame, the spatio-temporal dependence of these quantities must be expressed in terms of the primed variables, found by substitution obtained from the inverse Lorentz transformation

$$x = x', \quad y = y', \quad z = \gamma_f(z' + \beta_f ct'), \quad t = \gamma_f(ct' + \beta_f z'). \tag{1.56}$$

The fields are obtained from this expression using the following relations:

$$\vec{E}' = -\vec{\nabla}'\phi'_e - \frac{\partial \vec{A}'}{\partial t'}, \quad \vec{B}' = \vec{\nabla}' \times \vec{A}'. \tag{1.57}$$

The transformation of the electromagnetic fields described by Eqs(1.55)–(1.57) are often written as

$$\vec{E}'_\perp = \gamma_f(\vec{E}_\perp + \vec{v}_f \times \vec{B}_\perp), \qquad \vec{E}'_{||} = \vec{E}_{||},$$

$$\vec{B}'_\perp = \gamma_f\left(\vec{B}_\perp - \frac{1}{c^2}\vec{v}_f \times \vec{E}_\perp\right), \qquad \vec{B}'_{||} = \vec{B}_{||}, \tag{1.58}$$

here the symbols $\parallel$ and $\perp$ indicate the components of the field parallel to and perpendicular to the direction of Lorentz transformation of the frame $\vec{v}_f$, respectively.

Now we can return to the derivation of relativistic Lagrangian and Hamiltonian mechanics with electromagnetic fields. With our definition of canonical momenta, we can proceed to construct the Lagrangian by integrating the expression that defines these momenta,

$$p_{c,i} = \frac{\partial L}{\partial \dot{x}_i} = \gamma m_0 \dot{x}_i + eA_i, \tag{1.59}$$

with respect to the spatial coordinates. We thus obtain

$$L(\vec{x}, \dot{\vec{x}}) = -\frac{m_0 c^2}{\gamma} + q\vec{A} \cdot \vec{v} - q\phi_e(\vec{x}), \tag{1.60}$$

where we have allowed the presence in the Lagrangian of a conservative potential dependent only on $\vec{x}$, and identified it as the negative of the electrostatic potential energy $-q\phi_e(\vec{x})$.

The relativistically correct Hamiltonian is obtained from Eq. (1.60) by use of the definition

$$H = \vec{p}_c \cdot \dot{\vec{x}} - L = \frac{\vec{p}_c \cdot (\vec{p}_c - q\vec{A})}{\gamma m_0} - L$$

$$= \frac{(\vec{p}_c - q\vec{A})^2}{\gamma m_0} + \frac{m_0 c^2}{\gamma} + q\phi_e(\vec{x}). \tag{1.61}$$

By multiplying this expression by $\gamma m_0 c^2$ and using $H - q\phi = \gamma m_0 c^2$, we arrive at

$$(H - q\phi_e)^2 = (\vec{p}_c - q\vec{A})^2 c^2 + (m_0 c^2)^2, \tag{1.62}$$

or

$$H = \sqrt{(\vec{p}_c - q\vec{A})^2 c^2 + (m_0 c^2)^2} + q\phi_e. \tag{1.63}$$

We note that Eq. (1.62) could have been obtained by direct substitution of canonical definitions into Eq. (1.41) governing the norm of the mechanical momentum–energy four-vector, that is,

$$(\vec{p}_c - q\vec{A})^2 c^2 - (H - q\phi_e)^2 = -(m_0 c^2)^2. \tag{1.64}$$

.5   Hierarchy of beam descriptions

The methods for analyzing single particle dynamics given in Section 1.4 represent the first step in understanding the physics of charged particle beams. A real beam is made up not of a single particle, however, but a collection of many (N) particles. The second step towards describing the dynamics of an actual beam, therefore, is to consider a collection of N points in phase space, as illustrated in the phase plane plot shown in Fig. 1.7. It is not obvious how to proceed with the description of such a system, where the phase space has $2NM$ variables. It is therefore now necessary to discuss a hierarchy of descriptions that begin with single particle dynamics.

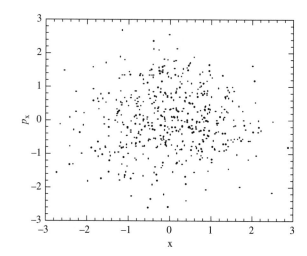

Fig. 1.7 Distribution of particles in (x, p_x) phase plane.

For many particle beams, the density of particles in phase space is sma enough that the particles are essentially a non-interacting ensemble, with bo macroscopic and microscopic electromagnetic fields created by the particle themselves contributing insignificantly to the motion. In this case, one on needs to solve the single particle equations of motion in the presence applied forces, and then proceed to produce a collective description of t beam ensembles' evolution based on the known single-particle orbits. Th book assumes the validity of this type of description, which straightfo wardly leads to analyses based the beam's **distribution**. In real charged partic beams, as well as in the modeling of such beams in multi-particle comp tations, this distribution is discrete, as illustrated in Fig. 1.7. On the oth hand, for analytical approaches, the distribution is viewed as a smooth pro ability function[7] in a $2M$-dimensional phase space $f(\vec{x}, \vec{p}, t)$, that is, th number of particles found in a differential phase space volume $\mathrm{d}V = \mathrm{d}^3\vec{x}\,\mathrm{d}^3$ in the neighborhood of a phase space location $\vec{x}, \vec{p}$ at a time t is simp given by $f(\vec{x}, \vec{p}, t)\,\mathrm{d}V$. While the computational approach to multi-partic dynamics is beyond the scope of this book, analytical approaches base on the **distribution function** $f(\vec{x}, \vec{p}, t)$ will be introduced in Chapter 5. O result concerning phase space distributions deserves prominent discussion this point, however, the conservation of phase space density, or **Liouville theorem**.

To begin, we write the total time derivative of the phase space distributic function,

$$\frac{\mathrm{d}f}{\mathrm{d}t} = \frac{\partial f}{\partial t} + \dot{\vec{x}} \cdot \vec{\nabla}_{\vec{x}} f + \dot{\vec{p}} \cdot \vec{\nabla}_{\vec{p}} f, \tag{1.6}$$

where the second and third terms on the right-hand side of Eq. (1.62) are th convective derivatives in phase space, derived simply by the chain rule. (Not the subscript on the gradient operators indicates differentiation with respect either coordinate or momentum components.) If the forces are derivable fro a Hamiltonian, then

$$
\begin{aligned}
\frac{\mathrm{d}f}{\mathrm{d}t} &= \frac{\partial f}{\partial t} + \sum_i \left(\frac{\mathrm{d}x_i}{\mathrm{d}t} \frac{\partial f}{\partial x_i} + \frac{\mathrm{d}p_i}{\mathrm{d}t} \frac{\partial f}{\partial p_i} \right) \\
&= \frac{\partial f}{\partial t} + \sum_i \left(\frac{\partial H}{\partial p_i} \frac{\partial f}{\partial x_i} - \frac{\partial H}{\partial x_i} \frac{\partial f}{\partial p_i} \right) \\
&= \frac{\partial f}{\partial t} + \sum_i \left(\frac{\partial H}{\partial p_i} \frac{\partial H}{\partial x_i} \frac{\mathrm{d}f}{\mathrm{d}H} - \frac{\partial H}{\partial x_i} \frac{\partial H}{\partial p_i} \frac{\mathrm{d}f}{\mathrm{d}H} \right) = \frac{\partial f}{\partial t}.
\end{aligned} \tag{1.66}
$$

Because we have assumed that the forces are derived from a Hamiltonian, it can be seen that the summation in Eq. (1.66) vanishes, and $\mathrm{d}f/\mathrm{d}t = \partial f/\partial t$. In Eq. (1.65), it should be understood that the total derivative is taken by moving about phase space according to the particle's equations of motion at a given phase space point $(\vec{x}, \vec{p})$. If neither creation nor destruction of the particles are allowed, we also have, by drawing a small constant differential volume $\mathrm{d}V$ around a given particle, $f(\vec{x}, \vec{p}, t) = \mathrm{d}V^{-1}$ in the neighborhood of this point. Then we have the result that $\partial f/\partial t = 0$ and

$$
\frac{\mathrm{d}f}{\mathrm{d}t} = 0. \tag{1.67}
$$

This result is termed Liouville's theorem, and it states that the phase space density encountered as one travels with a particle in a Hamiltonian system is conserved. The derivation of Eq. (1.67) may seem to be tautological, since we stated that $\partial f/\partial t = 0$ without referring to any property of Hamiltonian. Equation (1.67) is more illustrative than it seems, however, because the properties of the Hamilton equations guarantee that the density of **any** volume of phase space whose boundary follows these equations is constant. An alternative point of view, suggested by Eq. (1.65), is that phase space itself is incompressible. Perhaps the most common statement derived from interpretation of Eq. (1.67) is that the volume occupied by particles in phase space is conserved. For systems in which the motion of all phase planes is uncorrelated, the motion in the separate phase planes is independent. Then one can state that the **emittance** or area occupied by particles in a phase plane is conserved. The conservation of emittance, which is discussed further in Chapter 5, plays a very important role in the theory and design of particle accelerators.

Here, we concentrate on cases where Liouville's theorem holds, since we will almost exclusively consider motion due to applied forces that are derivable from a Hamiltonian. Deviations from this physical scenario form much of the following volume on advanced subjects in beam physics so we restrict comment on such topics to a few general statements. When multi-particle interaction effects become important, there are two distinct regimes to consider. The first can be described as one in which the collective macroscopic fields arising from the bulk beam charge and current density is the dominant self-interaction mechanism of the beam. These self-fields are most often termed **space-charge** when they arise from the near-field of the beam's charge distribution and **wake-fields** when they arise from the beam's collectively radiated fields. The evolution of the beam distribution can be dominated by space-charge fields, in which case one can use a description that is based on the notions of cold fluid motion developed in the field of plasma physics. In this scenario, simple collective (plasma) oscillations are encountered. When only macroscopic beam fields are important, these fields may, in principle, be included into the Hamiltonian; Eq. (1.67) is unchanged, and the implications of Liouville's theorem still hold.

On the other hand, microscopic binary collisions may strongly affect the beam distribution evolution, and diffusion (heating) may occur in phase space. There are a number of methods for dealing with this complication, but most are beyond the scope of the present discussion. However, one method is relatively easy to appreciate in the context of this section. An explicit non-Hamiltonian term may be introduced on the right-hand side of Eq. (1.67), which accounts for the time derivative of the distribution due to collisions (the effects of microscopic fields), so that $df/dt = \partial f/\partial t|_{\text{coll}}$. In this generalized form, the revised version of Eq. (1.67) is termed the **Boltzmann equation**. Thus, Eq. (1.67), which is commonly known (when the full derivative is written as in Eq. (1.65)) as the **Vlasov equation**, is also referred to as the **collisionless Boltzmann equation**. The non-macroscopic-Hamiltonian physics arising from emission of radiation by particle beams gives rise to dissipation and damping of phase space trajectories. The description of the effect of such phenomena on the distribution function may be treated with a full Boltzmann equation approach, or though an analysis based on the Fokker–Planck equation.

1.6   The design trajectory, paraxial rays, and change of independent variable

While a rigorous analysis of classical motion can be performed, as discussed above, using canonical momenta and coordinates within the confines of the Hamiltonian formalism, one often finds it useful in practice to use a more physically transparent description. This is obtained by use of **paraxial rays**, vector representations of the local trajectory which, by definition, have an angle with respect to a **design trajectory** that is much smaller than unity. Trajectories of interest in beam physics are always paraxial—one must confine the beam inside of small, near-axis regions, such as the drift tubes shown in Fig. 1.1.

Both the paraxial ray and the design trajectory are illustrated in Fig. 1.8. In this figure, an example of a design trajectory, defined as the ideally preferred trajectory—a locally straight or curved line— through the system, is displayed. As we shall see, one defines the coordinate system for analyzing beam physics problems locally by use of the design trajectory. As a result, the coordinate systems we encounter in this text may naturally be locally Cartesian, or locally curvilinear (i.e. inside of bend magnets). Curvilinear coordinate systems are obviously found in circular accelerators where the design trajectory is closed upon itself and is often referred to as the **design orbit**.

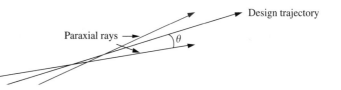

Fig. 1.8 Design trajectory and examples of small angle paraxial rays.

Figure 1.8 also displays examples of paraxial rays. The ray is a useful visualization tool and can be described mathematically by a coordinate offset, and an angle θ. In a locally Cartesian coordinate system,[8] we take the distance along the design trajectory to be z. In this system, a horizontal offset is designated by and the horizontal projection of the angle is θ_x. This angle is given in terms of the momenta as

$$\tan\theta_x = \frac{p_x}{p_z} = \frac{v_x}{v_z}. \tag{1.68}$$

Analogous definitions hold for the other transverse offset and angle, which we term vertical and indicated by the variable y. We are ultimately interested in a description of particle dynamics that uses the distance along the design trajectory as the independent variable. The reason for this is straightforward; in an optics system (for charged particles or photons) the forces encountered are always specified in space and not in time. Thus, a spatial description is more efficient and natural.

In order to use z as the independent variable, we must be able to write equations of motion in terms of z. This is accomplished by writing total time derivatives (written in compact notation as $(\cdot) \equiv d/dt$) in terms of total spatial derivative in z,

$$(\ )' \equiv \frac{d}{dz} = \frac{1}{v_z}\frac{d}{dt}. \tag{1.69}$$

The derivative of a horizontal offset with respect to z is given by

$$x' = \frac{dx}{dz} = \tan(\theta_x). \tag{1.70}$$

When analyzing beams, which are by definition are collections of rays localized in offset and angle near the design orbit, it can be seen from Eq. (1.70) that the transverse momentum (normal to the design orbit) is much smaller than the longitudinal momentum, $p_{x,y} \ll p_z \cong |\vec{p}|$. As a consequence, we may in most cases use the small angle, or paraxial, approximation, which allows us to write a series of useful approximate expressions,

$$x' = \tan(\theta_x) \cong \theta_x \cong \sin(\theta_x) \ll 1. \tag{1.71}$$

It is not immediately apparent how this assumption restricts the offset from the design orbit. As we shall see, all particle beam optics systems that focus and control the rays fundamentally resemble simple harmonic oscillators, where the transverse force is linear in offset, $F_x = -Kx$. Using this analogy, we can write a model equation for the transverse offset

$$\ddot{x} + \omega^2 x = 0, \tag{1.72}$$

where $\omega^2 = K/\gamma m_0$. Further, Eq. (1.72) can be cast in terms of a ray description, using Eq. (1.70), to obtain

$$x'' + k^2 x = 0. \tag{1.73}$$

Here, $k \equiv \omega/v_z$ is the characteristic oscillation wavenumber of the optics system. The solutions of Eq. (1.73) are of the form $x = x_m \cos(kz + \phi)$, yielding an approximate angle of $kx_m \sin(kz + \phi)$, and so the paraxial approximation is obeyed for offsets $x_m \ll k^{-1}$. Another restriction on the validity of a paraxial description is relevant in the case of a curvilinear design orbit. In this case, we

[8] When the design orbit is derived from rectilinear motion, we will choose z as the distance along the design orbit, in part to easily connect to field descriptions written in cylindrical coordinates. When the design orbit is bent, we will emphasize the curvilinear nature of the coordinate system by using the variable s as the distance along the design orbit.

must require that $x_m \ll R$ (where R is the local radius of curvature of the design trajectory) for the paraxial approximation to hold.

When one uses paraxial equations of the physically transparent form shown in Eqs (1.72) and (1.73), the analysis strays somewhat from a rigorous phase space description. In fact, one often uses this paraxial formalism instead of a canonically correct approach, even to the point of replacing a momentum (e.g. p_x) with the angle (x') in phase plane plots. In this case, when we make a plot of the trajectory in, for example, the (x, x') plane, which is an example of a **trace space**, we construct what is termed a **trace space plot**. This does not introduce complications in understanding the motion at constant values of p_z, because we can always recover the transverse momentum by using $p_x = p_z x' \cong \beta \gamma m_0 c x'$. If longitudinal acceleration occurs, however, the angle is diminished and an apparent damping (so-called **adiabatic damping**, see Section 2.6) of the motion is observed.

With a change of independent variable evidently of high interest to particle beam physics, we must revisit the question of Hamiltonian analysis in this context. In order to proceed with transformation of the Hamiltonian into one with a new independent variable, we first note that the symmetry of Eqs (1.62) and (1.64) (in which the energy and momenta are on equal footing) is suggestive of a notion—the choice of the energy function as the Hamiltonian is a bit arbitrary as well. We could just as naturally have chosen one of the momenta as the Hamiltonian from which we derive equations of motion, and the form of Eq. (1.62) would remain the same, except for a minus sign. The process of changing the Hamiltonian from an energy-based function to a momentum based function also requires that the independent variable be changed from t to the coordinate canonical to the new Hamiltonian. This can be illustrated by referring to the stationary property of the action integral (Eq. (1.11)) which gives rise to the equations of motion,

$$\delta \int_{t_1}^{t_2} L \, dt = \delta \int_{t_1}^{t_2} (\vec{p}_c \cdot \dot{\vec{x}} - H) \, dt = 0. \tag{1.74}$$

We can accomplish the change of independent variable, from t to z, by rewriting Eq. (1.74) as follows:

$$\delta \int_{z(t_1)}^{z(t_2)} (p_{c,x} x' + p_{c,y} y' - H t' + p_{c,z}) \, dz = 0, \tag{1.75}$$

where again the prime indicates differentiation with respect to the new independent variable, $(\,)' \equiv d/dz$. In Eq. (1.64), the role of the new Hamiltonian is obviously played by $-p_{c,z} \equiv G$,

$$G = -p_{c,z} = \sqrt{(H - q\phi_e)^2 - (p_{c,y} - qA_y)^2 c^2 - (p_{c,x} - qA_x)^2 c^2 - (m_0 c^2)^2}, \tag{1.76}$$

the role of the third coordinate is taken by $-t$, and the role of the third momentum by H. Note that this statement implies that H and $-t$ are canonically conjugate.

This type of independent variable transformation is a canonical transformation by design, and is used in applications, such as particle beam dynamics, where the applied forces are described more naturally by functions of a spatial coordinate than by functions of time. Because the transformation is canonical, the Hamilton's equations of motion are obtained in the usual fashion,

$$
p'_{c,x} = -\frac{\partial G}{\partial x} = \frac{\partial p_{c,z}}{\partial x}, \quad p'_{c,y} = -\frac{\partial G}{\partial y} = \frac{\partial p_{c,z}}{\partial y}, \quad H' = \frac{\partial G}{\partial t} = -\frac{\partial p_{c,z}}{\partial t},
$$

$$
x' = \frac{\partial G}{\partial p_{c,x}} = -\frac{\partial p_{c,z}}{\partial p_{c,x}}, \quad y' = \frac{\partial G}{\partial p_{c,y}} = -\frac{\partial p_{c,z}}{\partial p_{c,y}}, \quad t' = -\frac{\partial G}{\partial H} = \frac{\partial p_{c,z}}{\partial H}.
$$

$$
(1.77)
$$

For problems in which one of the spatial coordinates (z in particle beam physics) is taken to be the independent variable, it is assumed that the motion can be followed monotonically in this coordinate. Note that this is always the case for time as an independent variable (as far as we know!). If the trajectory describing the motion is not a monotonic function of the independent variable, then it is not uniquely described by this variable. Fortunately, in particle beams this assumption is also always correct.

1.7   Summary and suggested reading

In this chapter, we have motivated much of the contents of this book by introducing particle accelerators in their scientific and historical context. In order to build up the tools needed to analyze the dynamics of charged particle beams, we have reviewed methods in Lagrangian and Hamiltonian dynamics as well as special relativity, in a unified way. These general subjects gave way to concepts more specific to describing the motion charged particles in beams. We have examined the notion of phase space, and the conservation of its density in Hamiltonian systems—the Liouville theorem. We introduced the concept of the design trajectory, an ideal trajectory through an accelerator or transport system, which allows nearby trajectories (paraxial rays) to be defined. The design trajectory also gives one the freedom to analyze the motion using distance along such a trajectory as the independent variable, instead of time. This way of approaching description of trajectories is natural in beam optics and other problems in beam physics.

The subject of classical mechanics can be reviewed in a recommended number of texts:

1. K.R. Symon, *Mechanics* (Addison-Wesley, 1971). A classic and readable undergraduate text that introduces variational methods.
2. J.B. Marion and S.T. Thornton, *Classical Dynamics of Particles and Systems* (Harcourt Brace & Company, 1995). A more expansive treatment of mechanics for advanced undergraduate readers.
3. H. Goldstein, C. Poole, and J. Safko, *Classical Mechanics* (Addison-Wesley, 2002). One of the best graduate treatments of variational principles in mechanics.
4. L. Michelotti, *Intermediate Classical Mechanics with Applications to Beam Physics* (Wiley, 1995). A fairly complete introduction to modern methods in nonlinear dynamics, with an excellent array of problems and, as a bonus, has detailed examples in beam physics. Written at the graduate level.

Electromagnetic theory is central to this book, and provides the natural context for introducing special relativity. A selection of useful texts includes:

5. R. Wangsness, *Electromagnetic Fields* (Wiley, 1986). A complete, junior level introductory text.
6. R. Resnick, *Introduction to Special Relativity* (Wiley, 1968). A nice primer on special relativity.
7. M.A. Heald and J.B. Marion, *Classical Electromagnetic Radiation* (Harcourt Brace & Company, 1995). An advanced undergraduate course book.
8. J.D. Jackson, *Classical Electrodynamics* (Wiley 1975). The standard in graduate texts.
9. L.D. Landau and E.M. Lifschitz, *The Classical Theory of Fields* (Addison-Wesley, 1971). A dense, deep investigation for graduate students.
10. J. Schwinger, L.L. Deraad, and K. Milton, *Classical Electrodynamics* (Perseus, 1998). This text is a fine alternative to Jackson's, and a bit more modern.

This chapter has introduced some of the more basic notions of particle accelerators, including phase space descriptions and the design trajectory. Other recommended introductory texts, which will also serve as references to the following chapters, may include:

11. M.S. Livingston, *Particle Accelerators, Advances in Electronics I*, pp. 269–331 (Academic Press, 1948). This monograph gives an overview of the history of particle accelerators until 1948, and has an excellent reference list pointing to the primary articles in early accelerator scient. It anticipates the rise in importance of the proton synchrotron.
12. D. Edwards and M. Syphers, *An Introduction to the Physics of High Energy Accelerators* (Wiley, 1993). A fine primer on high-energy devices, and good reference book, especially on circular accelerators. Written at senior undergraduate level.
13. H. Wiedemann, *Particle Accelerator Physics I: Basic Principles and Linear Beam Dynamics* (Springer-Verlag, 1993). Very comprehensive and rigorous book, concentrating on circular accelerators. Written at the graduate level.
14. S.Y. Lee, *Accelerator Physics* (World Scientific, 1996). A comprehensive and sweeping introduction, especially in the areas of circular accelerators. Written at the graduate level.
15. M. Reiser, *Theory and Design of Charged Particle Beams* (Wiley, 1994). An in-depth treatment of beam optics and transverse collective effects, with emphasis on rigorous analytical methods. Written at the graduate level.
16. S. Humphries, Jr., *Principles of Charged Particle Acceleration* (Wiley, 1986). This is a strongly pedagogical text with good physics underpinning.
17. M. Sands, *The Physics of Electron Storage Rings: an Introduction* (Stanford, 1970). A ground-breaking book, with clear explanations of basic storage ring physics and radiation damping effects. Written at the graduate level.

There are a number of books one may examine to get a flavor for collective effects in charged particle beams, subjects that have been deemed to lie outside of this text's scope:

18. J.P. Lawson, *The Physics of Charged Particle Beams* (Clarendon Press, 1977). Very physics oriented, with excellent melding of plasma physics notions into the text. Written at the graduate level.
19. S. Humphries, Jr., *Charged Particle Beams* (Wiley, 1990). A unified presentation of the physics of high power and high brightness beams. Written at the graduate level, it is a companion to Ref. 16, by the same author.
20. R. B. Miller, *Intense Charged Particle Beams* (Plenum, 1982). A good first look at very high current beams, with collective forces emphasized. Written at the graduate level.
21. A. Chao, *Physics of Collective Beam Instabilities in High Energy Accelerators* (Wiley, 1993). A classic treatment of instabilities, very broad in scope and rigorous in approach. Written at the graduate level.

The are some general references that will aid a student or practitioner in the field:

22. A. Chao and M. Tigner, editors, *Handbook of Accelerator Physics and Engineering* (World Scientific, 1999). Not a text, but a vade mecum for the accelerator field, with summaries of basic principles, and useful formulae covering almost every conceivable aspect of accelerators. For professional use.
23. *The Particle Data Book*, an annually-updated updated reference for high-energy physics, available from Lawrence Berkeley National Laboratory at http://pdg.lbl.gov/.

Exercises

(1.1) Consider the three-dimensional simple harmonic oscillator where $T = \frac{1}{2}m\dot{\vec{x}}^2$ and $V = \frac{1}{2}K\vec{x}^2$.

    (a) Construct the Lagrangian of this system in Cartesian coordinates, $\vec{x} = (x, y, z)$.

    (b) Derive the canonical momenta and construct the Hamiltonian for this system. Show that the total energy of the system is conserved, along with the energy associated with motion in each phase plane.

    (c) Construct the Lagrangian of this system in cylindrical coordinates (r, z, ϕ), noting that $\dot{\vec{x}} = \dot{\rho}\hat{\rho} + \dot{z}\hat{z} + \rho\dot{\phi}\hat{\phi}$ and that $\dot{\vec{x}}^2 = \rho^2$.

    (d) Construct the Hamilton for this system in cylindrical coordinates. From inspection of the Hamiltonian, deduce any constants of the motion associated with the angular momentum.

    (e) Construct the Lagrangian of this system in spherical polar coordinates (r, θ, ϕ), noting that $\dot{\vec{x}} = \dot{r}\hat{r} + r\dot{\theta}\hat{\theta} + r\sin\theta\dot{\phi}\hat{\phi}$ and that $\vec{x}^2 = r^2$.

    (f) Construct the Hamilton for this system in spherical polar coordinates. From inspection of the Hamiltonian, deduce any constants of the motion associated with angular momenta.

(1.2) In order for the motion to be stable in the simple harmonic oscillator case illustrated in Fig. 1.4, the force must be restoring or $K > 0$. Assuming an unstable system, however, we have $K < 0$. Plot the curves corresponding to $H = -\frac{1}{8}, -\frac{1}{2}$, and -2 in (x, p_x) phase space. Note that the curves are not closed, indicating unbounded motion.

(1.3) While the forces most commonly associated with charged particle motion in accelerators are of the form of a simple harmonic oscillator, meaning they are linearly proportional to distance from an equilibrium point (i.e. $F_x = -Kx$), other types of forces may be present. For instance, in a sextupole magnet, the force is of the form, $F_x = -ax^2$, where a is a constant.

(a) What is the Hamiltonian associated with one-dimensional motion under this applied field?

(b) Draw some representative constant H curves in (x, p_x) phase space for this Hamiltonian. Comment on whether the motion is bounded or unbounded (see Example 1.2).

(c) In an octupole-like field, the force is of the form $F_x = -ax^3$. Construct the Hamiltonian and plot some constant H curves. Consider the effect of changing the sign of the constant a and discuss whether the motion is bounded or unbounded.

(1.4) Consider an undamped oscillator consisting of a weight hanging from a spring. This spring is set in motion and has a certain action J and frequency ω. If a window to the cool outside air is opened in the room containing this oscillator, the spring becomes colder and more rigid, causing ω to slowly rise. As J is an adiabatic invariant, the total energy in the oscillator grows as the thermal energy is removed from the spring! Explain. (Hint: consider the microscopic internal degrees of freedom of the spring.)

(1.5) Show, by Taylor expansion of the particle mechanical energy, $U = \gamma m_0 c^2$ in terms of the velocity when $v \ll c$, that this total energy is approximately the sum of the particle rest energy and the non-relativistic expression for the kinetic energy.

(1.6) The Thomson backscattering analysis of photons by electrons given above is only approximate and applies when the incident photon energy is very small compared to the rest energy of the electron, $m_0 c^2 = 0.511 \,\text{MeV}$. This analysis, referred as **Compton scattering**, should be familiar to the reader. One can thus recognize that Thomson scattering, where the scattered photon has the same frequency as the incident photon in the electron rest frame, is only an approximate description—a limit of the general Compton scattering phenomenon for low-energy photons.

(a) The usual Compton scattering analysis is performed in the electron rest frame. From the Lorentz transformation above, find an expression of electron energy $\gamma m_e c^2$ and incident (counter-propagating) photon energy $\hbar\omega$ such that the photon energy in the electron rest frame is less than 10 per cent of the electron rest energy, $\hbar\omega' < m_e c^2/10$.

(b) By first performing (in the electron rest frame) the usual Compton scattering analysis of scattered wavelength λ' as a function of angle θ', and then performing a Lorentz transformation on the results back to the lab frame, find the scattered wavelength as a function of laboratory angle.

(c) Now consider the photon to be incident in the lab frame at some arbitrary angle θ_i, where $\theta_i = 0$ is defined as the counter-propagating case. Find the angle and energy of the incident photon in the electron rest frame by Lorentz transformation. Find the energy of the backscattered light (copropagating with the electron velocity) as a function of θ_i, using a full Compton analysis.

(d) With the Thomson assumption that (in the electron rest frame) the frequency does not change during the scattering event, perform the same analysis as in part (c). For an energy in the electron rest frame of $\hbar\omega' = m_e c^2/10$, compare the backscattered energies given by the Thomson and Compton analyses.

(1.7) Using the Lorentz transformation of the $\vec{P}$ four-vector, generalize the addition of velocities expression given by Eq. (1.51) to account for any arbitrary angle (set to zero in Eq. (1.51)) between the particle and frame velocities.

(1.8) Consider a stationary uniform cylinder of charge that has charge density ρ_e up to a radius a, and then vanishes outside of this radius.

(a) From Gauss' law, find the radial electric field associated with this charge distribution.

(b) Now assume that this distribution is in motion along its symmetry axis with speed v and with respect to the lab frame. From Lorentz transformation-derived rules for determining the fields (Eq. (1.58)), find the electric and magnetic fields in the lab frame associated with the moving charge distribution.

(c) From Lorentz transformation of the charge–current four-vector, find the density ρ_e' and current density $J_{e,z}'$ associated with the moving charge distribution.

(d) Using Gauss' and Ampere's law directly, calculate the radial electric field, and the azimuthal magnetic field associated with the moving charge distribution. Compare with your answer in part (b).

(e) What is the net radial force on a particle inside of the beam (that has velocity v)?

(1.9) For a beam particle acted upon by the forces given in Exercise 1.8:

(a) Determine the scalar and vector potentials. (Hint: find the scalar potential in the rest frame of the beam first.)

(b) Construct the Lagrangian for this system.

(c) Derive the Hamiltonian for the system in Exercise 1.8(c).

(d) Derive the radial equations of motion for this particle.

(e) If you change the sign of the particle but not its velocity, the radial motion of the particle is approximately simple harmonic for motion for $\rho < a$. Under what conditions on ρ_e, v, and a is this approximation valid? Assuming these conditions, plot the phase space using constant H curves both inside and outside of the beam.

(1.10) Non-equilibrium solutions to the Vlasov equation are generally quite difficult, but it is possible to make a very powerful statement about solutions for cases in equilibrium, where $\partial f / \partial t = 0$. As an example of the analysis of a Vlasov equilibrium, show that for a time-independent Hamiltonian, $H(\vec{x}, \vec{p})$,

$$f(\vec{x}, \vec{p}) = g[H(\vec{x}, \vec{p})],$$

is an equilibrium solution to the Vlasov equation, where g is any differentiable function of the Hamiltonian. This result will be important to the discussion of equilibrium distributions in Chapter 5 (cf. Ex. 5.2).

(1.11) A practical check on the paraxial approximation is to see that the error made in assuming that the longitudinal velocity is approximately the total velocity, $v_z \cong v$, is small.

(a) Assuming $p_{x,y} \ll p_z$, write v as a function of p_x, p_y, and p_z. Taylor expand this expression to second order in p_x/p_z and p_x/p_z (x' and y').

(b) For what ratio of $\sqrt{p_x^2 + p_y^2}/p_z$ is the error in the expression $v_z \cong v$ kept to 0.1 per cent?

(1.12) For the Hamiltonian derived in Exercise 1.9 for the unlike-sign particle case:

(a) Perform a canonical transformation to use z as the independent variable.

(b) Derive the equations of motion for the radial and angular coordinates and momenta with z as the independent variable.

(c) Assuming the angular momentum of the particle is zero, construct (r, r') trace space plots for motion both inside and outside of the beam.

(d) What is the equation of motion for H? What about t?

2 Charged particle motion in static fields

Many of the physical scenarios that unfold in the sometimes complex-looking world of particle beams and accelerators can be understood by first referring to a simplified situation, from which the essential physics can be extracted. The purpose of this chapter is to survey the most illustrative of these situations, where only static fields play a role. We thus examine the motion of charged particles in: uniform (dipole) magnetic and electric fields, both separately and in combination; magnetic dipole fields that have periodic longitudinal variation of the dipole orientation (undulators); and magnetic and electric quadrupole fields. These analyses will serve to both introduce physical concepts, and to develop mathematical techniques which are needed for the discussions in following chapters.

2.1 Charged particle motion in a uniform magnetic field

In a uniform (dipole) magnetostatic field, with field direction chosen along the z-axis, $\vec{B} = B_0 \hat{z}$, the Lorentz force equation governing the time evolution of the momentum can be written in two components, one parallel and one perpendicular to the field,

$$\frac{dp_z}{dt} = 0, \qquad \frac{d\vec{p}_\perp}{dt} = q(\vec{v}_\perp \times \vec{B}) = \frac{qB_0}{\gamma m_0}(\vec{p}_\perp \times \hat{z}). \qquad (2.1)$$

One immediately sees from Eq. (2.1) that p_z is a constant of the motion and, further, since the magnetic force does no work on the particle, the total momentum $p = \sqrt{p_z^2 + p_\perp^2}$ (and therefore the total energy $\gamma m_0 c^2$) is also constant. Thus the transverse momentum must also be constant in amplitude, but changing in direction. It is well known that this implies circular motion in the plane perpendicular to the magnetic field. In general, therefore, the motion is the superposition of a circular orbit and a uniform drift normal to the circle—in other words, it is helical. Although the circular orbit traced by the particle in the plane perpendicular to the field is familiar from non-relativistic mechanics, we now examine this motion in more detail in the context of relativistic particles.

One approach is to write the differential equations governing the transverse velocities,

$$\frac{dv_x}{dt} = \frac{qB_0}{\gamma m_0} v_y, \qquad \frac{dv_y}{dt} = -\frac{qB_0}{\gamma m_0} v_x. \qquad (2.2)$$

Taking the time derivative of Eq. (2.2) and combining the results with the original expressions yields simple harmonic oscillator equations,

$$\frac{d^2v_x}{dt^2} + \omega_c^2 v_x = 0, \qquad \frac{d^2v_y}{dt^2} + \omega_c^2 v_y = 0, \tag{2.3}$$

where we have substituted in the relativistic cyclotron frequency,

$$\omega_c \equiv \frac{qE_0}{\gamma m_0}. \tag{2.4}$$

The relativistically correct transverse velocities are, from Eq. (2.3), obviously harmonic functions of time, having angular frequency ω_c. This definition of the cyclotron frequency differs from its non-relativistic analogue by the presence of the factor of γ in the denominator. This factor of γ is due to the relativistic change to the inertial mass (cf. Eq. 1.50)—the particle appears to be heavier, and the "fictitious force" associated with centripetal acceleration becomes larger, $mv^2/R \rightarrow \gamma mv^2/R$. This point is explored further in the exercises.

Equations (2.2) and (2.3) taken together indicate harmonic oscillations of equal amplitude in x and y, which are 90° apart, having the general solution

$$v_x = -v_m \sin(\omega_c t + \phi), \qquad v_y = v_m \cos(\omega_c t + \phi). \tag{2.5}$$

Equation (2.5) can be integrated to find the transverse motion in its most general form:

$$x = R\cos(\omega_c t + \phi) + x_0, \qquad y = R\sin(\omega_c t + \phi) + y_0. \tag{2.6}$$

One can easily deduce from Eq. (2.6) that the particle orbit is a circle of radius $R = v_m/\omega_c$ centered at (x_0, y_0).

We have not yet established the dependence of R on other physical parameters of the system. To do this, we note that balancing of radial force and centripetal acceleration implies

$$\frac{\gamma mv_\perp^2}{R} = qv_\perp B_0, \quad \text{or} \quad p_\perp = qB_0 R. \tag{2.7}$$

This relationship between the transverse momentum component and the magnetic field is often written in a form which is useful for easy calculations,

$$p_\perp(\text{MeV}/c) = 299.8 \cdot B_0(\text{T})R(\text{m}). \tag{2.8}$$

Equation (2.8) recasts the second of Eq. (2.7) in the "engineering" units of high-energy accelerators.

In this discussion we have described the relativistically correct helical motion of a charged particle in a uniform magnetic field, which is the spring-board for examining two situations of present interest: the circular accelerator, and the focusing solenoid.

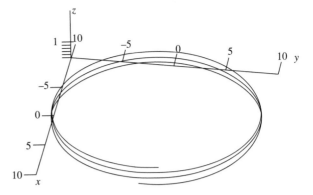

Fig. 2.1 Helical orbit with small pitch angle (0.1) in uniform magnetic field.

2.2   Circular accelerator

The results of the previous section can be applied to discussion of the circular accelerator if the pitch angle ($\theta_p = \tan^{-1}(p_z/p_\perp)$) associated with the motion is small, as is shown in Fig. 2.1. In fact, for simplicity of analysis, we must begin the discussion by assuming that the pitch angle vanishes, thus allowing us to initially ignore the out of bend plane motion. To see why this is so, we reintroduce in this context the notion of the circular design orbit, the trajectory in the accelerator that the ideal particle—the one tracing the exact trajectory that the designer desires—follows. Because this orbit must be closed in the circular accelerator, we are clearly restricted to in-plane motion, with $p_z = 0$. In the example of Fig. 2.1, we may think of the design orbit as being in the $z = 0$ plane (the bend plane), with radius $R = 5$ centered on the point $x = 0$, $y = 0$.

The motion of charged particle trajectories near the design orbit may be stable (tending to remain near the design orbit) or unstable (tending to diverge from the design orbit). In some circular accelerators, it is necessary that the charged particles be stored in the accelerator near the design orbit for more than 10^{10} revolutions, or turns, so it is obviously of paramount importance that motion near the design orbit be stable.

It is obvious from Fig. 2.1 that the vertical motion (along the z-axis, out of the bend plane) is not at all stable—the motion in this system is unbounded in the z-dimension. We shall return to this point in Section 3.1 when we discuss motion in a nonuniform magnetic field, and resolve the problem of unbounded motion in the dimension normal to the plane containing the design orbit.

Concentrating now on the motion in the bend plane, there are two possible ways in which this motion can be perturbed from the design orbit—first, by an error in center of curvature of the orbit, and second, by an error in the value of the radius of curvature, which is, of course, equivalent to a deviation in particle momentum from that of the ideal (design) particle. The first type of perturbation is illustrated in Fig. 2.2, which shows two orbits, with the second (perturbed) orbit having the same radius, but with center offset from the design orbit center by a small amount in y. It can be clearly seen that this perturbed orbit displays stability, as it passes through the design orbit twice per turn around the circle. Two different classes of perturbations, both leading to an error in the center of curvature, are displayed in Fig. 2.2: pure angle errors at $x = 0$, 10 and $y = 5$; pure offset errors at $x = 5$ and $y = 0$, 10. The errors at all other points

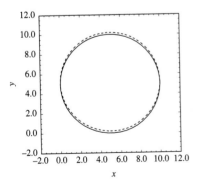

Fig. 2.2 Design (solid line) and transversely perturbed (dashed line) orbits in uniform magnetic field.

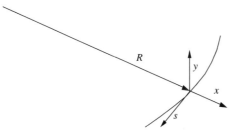

Fig. 2.3 Locally defined right-handed coordinate system used with bending design. Here s is the distance along the design orbit, x is the distance from this orbit along the radius of curvature, and y is the distance from the design orbit out of the bend plane.

along the trajectory are a superposition of angle and offset perturbations. From these observations, we may strongly suspect that the perturbed trajectory is a harmonic oscillation, but we must first develop the proper analysis tools to verify this suspicion.

As stated in Chapter 1, as we will typically take the distance along the design orbit to be the independent variable (in this case indicated by s), we implicitly wish to analyze the charged particle dynamics near the design orbit. In the present case, this orbit is specified by a certain radius of curvature R (and thus a certain momentum $p_0 = qB_0R$), and center of curvature, (x_0, y_0). With this choice of analysis geometry, we can locally define a new right-handed coordinate system (x, y, s), as shown in Fig. 2.3. In this coordinate system, x is the distance of the orbit under consideration from the design orbit, in the direction measured along the radius and normal to s. The distance y (formerly indicated by the coordinate z in Section 2.1) is measured from the design orbit to the particle orbit under consideration, in the direction out of the bend plane.[1] The choice of a right-handed system in this case is a function of the direction of the bend, and in simple circular accelerators, one is free to construct the curvilinear coordinates once and for all. On the other hand, when we encounter bends in the opposing direction, as in chicane systems (see Chapter 3), we will choose to consistently define the coordinate x, so that it is positive along the direction away from the origin of the bend. As we will also choose to leave the vertical direction unchanged in this transformation, a left-handed coordinate system will result when the bend direction is changed.

The coordinate system shown in Fig. 2.3 is quite similar to a cylindrical coordinate system, with x related to the radial variable ρ by the definition $x \equiv \rho - R$, s replacing the azimuthal angle ϕ ($ds = R\,d\phi$), and y, as previously noted, replacing z. Thus we can write the equations of motion for orbits in this system by using the Lagrange–Euler formulation (see Problem 2.1), as

$$\frac{dp_\rho}{dt} = \frac{\gamma m_0 v_\phi^2}{\rho} - qv_\phi B_0, \tag{2.9}$$

where $v_\phi = \rho\dot\phi$ is the azimuthal velocity.

Equation (2.9) can be cast as a familiar differential equation by using x as a small variable ($x \ll R$, which is also equivalent, as will be seen below, to the paraxial ray approximation) to linearize the relation. This is accomplished through use of a lowest order Taylor series expansion of the motion about the design orbit equilibrium ($p_x = p_\rho = 0$) at $\rho = R$,

$$\frac{dp_x}{dt} \cong -\frac{\gamma_0 m_0 v_0^2}{R^2}x. \tag{2.10}$$

[1]This convention, in which the symbol y is defined to be the distance out of the bend plane is typical of the American literature. European beam physicists more often use the symbol z instead, but we do not follow this convention even though it connects more naturally to our previous discussion. This is because our adopted convention makes subsequent derivations somewhat easier to understand, and also because it allows the connection between linear accelerator and circular accelerator coordinate systems to become more obvious.

Using Eq. (2.7), we write the design radius as $R = \gamma_0 m_0 v_0 / q B_0$, to obtain

$$\frac{d^2 x}{dt^2} + \omega_c^2 x = 0, \tag{2.11}$$

where $p_0 = \gamma_0 m_0 v_0$ is the design momentum. Equation (2.11) indicates simple harmonic motion with the same frequency as that of the cyclotron motion describing the design orbit.

Equation (2.11) is written in more standard form by using s as the independent variable, with shorthand designation of differentiation $(\ )' \equiv d/ds = (1/v_0)\, d/dt$, as

$$x'' + \left(\frac{1}{R}\right)^2 x = 0. \tag{2.12}$$

The simple harmonic oscillations about the design orbit described by Eq. (2.12), associated with the perturbed orbits of particles having the same momentum as the design particle, are termed **betatron oscillations**, because they were first described in the context of the betatron.

These oscillations may seem to be a bit of a mystery at this point, even though the mathematical derivation leading to Eqs (2.11) and (2.12) is straightforward. The question may be asked: if the force and effective mass of the charged particles on and off of the design orbit are the same, how does the "restoring force" arise? This question can be answered most easily by looking at the description of the motion using s as the independent variable, and using the picture shown in Fig. 2.4.

The total momentum transfer of the particle on an arbitrary offset orbit, as shown in Fig. 2.4, is calculated as follows:

$$\Delta p_x = -q \int_{t_1}^{t_2} v_0 B_0 \, dt = -q \int_{s_1}^{s_2} B_0 \left(1 + \frac{x}{R}\right) ds. \tag{2.13}$$

The momentum transfer for offset orbits is different in this case because the integration path length is different—integration of the force equation with s as the independent variable is equivalent to using the angle along the design orbit, as parameterized by $ds = R\, d\theta$. Integration of the force over the offset path length for a given angular increment covers a larger differential length $ds_x = (R + x)\, d\theta$. For this reason, the focusing effect described by Eqs (2.11) and (2.12) is sometimes termed **path length focusing**. This type of focusing

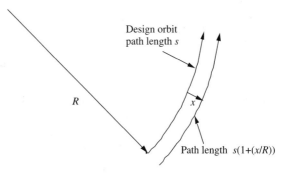

Fig. 2.4 Path length difference between design orbit and offset betatron orbit.

forms the basis of so-called **weak focusing** systems, which are discussed in the next chapter.

As a final comment in this cursory introduction to circular accelerators, we remind the reader that the betatron motion (which we indicate by the subscripted variable x_β) is that due only to trajectory errors for particles that have the design momentum. The displacement of an arbitrary particle from the design orbit has another important component, that due to deviations from the design momentum $\delta p \equiv p - p_0$. An analysis which treats the particle dynamics only in a first-order Taylor series in both betatron (angle and offset) and momentum errors (which requires both $x_\beta \ll R$ and $\delta p/p_0 \ll 1$) is by assumption a description which is additive in these quantities, that is,

$$x = x_\beta + \eta_x \frac{\delta p}{p_0}. \tag{2.14}$$

The coefficient η_x is termed the **momentum dispersion** (in this case in the x-direction, or **horizontal momentum dispersion**), and is, as we shall see, generally a detailed function of the magnetic field profile, with variation in s. In the case of the uniform magnetic field that we have been studying, however, it is a constant, as can be seen by the schematic shown in Fig. 2.5. Because the radius of curvature of each momentum component is linear in p, the momentum dispersion function in this case is constant,

$$\eta_x = \frac{\partial x}{\partial \left[\frac{\delta p}{p_0} \right]} = p_0 \frac{\partial R}{\partial p} = R(p_0). \tag{2.15}$$

The momentum dispersion about the design orbit is simply the radius of curvature of the design orbit (at the design momentum) in the case of a uniform magnetic field.

Even though this section is devoted to discussion of charged particle motion under the influence of static magnetic fields, because we have discussed circular design orbits we refer the reader to an exercise at this point which illustrates acceleration in a time-varying magnetic field in the betatron, Exercise 2.2.

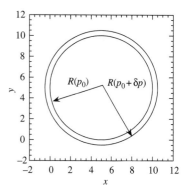

Fig. 2.5 Momentum dispersion in a uniform magnetic field.

2.3 Focusing in solenoids

The motion of a charged particle in a focusing solenoid magnet is conceptually the complement of that in the circular accelerator. This can be seen by noting that in the solenoid, the design orbit is that which travels straight down the longitudinal axis of the device (the z-direction), the direction parallel to the uniform magnetic field. Thus only off-axis orbits with non-zero angular momentum—defined about the z-axis—are deflected by this solenoid field. The assumption of paraxial orbits means that $p_z \gg p_\perp$, and the pitch angle of the helical orbit is very large, as illustrated by Fig. 2.6. The key to understanding the motion of a charged particle in a focusing solenoid is to recognize how the angular momentum, which drives this helical motion, arises. To do this we must violate the assumptions of the previous two sections slightly, and ask what happens when the charged particle moves from a region where the magnetic field vanishes to one where it is uniform.

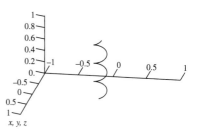

Fig. 2.6 Helical orbit typical of a solenoid, where the pitch angle is large.

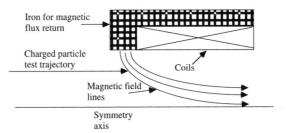

Iron for magnetic
flux return

Charged particle
test trajectory

Coils

Magnetic field
lines

Symmetry
axis

Fig. 2.7 Fringe field region of solenoid, with
initially offset particle trajectory encountering
radial component of magnetic field.

Fig. 2.7 Fringe field region of solenoid, with
initially offset particle trajectory encountering
radial component of magnetic field.

In this transitional region, the magnetic field must "fringe" to satisfy the divergence-free criterion, $\vec{\nabla} \cdot \vec{B} = 0$. Assuming a cylindrically symmetric geometry, this criterion becomes

$$\frac{1}{\rho} \frac{\partial}{\partial \rho} \rho B_\rho = -\frac{\partial B_z}{\partial z}. \tag{2.16}$$

Thus, as the charged particle enters the region where the solenoid field B_z rises and $\partial B_z/\partial z$ is non-vanishing, a radial component of the magnetic field is encountered. This is illustrated in Fig. 2.7, in which the fringe-field region of a solenoid is schematically shown.

In order to integrate Eq. (2.16), we assume the lowest order approximation on the form of the longitudinal field, that it is independent of radius. Because of symmetry, this approximation is good to second order in ρ, that is $B_z(\rho, z) \cong B(0, z)[1 + \alpha\rho^2 + \cdots]$, where α is a constant. Assuming in the region of interest that $|\alpha\rho^2| \ll 1$, we may write

$$B_\rho \cong -\frac{\rho}{2} \frac{\partial B_z}{\partial z}\bigg|_{\rho=0}. \tag{2.17}$$

Further, using the paraxial approximation, we may obtain an excellent estimate of the total angular momentum imparted to a charged particle as it passes through the fringe field region of a solenoid by evaluating the force integral assuming a constant radial offset ρ_0,

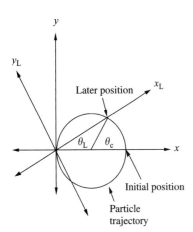

Fig. 2.8 Transverse trajectory of charged particle with no initial transverse momentum (before magnet) in a solenoid. The rotating Larmor frame is defined by the x_L-axis passing through the charged particle that begins its motion on the x-axis.

$$\Delta p_\phi \cong q \int_{t_1}^{t_2} v_z B_\rho \, dt = q \int_{z_1}^{z_2} B_\rho \, dz = -q\frac{\rho_0}{2} \int_{z_1}^{z_2} \frac{\partial B_z}{\partial z}\bigg|_{\rho=0} dz$$

$$= -q\frac{\rho_0}{2} \int_{z_1}^{z_2} \frac{dB_z}{dz}\bigg|_{\rho=0} dz = -q\frac{\rho_0}{2}[B_z(z_2) - B_z(z_1)] = -q\frac{\rho_0}{2}B_0. \tag{2.18}$$

Here we have explicitly assumed that the particle velocity does not change significantly in direction or magnitude while passing through the magnetostatic fringe field region enclosed in the interval (z_1, z_2). Equation (2.18) is sometimes known as **Busch's theorem**.

It is important to note that the total transverse momentum[2] "kick" imparted to a particle entering the solenoid, $\Delta p_\phi \cong q\rho_0 B_0/2$, gives rise to subsequent transverse motion with radius of curvature $R = \Delta p_\phi/qB_0 \cong \rho_0/2$. Therefore a charged particle with no initial transverse motion displays helical motion inside of the solenoid, with radius of curvature such that the particle orbit passes through the axis ($\rho = 0$). This somewhat surprising result is illustrated in Fig. 2.8, where both the particle trajectory in this case, and the proper coordinate system for further analysis of the problem, are shown.

[2] Note that here we are examining the transverse momentum "kick" in the angular direction. This should not be confused with the **angular momentum**, which has a specific meaning in the context of Hamiltonian dynamics—it is the momentum which is canonical with the azimuthal coordinate ϕ (see Ex. 2.4), and of course has units of momentum times length.

If the particle begins its trajectory offset in x ($x = x_0$), but not in y, and with no transverse momentum before the magnetic field region, the angle that this particle's trajectory makes with respect to the x-axis is the **Larmor angle** θ_L. As can be seen from Fig. 2.8, the Larmor angle is simply related to the cyclotron angle of the motion (centered on $x = x_0/2$) by $2\theta_L - \theta_c = 0$, or $\theta_L = \theta_c/2$. Since the cyclotron angle grows linearly in time, so does the Larmor angle, and thus we can define the **Larmor frequency**,

$$\omega_L \equiv \frac{d\theta_L}{dt} = \frac{\omega_c}{2} = \frac{qB_0}{2\gamma m_0}. \tag{2.19}$$

The **Larmor frame** can now be introduced—it is simply the frame that rotates about the z-axis with the Larmor frequency, as illustrated in Fig. 2.8.

In the Larmor frame, the motion in x_L is simple harmonic with the Larmor frequency,

$$x_L = x_0 \cos(\theta_L) = x_0 \cos(\omega_L t). \tag{2.20}$$

The central advantage of introducing the Larmor frame is that the motion is simple harmonic for both of the transverse dimensions in this frame. To see this, we note that the coordinates in the Larmor frame can be written in compact form as

$$\begin{pmatrix} x_L \\ y_L \end{pmatrix} = \begin{pmatrix} \cos(\omega_L t) & \sin(\omega_L t) \\ -\sin(\omega_L t) & \cos(\omega_L t) \end{pmatrix} \begin{pmatrix} x \\ y \end{pmatrix}. \tag{2.21}$$

Thus, we have, using Eqs (2.2) and (2.21),

$$\ddot{x}_L + \omega_L^2 x_L = 0 \quad \text{and} \quad \ddot{y}_L + \omega_L^2 y_L = 0. \tag{2.22}$$

These equations describe simple harmonic oscillations in both x_L and y_L as observed in the Larmor frame. Thus, the motion in a solenoid can be described as a rotation about the origin that proceeds at the Larmor frequency, with simple harmonic (betatron) oscillations occurring in the rotating frame, also proceeding with angular frequency ω_L.

The last point to be made in discussing solenoid focusing concerns casting of the problem in standard accelerator physics form, with the distance z down the axis of the solenoid used as the independent variable. In the paraxial approximation, we take $v_z \cong v$, and thus

$$x_L'' + k_L^2 x_L = 0, \tag{2.23}$$

$$y_L'' + k_L^2 y_L = 0, \quad \text{with} \quad k_L = \frac{\omega_L}{v_z} = \frac{qB_0}{2p_z} \cong \frac{qB_0}{2p}. \tag{2.24}$$

The inverse of the wave number associated with the betatron harmonic motion in the Larmor frame is approximately twice the radius of curvature $R = p/qB_0$ of a charged particle of the same momentum traveling normal to the solenoid field.

2.4 Motion in a uniform electric field

The substitution of a uniform electric field $\vec{E} = E_0\hat{z}$ for the uniform magnetic field changes our point of view quite dramatically, as the energy of the particle is no longer constant, but increases due to longitudinal acceleration. The equations of motion in this case, with $\vec{B} = 0$, read

$$\frac{dp_z}{dt} = qE_0, \qquad \frac{d\vec{p}_\perp}{dt} = 0. \tag{2.25}$$

Since the momentum transverse to the electric field is conserved we may, for the moment, ignore its effects and concentrate on the one-dimensional problem of the motion in z.

The most straightforward way to accomplish this is to note that the electro-static field $\vec{E} = E_0\hat{z}$ may be derived from a potential, and thus the potential energy is

$$q\phi_e = -qE_0 z. \tag{2.26}$$

The Hamiltonian, which in this case is the total energy (see Section 1.4), is

$$H = \gamma m_0 c^2 + q\phi_e = \gamma m_0 c^2 - qE_0 z. \tag{2.27}$$

Because the Hamiltonian is independent of t, it is a constant of the motion, and it can be evaluated at given initial conditions, $H|_{z=0}$. The particle's mechanical (rest plus kinetic) energy is therefore given by

$$\gamma m_0 c^2 = H|_{z=0} + qE_0 z, \tag{2.28}$$

or in normalized form,

$$\gamma(z) = \frac{H|_{z=0}}{m_0 c^2} + \gamma' z, \tag{2.29}$$

where $\gamma' \equiv qE_0/m_0 c^2$, and the initial conditions-derived constant $H|_{z=0} = \gamma|_{z=0} m_0 c^2$. Note that the energy increases linearly in z, as $dU = \vec{v} \cdot d\vec{p}$, or for one-dimensional motion

$$\frac{dU}{dz} = \frac{dp}{dt} = qE_0. \tag{2.30}$$

This simple relation helps clarify some jargon encountered in the particle beam physics field, that the amplitude of the accelerating force (change in momentum per unit time) is referred to as an **acceleration gradient**, or spatial energy gradient, commonly quoted in units of MeV/m. The normalized acceleration gradient γ' defines a **scale length** for acceleration $L_{acc} = \gamma'^{-1}$, over which the particle gains one unit of rest energy.

In this simple scenario, other relevant dynamical quantities can be derived from knowledge of $\gamma(z)$,

$$p(z) = \beta\gamma m_0 c = \sqrt{\gamma^2(z) - 1}\, m_0 c, \tag{2.31}$$

and

$$v(z) = \beta c = \frac{p(z)c^2}{U(z)} = c\sqrt{1 - \frac{1}{\gamma^2(z)}}. \tag{2.32}$$

The deduction of the velocity from γ is of prime importance because it allows the transformation of independent variable from t, which enters naturally into the equations of motion, to z.

In the analysis given in this section, we have effectively adopted z as the independent variable. It is also interesting to explore acceleration from the point of view of explicit time dependence. In this case, we can write the first of Eq. (2.25) as

$$\frac{dp_z}{dt} = m_0 c \frac{d(\beta_z \gamma)}{dt} = m_0 c [\gamma + \beta_z^2 \gamma^3] \frac{d\beta_z}{dt} \cong \gamma^3 m_0 c \frac{d\beta_z}{dt} = qE_0, \quad (2.33)$$

or

$$\frac{dv_z}{dt} = \frac{qE_0}{\gamma^3 m_0}, \quad (2.34)$$

where we have assumed in Eq. (2.33) that $\beta_z \cong \beta$, the motion is predominantly along the z direction. Equation (2.34) seems to indicate that the effective mass of a relativistic particle in accelerating parallel to its velocity vector is $\gamma^3 m_0$. This longitudinal mass effect will be revisited when we discuss charged particles' longitudinal oscillations in accelerators.

Let us now examine a phase plane plot of uniform acceleration, as illustrated in Fig. 2.9. The curves of constant H are hyperbolae, which look locally parabolic near $p_z = 0$, the region of the non-relativistic limit. Asymptotically in large $p_z/m_0 c$, the ultra-relativistic region, the curves approach straight lines.

In Section 2.3, we found that edge effects—the effects of moving from a field-free region to one with field present—were very important in understanding the transverse motion of a charged particle in a longitudinal magnetic field. This is also true for the case of entry into a uniform electric field. The analysis proceeds just as in Eqs (2.16)–(2.18), and begins with an expansion near the axis of Eq. (1.17),

$$\frac{1}{\rho} \frac{\partial}{\partial \rho} \rho E_\rho = -\frac{\partial E_z}{\partial z}. \quad (2.35)$$

Integration of Eq. (2.35) outward from the axis yields

$$E_\rho \cong -\frac{\rho}{2} \frac{\partial E_z}{\partial z} \Big|_{\rho=0}, \quad (2.36)$$

which can be further used to find the radial momentum kick imparted to the particle as in passes through the fringing field region where E_ρ is non-zero,

$$\Delta p_\rho = q \int_{t_1}^{t_2} E_\rho \, dt = \frac{q}{v} \int_{z_1}^{z_2} E_\rho \, dz = -\frac{\rho q}{2v} \int_{z_1}^{z_2} \frac{\partial E_z}{\partial z} \Big|_{\rho=0} dz$$

$$= -\frac{\rho q}{2v} \int_{z_1}^{z_2} \frac{dE_z}{dz} \Big|_{\rho=0} dz = -\frac{\rho q}{2v} [E_z(z_2) - E_z(z_1)] = -\frac{\rho q}{2v} E_0. \quad (2.37)$$

We have assumed in Eq. (2.37) that the particle moves from a field-free region to one with uniform longitudinal electric field E_0. For an accelerating field ($qE_0 > 0$), this effect is obviously focusing, with the momentum kick tending to push the particle towards the axis. When a particle leaves the uniform accelerating field region, however, the lines of force fringe outward instead of inward and correct evaluation of Eq. (2.37) in this case yields a defocusing momentum kick.

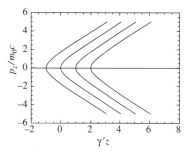

Fig. 2.9 Longitudinal phase plane $(\gamma'z, p_z/m_0 c)$ trajectories for uniform acceleration.

2.5   Motion in quadrupole electric and magnetic fields

In our discussion of beam optics thus far we have seen that one can use solenoidal magnetic fields for focusing particles during linear transport, and path length focusing in circular accelerators. These forms of focusing have introduced some basic concepts, such as harmonic betatron oscillations about the design orbit, but do not include the most widely used magnetic focusing scheme, that which employs transverse quadrupole fields.

Since we are using a coordinate system with one preferred longitudinal axis (z), we use cylindrical coordinates (ρ, ϕ, z) to evaluate the form of the scalar potentials from which either static transverse electric or magnetic[3] fields can be derived. Let ψ be such a potential, which, in the limit of a device long compared to its transverse dimensions, approximately obeys the two-dimensional Laplace equation

$$\vec{\nabla}_{\perp}^2 \psi = 0. \tag{2.38}$$

The solutions to this equation are of a form that is well behaved on axis ($\rho = 0$)

$$\psi = \sum_{n=1}^{\infty} a_n \rho^n \cos(n\phi) + b_n \rho^n \sin(n\phi), \tag{2.39}$$

Let us discuss the first few of the n multipole forms of the solution. For $n = 1$, we have

$$\psi_1 = a_1 \rho \, \cos(\phi) + b_1 \rho \, \sin(\phi) = a_1 x + b_1 y \tag{2.40}$$

and the fields are of the **dipole** form. Let us now assume we are talking about magnetic fields, so $\vec{B} = -\vec{\nabla}\psi$, and

$$\vec{B}_1 = -\vec{\nabla}\psi_1 = -a_1\hat{x} - b_1\hat{y}. \tag{2.41}$$

The form given in Eq. (2.41) is referred to as a dipole field because it can be formed with a magnet possessing only two poles. Such magnets will be discussed further in Chapter 6. They are the devices that yield the uniform fields that we have analyzed in Sections 2.1–2.3.

For $n = 2$, we have

$$\psi_2 = a_2 \rho^2 \cos(2\phi) + b_2 \rho^2 \sin(2\phi) = a_2(x^2 - y^2) + 2b_2 xy \tag{2.42}$$

and the associated magnetic field is

$$\vec{B}_2 = 2a_2(-x\hat{x} + y\hat{y}) - 2b_2(y\hat{x} + x\hat{y}), \tag{2.43}$$

If the potential coefficient a_2 is non-vanishing, there is a force on a charged particle traveling in the z direction, directed in the x dimension and which is proportional to y, and vice versa. This type of force is called **skew quadrupole**, and gives rise to generally undesirable coupling between the x and y phase planes. The coefficient b_2 indicates the presence of **normal quadrupole** fields, from which one obtains a force in the x dimension proportional to x, and a force in y that is proportional to y,

$$\vec{F}_{\perp} = qv_z\vec{z} \times \vec{B}_2 = 2qv_z b_2(y\hat{y} - x\hat{x}). \tag{2.44}$$

Assuming qb_2 is positive, this force is focusing in the x dimension and defocusing in y. Obviously if qb_2 is negative, these focusing/defocusing roles are

[3] If there is no current density $\vec{J} = 0$ in the region of interest, the magnetic field can be derived from a scalar potential. The case where the current density is non-vanishing is treated in Chapter 6.

reversed. Since the coefficient b_2 measures the gradient in magnetic field away from the axis, one often gives its strength in terms of this gradient $b_2 = -\partial_x B_x/\partial y \equiv B'/2$. The term "quadrupole" can be understood by looking at Fig. 2.10, which shows a design for a quadrupole built at UCLA, with its four poles excited by current-carrying coils in alternating polarity. The design principles of this magnet are discussed in Chapter 6. Let us content ourselves at the present to point out that the iron poles, whose surfaces form magnetic equipotentials, are hyperbolae, as suggested by Eq. (2.42).

If we assume paraxial motion near the z-axis, we may write the transverse equations of motion for a particle of charge q and momentum p_0 as

$$x'' = \frac{F_x}{\gamma m_0 v_0^2} = -\frac{qB'}{p_0}x, \tag{2.45}$$

and

$$y' = \frac{F_y}{\gamma m_0 v_0^2} = \frac{qB'}{p_0}y. \tag{2.46}$$

These equations are written in standard oscillator form as

$$x'' + \kappa_0^2 x = 0 \tag{2.47}$$

and

$$y'' - \kappa_0^2 y = 0. \tag{2.48}$$

The square wave number $\kappa_0^2 \equiv qB'/p_0$ is sometimes known as the focusing strength K. It can be simply calculated by using the handy shortcut $qB'/p_0 = B'/BR$ where $BR(\text{Tm}) = p_0(\text{MeV}/c)/299.8$ (cf. Eq. 2.8) is known as the magnetic rigidity of the particle.

Assuming $\kappa_0^2 > 0$, one has simple harmonic oscillations in x,

$$x = x_0 \cos[\kappa_0(z - z_0)] + \frac{x_0'}{\kappa_0} \sin[\kappa_0(z - z_0)], \tag{2.49}$$

and the motion in y is hyperbolic,

$$y = y_0 \cosh[\kappa_0(z - z_0)] + \frac{x_0'}{\kappa_0} \sinh[\kappa_0(z - z_0)]. \tag{2.50}$$

If $\kappa_0^2 < 0$, the motion is simple harmonic (oscillatory) in y, and hyperbolic (unbounded) in x. Focusing with quadrupoles alone can only be accomplished in one transverse direction at a time. Ways of circumventing this apparent limitation in achieving transverse stability, by use of alternating gradient focusing, are discussed in the Chapter 3.

As noted in Chapter 1, it is much easier to build transverse field magnets than transverse electric field-supporting electrode arrays that impart equal force upon charged particles traveling nearly the speed of light. Therefore, the transverse electric field quadrupole is found mainly in very low energy applications such as the electron microscope. Note that an electric quadrupole that produces decoupled forces in analogy with the magnetic forces of Eq. (2.41) has an electrode array with hyperbolic surfaces rotated by 45° from that shown in Fig. 2.10. In this case we may say that from the viewpoint of electric forces, if

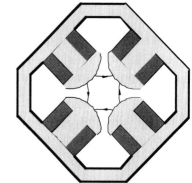

Fig. 2.10 Quadrupole magnet showing iron pole tips and yoke (hatching), current windings (gray), and orientation of magnetic field lines.

the same form of the solution for the electrostatic potential ϕ_e is used as was taken for ψ (Eq. 2.3), then a_2 is the coefficient of normal electric quadrupole, and b_2 is the coefficient of the skewed electric quadrupole.

The paraxial ray equations in the case of an electric quadrupole are identical to those given in Eqs (2.47) and (2.48), with $\kappa_0^2 \equiv qE'/p_0v_0$. When the motion is focusing in x it is defocusing in y, as before. The differences between the two cases lie in the deviations from the assumptions of the paraxial approximation. In the case of magnetic forces, the energy and total momentum are constant, but the longitudinal momentum p_z, and thus the longitudinal velocity v_z may change, which in turn forces a small modification of Eqs (2.47) and (2.48). With electric forces the situation is slightly different—the energy is not constant, as work may be performed on a particle with transverse motion, but the longitudinal momentum p_z is now constant. In this case the longitudinal velocity v_z may also change non-negligibly when the paraxial approximation is not valid.

2.6   Motion in parallel, uniform electric and magnetic fields

In the case of a uniform electric field, $\vec{E} = E_0\hat{z}$, and a parallel uniform magnetic field, $\vec{B} = B_0\hat{z}$, the electric field provides uniform acceleration while the magnetic field provides solenoidal focusing. This example scenario shows the combination of acceleration and focusing well, nicely illustrating the phenomenon of adiabatic damping of transverse oscillations. The equations of motion in this field configuration are

$$\frac{dp_z}{dt} = qE_0 \qquad \frac{d\vec{p}_\perp}{dt} = q(\vec{v}_\perp \times \vec{B}) = \frac{qB_0}{\gamma m_0}(\vec{p}_\perp \times \hat{z}). \qquad (2.51)$$

Equation (2.51) are coupled by the presence of γ in the second equation, and so the method of solution offered previously for the transverse motion must be re-examined.

On the other hand, as the amplitude of $\vec{p}_\perp$ is invariant, the energy may still be trivially found as a function of z,

$$U(z) = \gamma(z)mc^2 = \sqrt{(p_\perp^2 + p_z^2|_{z=0})c^2 + (m_0c^2)^2} + qE_0z. \qquad (2.52)$$

In the paraxial limit, one may ignore the transverse momentum contribution to the energy, and Eq. (2.52) gives

$$U(z) = \gamma(z)mc^2 = \sqrt{p_z^2|_{z=0}c^2 + (m_0c^2)^2} + qE_0z. \qquad (2.53)$$

After the energy and the longitudinal momentum are determined by Eqs (2.53) and (2.31), one can begin to write the transverse equations of motion in the Larmor frame (rotating with local frequency $\omega_L(z) = qB_0/2\gamma(z)m_0$) through the definition

$$x'_L \equiv \frac{p_{x_L}}{p_z}, \qquad (2.54)$$

with an analogous expression in y_L. Since p_{x_L} is a Cartesian projection of the constant amplitude $\vec{p}_\perp$, we expect the maximum angle in x_L (and y_L) to be secularly—meaning on a time scale longer than the relevant (Larmor) oscillation period—damped by the acceleration, as $x'_L \propto p_z^{-1}$.

Concentrating on the motion in x_L, we differentiate Eq. (2.54) to obtain

$$x''_L + \frac{p_{x_L} p'_z}{p_z^2} - \frac{p'_{x_L}}{p_z} = 0, \tag{2.55}$$

or using the paraxial approximation $p_z \cong p$,

$$x''_L + \frac{(\beta\gamma)'}{\beta\gamma} x'_L + \left(\frac{qB_0}{2\beta\gamma m_0 c}\right)^2 x_L = 0. \tag{2.56}$$

If we look only at highly relativistic motion, $\beta \cong 1$, we can further approximate Eq. (2.56),

$$x''_L + \frac{\gamma'}{\gamma} x'_L + \left(\frac{b\gamma'}{2\gamma}\right)^2 x_L = 0. \tag{2.57}$$

The solution to this homogeneous equation is of the form

$$x_L(z) = x_{L,0} \cos\left[\frac{b}{2} \ln\left(\frac{\gamma(z)}{\gamma_0}\right)\right] + \frac{2\gamma_0}{b\gamma'} x'_{L,0} \sin\left[\frac{b}{2} \ln\left(\frac{\gamma(z)}{\gamma_0}\right)\right], \tag{2.58}$$

where $b = B_0 c/E_0$ and the initial offset, angle, and Lorentz factor are $x_{L,0}, x'_{L,0}$, and γ_0, respectively. Thus the solution appears as a harmonic oscillator with a logarithmically (as opposed to linearly) increasing argument in the oscillatory trigonometric functions. We now compare this system to that of the simple harmonic osciallator.

Simple harmonic motion is associated with two invariants: the angular frequency ω and the value of the Hamiltonian (the total oscillator energy). As discussed in Sections 1.3 and 2.3, these quantities are related through

$$H = \frac{1}{2m}[p_x^2 + m^2\omega^2 x^2] = J_x\omega, \tag{2.59}$$

where the action $J_x = \oint p_x \, dx/2\pi = A/2\pi = x_{max} p_{x,max}/2$, is the area in the phase plane over 2π, enclosed by the elliptical trajectory of the oscillator (see Fig. 1.4).

An analogue to the action in trace space can be defined as $J_{x,\text{trace}} \equiv x_{max} x'_{max}/2$. For the system under study we obtain the angle in x_L from Eq. (2.58) as,

$$x'_L(z) = -x_{L,0}\frac{b\gamma'}{2\gamma} \sin\left[\frac{b}{2} \ln\left(\frac{\gamma(z)}{\gamma_0}\right)\right] + \frac{\gamma_0}{\gamma} x'_{L,0} \cos\left[\frac{b}{2} \ln\left(\frac{\gamma(z)}{\gamma_0}\right)\right]. \tag{2.60}$$

Setting $x'_{L,0} = 0$ for simplicity, one can see that

$$J_{\text{trace}} \equiv \frac{x_{L,max} x'_{L,max}}{2} = x_{L,0}^2 \frac{\gamma'}{\gamma} \frac{b}{2}, \tag{2.61}$$

and the action in trace space damps as γ^{-1}. According to Eq. (2.60), it is clear that this apparent damping is due to the diminishing of x'_L with acceleration. This diminishing should have the functional dependence on momentum of $(\beta\gamma)^{-1}$. We have only approximately obtained this value by assuming the limit $\beta = 1$

for the purposes of solving the paraxial equation of motion, Eq. (2.56), exactly. It should be equally clear that the true phase plane action variable is constant in this case.

As mentioned in Section 1.6, the role of the action in the general theory of oscillators is that they are adiabatic invariants. This means that if the oscillator parameters are changed on a time scale that is slow compared to the oscillation period, then the action is conserved. The true phase plane action is an adiabatic invariant and is conserved, not damped, under slow acceleration. On the other hand, the apparent action in trace space is damped. Because of the guaranteed adiabatic invariance of the phase plane action under slow acceleration and concomitant changing of the oscillation frequency, the damping of the trace space action is termed adiabatic.

Use of the term adiabatic damping to describe this phenomenon is thus actually a bit of a misnomer. It is especially so in the context of the system described in this section, where the transverse motion under acceleration is not required to be an adiabatic process at all. It was only required to be one in which the transverse momentum is constant and the longitudinal momentum grows.

2.7   Motion in crossed uniform electric and magnetic fields*

Let us now consider the case where there are crossed uniform electric and magnetic fields, which we choose as $\vec{E} = E_0\hat{y}$, and $\vec{B} = B_0\hat{z}$. In this case, there is a particular solution to Eq. (1.11) obtained when the Lorentz force vanishes, given by

$$\vec{E} + \vec{v} \times \vec{B} = 0. \tag{2.62}$$

The velocity that is consistent with Eq. (2.62) is termed the drift velocity, and is found by taking the cross-product of Eq. (2.62) with $\vec{B}$, to give

$$\vec{v}_d = \frac{\vec{E} \times \vec{B}}{\vec{B}^2} = \frac{E_0}{B_0}\hat{x}. \tag{2.63}$$

The drift velocity is normal to both the magnetic and electric field directions.

The general motion in this case can be deduced by transforming the Lorentz frame traveling with the drift velocity using Eq. (1.58) to obtain

$$\vec{E}'_\perp = \gamma_d(\vec{E}_\perp + \vec{v}_d \times \vec{B}_\perp) = 0$$

$$\vec{B}' = \gamma_d\left(\vec{B}_\perp - \frac{1}{c^2}\vec{v}_d \times \vec{E}_\perp\right) = \gamma_d B_0\left(1 - \frac{\vec{v}_d^2}{c^2}\right)\hat{z} = \frac{B_0}{\gamma_d}\hat{z}. \tag{2.64}$$

The field is purely magnetic and uniform in the new frame. This means that any motion in this frame is simply that of the helical cyclotron form discussed in Section 2.1. Looking down the axis (z) of the magnetic field, a direction in which there is no net force, the motion appears cycloidal, in other words cyclic motion with a secular velocity (the drift velocity) superimposed, as shown in Figure 2.11.

The drift velocity and associated cycloid motion is not as important of an effect in accelerators as those encountered in previous sections. Such behavior

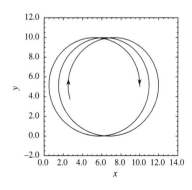

Fig. 2.11 Cycloid motion in crossed electric and magnetic fields: $\vec{E} = E_0\hat{y}, \vec{B} = B_0\hat{z}$.

s notably found in an analysis of intense beam rotation in solenoids due to the ongitudinal magnetic field and the radial self-electric field of the beam (see Ex. 2.14). The drift velocity is, however, quite important in plasma physics, and in eparation devices—a particle with the correct velocity passes through a crossed electric and magnetic field region, regardless of charge. This is an important echnique in experimental physics, and can be used for velocity identification of a particle. The cross-field configuration, in tandem with traversal of a region of uniform magnetic field, in which the momentum of the particle is determined, allows the mass and energy of the particle to also be determined.

2.8 Motion in a periodic magnetic field*

As a final scenario for this chapter, let us consider the case of a spatially periodic (in the z-direction) static magnetic field. There are two possible configurations of this **undulator** magnetic field, a **planar** or **helical polarization**, names that efer to analogous electromagnetic wave polarizations. In this section we will discuss the planar undulator, leaving analysis of the helical configuration for he exercises. Also, we will emphasize here the bending-plane motion of the particle, of the undulator, with the out-of-plane motion examined further in Chapter 3.

The periodic, vertically polarized magnetic field of interest here can be described mathematically by

$$\vec{B} = B_0 \sin(k_u z)\hat{y} \qquad (2.65)$$

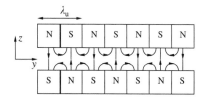

Fig. 2.12 Schematic representation of the magnetic field lines in a planar undulator configuration.

where $k_u = 2\pi/\lambda_u$ is the wave number associated with the **undulator** wavelength λ_u of the field. This expression is valid only in the symmetry plane of the field, which we take to be $y = 0$. For the planar configuration, there must be an additional longitudinal field component, as illustrated in Fig. 2.12, and analyzed in the Exercise 2.15.

As can be seen by the results of Exercise 2.15, Eq. (2.65) is approximately correct if the vertical offset from the symmetry plane is much smaller than an undulator wavelength, $k_u y \ll 1$. Let us now examine the electron beam dynamics in the undulator field under this additional constraint, which can be considered an extension of the paraxial approximation. We will formulate the problem in Hamiltonian style. To begin, we write the canonical momenta

$$p_{c,x} = \beta_x \gamma m_0 c + qA_x = \beta_x \gamma m_0 c - q\frac{B_0}{k_u}\cos(k_u z)\cosh(k_u y),$$

$$p_{c,y} = \beta_y \gamma m_0 c, \qquad (2.66)$$

$$p_{c,z} = \beta_z \gamma m_0 c,$$

from which we can derive the relativistically correct Hamiltonian,

$$H = \sqrt{\left(p_{c,x} + q\frac{B_0}{k_u}\cos(k_u z)\cosh(k_u y)\right)^2 c^2 + p_{c,y}^2 c^2 + p_{c,y}^2 c^2 + (m_0 c^2)^2},$$

$$(2.67)$$

where the electrostatic potential ϕ_e vanishes. Viewing z as the independent variable, we write the new form of the Hamiltonian function as

$$G = -p_{z,c} = \sqrt{H^2 - (p_{c,y} - qA_y)^2 c^2 - (p_{c,x} - qA_x)^2 c^2 - (m_0 c^2)^2}$$

$$= \sqrt{U^2 - (p_{c,y} - qA_y)^2 c^2 - \left(p_{c,x} + q\frac{B_0}{k_u}\cos(k_u z)\cosh(k_u y)\right)^2 c^2 - (m_0 c^2)^2}$$

$$(2.68)$$

where we have substituted the **numerical energy** U for the old Hamiltonian **functional energy** H. This new Hamiltonian is independent of x and t, and thus the canonical x component of the momentum $p_{c,x}$ is a constant of the motion, as is the total mechanical energy U (or equivalently, the total mechanical momentum p_0), which is always the case in magnetostatic systems.

The first integrals of the Hamiltonian system (the momenta) are thus as follows:

$$p_{c,x} = \text{constant} = p_{x0} \Rightarrow \beta_x \gamma m_0 c = -qA_x + p_{x0} = q\frac{B_0}{k_u}\cos(k_u z) + p_{x0},$$

$$(2.69)$$

$$p_{c,y} = \beta_y \gamma m_0 c = 0 \text{ (by assumption, taking } y = 0\text{)}, \qquad (2.70)$$

and

$$p_{c,z} = \sqrt{p_0^2 - \left(q\frac{B_0}{k_u}\cos(k_u z) + p_{x0}\right)^2}. \qquad (2.71)$$

This analysis has been done formally, with the aid of the Hamiltonian, in order to make more advanced analysis of the motion in undulators possible in following chapters.

The most important step remaining in the analysis is to find the second integral in x, which is obtained by integrating the paraxial equation

$$x' = -\frac{\partial G}{\partial p_{x,c}} \cong \frac{qB_0}{p_0 k_u}\cos(k_u z) + \frac{p_{x0}}{p_0} = \frac{qB_0}{p_0 k_u}\cos(k_u z) + x_0', \qquad (2.72)$$

to give, with initial (evaluated before entry into the undulator field) horizontal offset and angle (x_0, x_0'),

$$x \cong x_0 + x_0' z + \frac{qB_0}{p_0 k_u^2}\sin(k_u z). \qquad (2.73)$$

It can be seen from Eq. (2.73) that the amplitude of the undulating portion of the transverse motion is $qB_0/p_0 k_u^2$, and from Eq. (2.72) that the maximum angle associated with the undulating part of the motion is $qB_0/p_0 k_u$. This angle is typically much smaller than unity (the bends are paraxial) for magnets referred as undulators. Furthermore, we note that Eq. (2.73) shows that there is no restoring, or focusing force in the x-direction associated with this configuration of magnetic field—an initial error $x_0' \neq 0$ is not corrected, and leads eventually to a trajectory with large horizontal offset x.

The conservation of total momentum combined with Eq. (2.69) gives Eq. (2.71), which indicates that the longitudinal momentum, and therefore the longitudinal velocity, must diminish in the undulator, approximately as

$$p_z = \sqrt{p_0^2 - \left(q\frac{B_0}{k_{\mathrm{u}}}\cos(k_{\mathrm{u}}z)\right)^2} \cong p_0\left[1 - \frac{1}{2}\left(\frac{qB_0}{k_{\mathrm{u}}p_0}\cos(k_{\mathrm{u}}z)\right)^2\right], \quad (2.74)$$

for paraxial bends. Averaging Eq. (2.74) over a period of the motion, we have simply

$$\langle p_z \rangle \cong p_0\left[1 - \left(\frac{qB_0}{2k_{\mathrm{u}}p_0}\right)^2\right]. \qquad (2.75)$$

This "slowing" of the particle in its z-motion is an important effect in free-electron lasers, which are discussed in Chapter 8.

2.9   Summary and suggested reading

This chapter has been concerned with introducing a number of model problems, based on the relativistic motion of charged particles in static electric and magnetic field configurations. These configurations have included:

1. Uniform (dipole) magnetic fields, in which we deduce many aspects of the motion in both circular accelerators and focusing solenoids from the general case of helical motion. In both scenarios, we found variants of simple-harmonic betatron oscillations.
2. Uniform electric fields, in which acceleration of charged particles to relativistic energies are introduced.
3. Quadrupole magnetic and electric fields, where the motion is simple harmonic in one transverse dimension, and divergent in the other.
4. Superpositions of uniform electric and magnetic fields, which produce damped oscillations when the fields are parallel, and drift motion when the fields are crossed.
5. The periodic magnetic dipole field, or magnetic undulator. This device is shown to produce transverse undulating motion, which forms the basis of the free-electron laser.

These model problems also allowed us to introduce some rudimentary examples of analyses that are based on both relativistic and Hamiltonian formalisms. These analyses will help form the basis of more complex investigations of charged particle motion in the coming chapters.

Many texts in electromagnetism also introduce aspects of charged particle motion in electric and magnetic fields. Readers wishing to review the theory of electrostatic and magnetostatic fields may also wish to review such texts. The following texts may be recommended as a supplement to this chapter:

1. R. Wangsness, *Electromagnetic Fields* (Wiley, 1986). There are many examples of dynamics calculations in this book, including useful appendices.
2. J.R. Reitz, F.J. Milford, and R.W. Christy, *Foundations of Electromagnetic Theory* (Addison-Wesley, 1993). Another fine undergraduate electromagnetism text with examples.

3. J.D. Jackson, *Classical Electrodynamics* (Wiley, 1975). A wealth of assigned problems are relevant to our discussions here.

4. L.D. Landau and E.M. Lifschitz, *The Classical Theory of Fields* (Addison-Wesley, 1971).

5. S. Humphries, Jr., *Principles of Charged Particle Acceleration* (Wiley, 1986). This text also examines many model problems as a path to understanding accelerator behavior.

6. J.D. Lawson, *The Physics of Charged Particle Reams* (Clarendon Press, 1977). There are many clear explanations of basic problems in charged particle motion.

7. F.F. Chen, *Introduction to Plasma Physics and Controlled Fusion: Plasma Physics* (Plenum, 1984). The field of plasma physics has a wide variety of physical effects it addresses, and some differing dynamics problems are addressed in this excellent introduction.

8. J.B. Marion and S.T. Thornton, *Classical Dynamics of Particles and Systems* (Harcourt Brace & Company, 1995).

9. H. Goldstein, C. Poole and J. Safko, *Classical Mechanics* (Addison-Wesley, 2002). Dynamics in electromagnetic fields using modern mechanics at a sophisticated level.

Exercises

(2.1) The mass increase implied by Eq. (2.4) can be obtained by using the relativistically correct Lagrangian (Eq. 1.60) to write the problem in cylindrical coordinates,

$$L(\vec{x}, \dot{\vec{x}}) = -\frac{m_0 c^2}{\gamma} + q\vec{A} \cdot \dot{\vec{x}}.$$

(a) Show that the cylindrically symmetric vector potential for a uniform magnetic field is $\vec{A} = (B_0 \rho/2)\hat{\phi}$.

(b) Write the Lagrangian in cylindrical coordinates (ρ, ϕ, z).

(c) Derive the Lagrange–Euler equations to show that the transverse motion (centered at the z-axis) is circular, with appropriate angular frequency and radius of curvature.

(2.2) In the **betatron**, the design orbit of the electrons is circular, with constant radius R, but the vertical magnetic field at the design orbit increases in time. This is described by the relation

$$p_0(t) = eB_0(t)R \quad \text{(i)}$$

which indicates that the design momentum must be increased in proportion to the strength of the guiding field. The acceleration for electrons traveling in a strictly circular orbit is given by

$$\frac{dp_0}{dt} = -eE_\phi, \quad \text{(ii)}$$

where E_ϕ is the azimuthal electric field, tangential to the electron path.

The **betatron condition** occurs when these requirements are consistent with each other. In order for this to be true, the field normal to the bend plane (B_z in cylindrical coordinates) must not be constant, but it should be a function of ρ, as shown in a representative way in Fig. 2.13. To derive the betatron condition, employ the following steps:

(a) Differentiate the expression (i) with respect to time, and

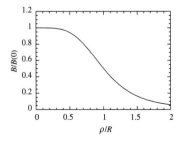

Fig. 2.13 Example of magnetic field profile as a function of radius in a betatron.

(b) Calculate the induced electric field from Stokes theorem applied to

$$\vec{\nabla} \times \vec{E} = -\frac{\partial \vec{B}}{\partial t}.$$

The betatron condition is often given as $B_0(t) = \bar{B}(t)/2$, where $\bar{B} \equiv (\int \vec{B} \cdot d\vec{A})/(\pi R^2)$ is the average normal magnetic field over the disk whose boundary is the design orbit. Is this the most general form of the betatron condition? Note that the betatron condition implies that the field is as shown in Fig. 2.13, having larger amplitude for $\rho < R$. This point is returned to in the next chapter.

(2.3) The cyclotron spoken of in Chapter 1 is schematically shown in Fig. 2.14. It consists of two opposing D-shaped iron magnet pole pairs (so-called Ds), giving rise to a roughly constant magnetic field B_0 normal to the diagram inside of the Ds. This field bends the particles inside of the D-regions in semi-circular trajectories. The Ds also are excited by a time-dependent voltage source of constant angular frequency ω_{rf} such that $V(t) = V_0 \sin(\omega_{rf} t)$, where rf stands for radio-frequency. Thus, when the particles cross the gap between the Ds they can be accelerated by the electric field between the two different potential regions, so that their energy is maximally incremented by $\Delta U = qV_0$ per crossing. This occurs only when the motion is **synchronous** with voltage wave-form, requiring $\omega_c \cong \omega_{rf}$. Note that the particles have larger radius of curvature as they gain energy and are eventually ejected from the machine.

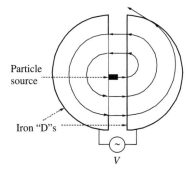

Fig. 2.14 Schematic diagram of magnetic field regions, voltage application, and design trajectory in a cyclotron. The particles are embedded in an approximately uniform magnetic field inside of the Ds, and accelerated by an electric field between the Ds.

(a) Assuming the particles are **non-relativistic** protons ($\gamma \cong 1$) and the magnetic field $B_0 = 1.5\,\mathrm{T}$, what is the rf frequency needed for the voltage supply? You may ignore the gap distance in this calculation.

(b) Assume the initial proton kinetic energy from the source is $150\,\mathrm{keV}$, $V_0 = 150\,\mathrm{kV}$, and the radius of curvature of the outside of the Ds is $0.75\,\mathrm{m}$. Approximately how many passes through

the system does a proton circulate for before it is ejected?

(c) What is the value of γ of the proton at the machine exit? Can you estimate the total phase slippage of a proton with respect to the applied voltage during acceleration due to relativistic changes in the cyclotron frequency?

(2.4) The Tevatron at Fermi National Accelerator Laboratory is a circular proton (colliding with counter-propagating antiprotons) accelerator that uses high field superconducting dipole magnets. It is approximately 1 km in radius, and accelerates protons to 900 GeV in total energy.

(a) What is the average magnetic field needed to keep 900 GeV protons on the design orbit?

(b) What is the circulation (cyclotron) frequency in this accelerator? The protons are injected into the Tevatron at 150 GeV. What is the circulation frequency at this energy? What is the circulation frequency at 900 GeV?

(c) If electrons were to be stored in the Tevatron for collision with the 900 GeV protons instead of antiprotons, what would their energy be? Be careful with round-off errors due to calculators in this problem.

(2.5) Busch's theorem can also be derived by use of the Hamiltonian formalism.

(a) Using the results of Exercise 2.1, write the Hamiltonian inside of a solenoid in terms of cylindrical coordinates. Hint: Be careful in handling the vector potential component of the angular momentum.

(b) Show that the canonical angular momentum is conserved in this case.

(c) Assuming the canonical angular momentum is zero (why?), what is the value of the mechanical momentum at radial offset ρ? Compare this to Eq. (2.18).

(2.6) If the transverse momentum is not small compared to the total momentum, we may not assume $p \cong p_z$ and the betatron wave number in the solenoid is not as given by Eq. (2.24).

(a) Derive a general expression for the wave number as a function of initial offset radius at the entrance to the solenoid. Hint: Normalizing the initial offset radius to p/qB_0 may make the answer more physically meaningful.

(b) At a certain radius, the particle cannot penetrate the interior of the solenoid. This is because all of its longitudinal momentum is given up to angular motion in the fringe field and the particle is reflected. For an electron with 100 keV kinetic energy, and $B_0 = 1$ T, what is this radius?

(2.7) By differentiation of Eq. (2.21) twice, derive Eq. (2.22).

(2.8) The high-energy linear collider at Stanford contains a linear accelerator of length $L = 3$ km, which uses electric fields in radio-frequency cavities to accelerate electrons (and positrons) to 50 GeV. The average longitudinal electric force performing work on the electrons in the accelerator is 16.7 MeV/m. The proper time τ, as observed in the frame of the accelerating electron, is related to the time measured in the laboratory t by $dt = \gamma \cdot d\tau$.

(a) Show that the time for the electrons to go from rest to final energy is

$$t = \frac{L}{c}\left[\sqrt{\frac{\gamma_f + 1}{\gamma_f - 1}}\right] \approx \frac{L}{c}$$

(b) Show that the proper time for the electrons to go from rest to final energy is

$$\tau = \frac{L}{c(\gamma_f - 1)}\cosh^{-1}(\gamma_f).$$

(c) What is the effective length of the linear accelerator as seen by the electrons at final energy? How long would it take to traverse this length at the final velocity of the electrons?

(2.9) Instead of writing the motion in terms of the variable v_z, consider the motion as a function of γ.

(a) Show that

$$\frac{d\gamma}{dt} = \frac{\sqrt{\gamma^2 - 1}}{\gamma}\frac{qE_0}{mc}.$$

(b) Integrate this equation and compare to the results of directly integrating the first of Eq (2.25).

(2.10) For magnetic focusing channels that use quadrupoles, the motion is the same for particles of like charge but different mass if the momentum of each species is the same. This can be seen by examination of Eqs (2.42) and (2.43). Now consider the motion of particles with equal charge but unequal mass in electric quadrupoles. What is the condition on dynamical quantities of the different species so that they can be focused identically in the electric quadrupole channel? Show, in the relevant non-relativistic limit, that this condition reduces to the equality of kinetic energies in the two species.

(2.11) For magnetic focusing channels, show that a particle traveling in a certain direction in z has the same equation of motion as its antiparticle of the same momentum traveling in the opposite direction in z. This is the basic optics principle behind particle/antiparticle colliding beam rings.

How do the paraxial optics change for these species i electric quadrupoles are used?

(2.12) In deriving Eq. (2.58) as a solution to Eq. (2.57), it i helpful to initially transform the independent variable t $u = \ln(\gamma(z))$, with inverse transformation $\gamma = \exp(u)$ This is known as a **Cauchy transformation**.

(a) Show that this transformation allows Eq. 2.58 to b written as

$$\frac{d^2 x_L}{du^2} + \left(\frac{b}{2}\right)^2 x_L = 0.$$

(b) Using the equation derived in part (a), construc Eq. (2.58).

(2.13) Can you solve the paraxial equation of motion (analogou to Eq. 2.55) for the case when the electric field is con stant, but the magnetic field is zero? You may start wit $x' \equiv p_x/p_z$ and $p_x = $ constant, then integrate.

(2.14) For cylindrically symmetric, intense beams propagating in equilibrium in solenoids there is an electric self-field i the radial direction that produces both an outward radia force and an azimuthal rotation of frequency ω_r. This fre quency may or may not be equal to the Larmor frequency and the related rotation may be viewed as an example o the $\vec{E} \times \vec{B}$ drift discussed previously.

(a) Show, for a uniform density (n_b) beam that thi radial space-charge electric force on a particle o charge q and rest mass m_0 is

$$F_e = qE_\rho = \frac{q^2 n_b}{2\varepsilon_0}\rho \equiv m_0 \frac{\omega_{p0}^2}{2}\rho,$$

where we have introduced the non-relativisti plasma frequency $\omega_{p0} = \sqrt{q^2 n_b/\varepsilon_0 m_0}$. Find th net (beam-induced) force, including the magneti force, on the particle, and write it in terms of th electric force and the particle's axial velocity v_0.

(b) When the beam rotates, there are two radial force that are introduced. One is due to centripetal accel eration, and the other is the radial component o the solenoidal force. Equilibrium (net radial forc equal to zero) can be established only when th beam rotates in a certain direction because bot the centripetal forces and the space-charge force are outward. Show that the condition on ω whic produces equilibrium can be written as

$$\omega_r^2 + \frac{\omega_p^2}{2} = \omega_r \omega_c,$$

where the relativistically correct plasma frequenc is given by $\omega_p^2 = \omega_{p0}^2/\gamma^3$.

(c) Illustrate this equilibrium condition by plotting ω_p/ω_c as a function of ω_r/ω_c. What rotation frequency ω_r maximizes the equilibrium density of the beam?

.15) The magnetic field given in Eq. (2.65) can be derived from a vector potential,

$$\vec{A} = -(B_0/k_u)\cos(k_u z)\hat{x},$$

in the $y = 0$ plane. Unfortunately, this vector potential does not satisfy the Laplace equation in the current-free region, $\nabla^2 \vec{A} = 0$, as required by the Maxwell relations, and so it cannot be valid.

(a) Show that the following appropriate potential (uniform in x, symmetric in y)

$$\vec{A} = -(B_0/k_u)\cos(k_u z)\cosh(k_u y)\hat{x}$$

obeys the Laplace equation.
(b) Find the field components associated with this vector potential.

.16) A **helically polarized** magnetic undulator has an on-axis field profile given by

$$\vec{B}(z) = \frac{B_0}{\sqrt{2}}[\sin(k_u z)\hat{x} + \cos(k_u z)\hat{y}].$$

It is typically constructed as shown schematically in the Fig. 2.15, with two main counter-rotating helical windings (bifilar helical undulator).

(a) The vector potential which gives rise to this field can be written in cylindrical coordinates as

$$A_\phi = \frac{B_0}{k_u}[I_0(k_u\rho) + I_2(k_u\rho)]\cos(\phi - k_u z)$$

and

$$A_\rho = -\frac{B_0}{k_u}[I_0(k_u\rho) - I_2(k_u\rho)]\sin(\phi - k_u z).$$

Here I_0 and I_2 are modified Bessel functions.

Show that these components satisfy the Laplace equation, $\nabla^2 \vec{A} = 0$.

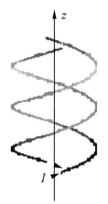

Fig. 2.15 Schematic of a bifilar helical undulator.

(b) Find the field components for this vector potential and verify that they give the correct field profile on axis.
(c) Write the Hamiltonian and equations of motion, using z as the independent variable, for the helical undulator. Show that in the limit $k_u\rho \ll 1$, there exists an orbit that is a perfect helix, in which the mechanical angular momentum is a constant. Note this is not true for the helix in the solenoid, where the **canonical angular momentum** is constant (the Hamiltonian is independent of azimuth).

3 Linear transverse motion

The phenomenon of stable transverse motion of charged particles near design trajectory has been introduced in Chapter 2. Simple harmonic betatro oscillations were encountered in the examples of the focusing solenoid, th quadrupole, and in the circular accelerator. The general physical and mathem atical tools for describing betatron oscillations are developed in this chapte The discussion begins, appropriately enough, within the historical context c the betatron, where the ideas of path length focusing and quadrupole focusin are used to provide simultaneous stability in both transverse dimensions. Afte introduction of this type of first-order (in the field strength), or **weak**, focusin we will move to the discussion of second-order, or **strong**, focusing. We end th chapter with a discussion of the first-order effects in transverse motion due to th dispersion of particles having total momenta deviating from the design value

3.1 Weak focusing in circular accelerators

The motion of a charged particle in a uniform magnetic field has been show to exhibit effective focusing in the bend (horizontal, x) plane. In a descrip tion based on examination of trajectories along the circular design orbit, th effective focusing is ascribed to differential path lengths taken by particles c differing betatron orbits. Such path length focusing is effective in stabilizin the horizontal motion, but not the vertical motion, which is completely uncoi strained in a uniform magnetic field. On the other hand, in our initial discussio of acceleration in the betatron (see Ex. 2.2), we have seen that the vertical com ponent of the magnetic field is not uniform, but diminishes with distance awa from the axis.

At the design orbit, which defines a curvilinear coordinate system, it is mo natural to adopt the lowest order (in $x \equiv \rho - R$) approximation of the variatio in the vertical component of the field as,

$$B_y(x) = B_0 + B'x + \cdots .$$ (3.1

The field gradient is negative, $B' < 0$, in the case of the betatron. Equation (3.1 implies, along with Ampere's law, that a horizontal component of the fiel exists,

$$B_x(x) = \int_0^y \frac{\partial B_y}{\partial x} d\tilde{y} \cong B'y.$$ (3.2

In this coordinate system, it is apparent that, near the design orbit, the mag netic field appears as a superposition of vertically oriented dipole and verticall focusing (horizontally defocusing) quadrupole fields. Since the field is ne

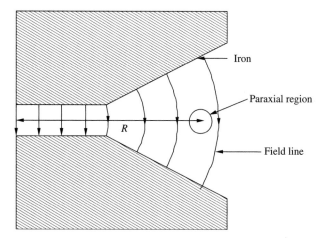

Fig. 3.1 Configuration of magnetic field in betatron, showing superposition of dipole and vertically focusing quadrupole components in vicinity of the design orbit.

designed to be a pure multipole in this case, but a superposition of two multipoles, it is termed a **combined-function** magnetic field (illustrated in Fig. 3.1) configuration.

Following the derivation of the horizontal equation of motion in the uniform magnetic field leading to Eq. (2.12), and using the equations of motion in a quadrupole (Eqs (2.42)–(2.43)), we can write the following equations describing paraxial motion in this combined function field:

$$x'' + \left[\left(\frac{1}{R}\right)^2 + \frac{B'}{B_0 R}\right] x = 0, \tag{3.3}$$

and

$$y'' - \frac{B'}{B_0 R} y = 0, \tag{3.4}$$

where we indicate differentiation with respect to the independent variable s as $' = d/ds$. If we normalize the focusing strengths in these equations to R^{-2}, we can write the resultant expressions in standard form,

$$x'' + \left(\frac{1}{R}\right)^2 [1 - n]x = 0, \tag{3.5}$$

and

$$y'' + \frac{n}{R^2} y = 0, \tag{3.6}$$

where the **field index** is given by $n \equiv -B'R/B_0$.

Equations (3.5) and (3.6) are often termed the **Kerst–Serber equations**, as they were deduced by D.W. Kerst and F. Serber during initial development of the betatron. From these relations, the condition for simultaneous stability in horizontal and vertical motion (both focusing strengths are positive) is simply given in terms of the field index,

$$0 < n < 1. \tag{3.7}$$

The simultaneous stability in x and y is clearly achieved by partially, but not fully, removing the natural path length focusing in the horizontal dimension through the introduction of a defocusing quadrupole field component. This

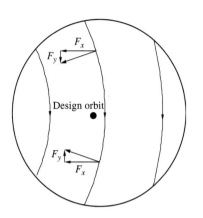

Fig. 3.2 Close-up view of paraxial region shown in Fig. 3.1, illustrating the vertically focusing components of the magnetic field at vertically offset positions.

component yields a force that is focusing in the vertical dimension, as can be seen in Fig. 3.2.

The Kerst–Serber equations are often written, especially in the context of simple magnetic optics systems, with the azimuthal angle $\theta = s/R$ around the design orbit as the independent variable,

$$\frac{d^2x}{d\theta^2} + [1 - n]x = 0 \quad \text{or} \quad \frac{d^2x}{d\theta^2} + \nu_x^2 x = 0, \tag{3.8}$$

and

$$\frac{d^2y}{d\theta^2} + ny = 0 \quad \text{or} \quad \frac{d^2y}{d\theta^2} + \nu_y^2 x = 0. \tag{3.9}$$

The forms of the Kerst–Serber equations displayed in Eqs (3.8) and (3.9) are illustrative, because they give a direct normalization of the betatron oscillation frequency in terms of the circulation frequency. The **tunes** (normalized frequencies), $\nu_x = \sqrt{1 - n}$ and $\nu_y = \sqrt{n}$, are the number of betatron oscillations per revolution in the horizontal and vertical dimensions, respectively. Assuming stability in both dimensions, the tunes are restricted to be smaller than unity. This restriction on the oscillation frequency is the source of the term "weak focusing." The weakness of the focusing becomes apparent when one attempts to scale the circular accelerator to larger design momenta and concomitant larger radii of curvature. For a given angular error of beam particle x' launched from the design orbit, the maximum offset found in the device is $x_m = Rx'/\nu_x$. Thus, the size of the beam that is contained in the machine scales with the radius of curvature. This scaling has serious implications in the design of the magnets and vacuum systems of the device. If the beam size, and thus the large clear aperture between magnetic poles becomes too large, the magnet becomes difficult, if not impossible, to build. Given a way to introduce stable oscillations with much higher tunes $\nu_{x,y} \gg 1$, the size of the beam could be greatly reduced. The discovery of such a method based on quadrupole magnets, termed **strong focusing**, has, therefore, allowed the development of very large radius of curvature circular accelerators. We will next introduce the tools needed to analyze strong focusing.

3.2   Matrix analysis of periodic focusing systems

The betatron oscillations of a particle in a weak focusing system can be intuitively understood, as they are merely examples of simple harmonic motion. This motion is characterized by having an oscillator (focusing) strength, κ_0^2, which is of second-order in field strength and/or the **field gradient** (parameterized by n in Section 3.1). Specifically, in a weak focusing circular accelerator, the path length focusing, when combined with a quadrupole gradient, allows simultaneous stability in x and y, yielding a focusing strength that is of second-order in the magnetic field amplitude $B_0, \kappa_0^2 \propto B_0^2$. In a simple quadrupole, however, the motion is not simultaneously stable in both transverse dimensions. Even though quadrupoles have a focusing strength that is of first-order in field gradient, **strong focusing** systems—based on arrays of quadrupoles and not dependent on combination with path length focusing—are also second-order focusing systems. We also note that systems based on solenoids have a focusing strength that is also of second-order in the field amplitude. We may infer from these examples that most focusing systems that are simultaneously stable in both transverse directions are of second order in field gradient or amplitude. In fact, this is characteristic of all focusing schemes that have no charge or current present within the beam channel.[1]

The purpose of this section, as well as in Section 3.3, is to introduce some analysis methods associated with strong focusing, in cases where the restoring force is periodic. We indicate this periodicity in the equation of motion by writing

$$x'' + \kappa_x^2(z)x = 0 \quad \text{with} \quad \kappa_x^2(z + L_p) = \kappa_x^2(z). \tag{3.10}$$

The assumed period length, L_p, must be, in the context of a circular accelerator, no larger than the circumference C of the accelerator. Most large accelerators are made up of several (M_p) identical modules and, therefore, have a periodicity of $L_p = C/M_p$. Also, in linear accelerators and transport lines, this periodic focusing (often termed **alternating gradient focusing**) is used by typically employing a very simple array of quadrupole magnets with differing sign field gradients. In fact, because the act of bending the design orbit introduces an asymmetry between the focusing in the bend (x) direction and non-bend (y) direction (as seen in Section 3.1), focusing in rectilinear—as opposed to curvilinear—transport is inherently simpler. For this reason, and to make the discussion as general as possible, we adopt z as the independent variable for the description of the motion in this section as well as in Sections 3.3 and 3.4.

There are two cases of interest that can be readily analyzed: (a) the focusing is piece-wise constant, as in magnetic quadrupoles, and (b) the focusing is a sinusoidally varying function in z. The first case is quite straightforward, as for a piece-wise constant value of the focusing strength, $\kappa_x^2(z) = \kappa_0^2$ in Eq. (3.10), the problem is reduced to that of the simple harmonic oscillator. The only generalization involved is that the focusing changes its oscillator characteristics in discrete steps at a finite number of points in z. Thus, the solution to the entire problem will involve "stitching together" a number of simple harmonic oscillator solutions. This is a straightforward process based on a **matrix description** of the dynamics.

The second case requires a perturbative analytical approach, which cannot be used for all conceivable physical parameters, so it is by nature a bit more problematic. In this section, we only discuss the first case and leave the second case for Section 3.4. The problem of the sinusoidally varying oscillator strength is encountered in more esoteric situations in charged particle dynamics (e.g. focusing of particles due to radio-frequency waves, ponderomotive forces in laser fields), but is of high conceptual value, since it illuminates the physical basis of the matrix results.

We now consider a piece-wise constant, periodic focusing described by a focusing strength (κ_0^2) that may be positive, negative, or vanishing. Physically, these cases may correspond, for example, to a focusing quadrupole, a defocusing quadrupole, and a force-free **drift**, respectively. For $\kappa_0^2 > 0$, Eq. (3.10) is a differential equation that describes a simple harmonic oscillator, where the solution can be written, with the constants of integration determined in terms of an **initial state vector** (a vector defined in trace space),

$$\vec{x}(z_0) \equiv \begin{pmatrix} x \\ x' \end{pmatrix}_{z=z_0} = \begin{pmatrix} x_i \\ x'_i \end{pmatrix},$$

[1] An electric focusing system based on a uniform column of charged particles (opposite in sign to the beam particles) lying along the beam axis produces a lens with cylindrically symmetric electric focusing fields (see Ex. 2.14), and is, therefore, stable in both transverse directions. This type of focusing (**ion focusing**) has a strength that is of first-order in the resultant electric field. A similar device based on magnetic fields can be created by using a current carrying plasma column (**plasma lens**).

as

$$x(z) = x_i \cos(\kappa_0(z - z_0)) + \frac{x_i'}{\kappa_0} \sin(\kappa_0(z - z_0)). \tag{3.11}$$

By differentiating Eq. (3.11), we also obtain the angle

$$x'(z) = -\kappa_0 x_i \sin(\kappa_0(z - z_0)) + x_i' \cos(\kappa_0(z - z_0)). \tag{3.12}$$

Equations (3.11) and (3.12) can be conveniently represented by a matrix expression, illustrating the relationship between the initial state vector $\vec{x}(z_0)$ and final state vector $\vec{x}(z)$ as

$$\vec{x}(z) = \mathbf{M_F} \cdot \vec{x}(z_0) \quad \text{with } \mathbf{M_F} = \begin{bmatrix} \cos(\kappa_0(z - z_0)) & \frac{1}{\kappa_0} \sin(\kappa_0(z - z_0)) \\ -\kappa_0 \sin(\kappa_0(z - z_0)) & \cos(\kappa_0(z - z_0)) \end{bmatrix}. \tag{3.13}$$

For the transformation of the vector $\vec{x}$ through a focusing section (lens) of length l, the matrix describing the transformation is simply

$$\mathbf{M_F} = \begin{bmatrix} \cos(\kappa_0 l) & \frac{1}{\kappa_0} \sin(\kappa_0 l) \\ -\kappa_0 \sin(\kappa_0 l) & \cos(\kappa_0 l) \end{bmatrix}. \tag{3.14}$$

In the case of a defocusing lens, we have $\kappa_0^2 < 0$ in Eq. (3.10), and the solution to this equation takes the form,

$$x(z) = x_i \cosh(|\kappa_0|(z - z_0)) + \frac{x_i'}{|\kappa_0|} \sinh(|\kappa_0|(z - z_0)), \tag{3.15}$$

with

$$x'(z) = |\kappa_0| x_i \sinh(|\kappa_0|(z - z_0)) + x_i' \cosh(|\kappa_0|(z - z_0)). \tag{3.16}$$

Thus, the transformation matrix describing the propagation of the vector $\vec{x}$ through a defocusing lens of length l is written, in analogy with Eq. (3.14), as

$$\mathbf{M_D} = \begin{bmatrix} \cosh(|\kappa_0|l) & \frac{1}{|\kappa_0|} \sinh(|\kappa_0|l) \\ |\kappa_0| \sinh(|\kappa_0|l) & \cosh(|\kappa_0|l) \end{bmatrix}. \tag{3.17}$$

Two interesting cases can be obtained by taking the limits of Eqs (3.14) and (3.17). The first is the force-free drift, in which case it is obvious that we are taking the limit that the force disappears, $\kappa_0 \to 0$. As a result, both Eqs (3.14) and (3.17) yield the same limit,

$$\mathbf{M_O} = \begin{bmatrix} 1 & L_d \\ 0 & 1 \end{bmatrix}, \tag{3.18}$$

where L_d is the length of the drift space. In a drift, the position x changes while the angle x' does not.

The other case of interest is the so-called thin-lens limit, where $\kappa_0^2 l$ is kept finite and constant, while $l \to 0$ in Eqs (3.14) and (3.17). As a result, we have

$$\mathbf{M_{F(D)}} = \begin{bmatrix} 1 & 0 \\ -\frac{1}{f} & 1 \end{bmatrix}, \tag{3.19}$$

where the lens focal length is defined as $f \equiv (\kappa_0^2 l)^{-1}$, and is positive for focusing lenses and negative for defocusing lenses. In the thin-lens limit, the change in position x is negligible and only the angle x' is transformed.

Two properties of the formalism developed so far should be noted. First, all of the transformation matrices—the focusing, defocusing, drift, and thin-lens matrices—have determinant equal to 1. This is a general property of **linear transformations**, which are precisely those that can be written as matrix transformations (see Ex. 3.8). The property of unit determinant transformation matrix is also a manifestation of Liouville's theorem, as is discussed later in this section.

The second notable property of the matrix formalism is that the full solution of the motion through a number of focusing elements (regions of constant κ_0^2) can be written in terms of the component element matrices' product. To clarify this, consider the example of a periodic system composed of a thin focusing lens, followed by a drift, a thin defocusing lens, and a final drift. In this case, the full transformation through one period of the system is written as

$$\vec{x}(L_p + z_0) = \vec{x}(2L_d + 2l + z_0) = \mathbf{M_O} \cdot \mathbf{M_D} \cdot \mathbf{M_O} \cdot \mathbf{M_F} \cdot \vec{x}(0) \equiv \mathbf{M_T} \cdot \vec{x}(z_0). \quad (3.20)$$

The total transformation matrix $\mathbf{M_T}$, being the product of matrices all of unit determinant, also has the property $\det(\mathbf{M_T}) = 1$. Note that the matrix product given in Eq. (3.20) is written in reverse order from that in which the component matrices are physically encountered in the beam line. Confusion on the ordering of matrices is the most common mistake made in the matrix analysis of beam dynamics!

For the example of Eq. (3.20), the total transformation matrix can be explicity written as

$$\mathbf{M_T} = \begin{bmatrix} \frac{\partial x}{\partial x_i} & \frac{\partial x}{\partial x_i'} \\ \frac{\partial x'}{\partial x_i} & \frac{\partial x'}{\partial x_i'} \end{bmatrix} = \begin{bmatrix} 1 - \frac{L_d}{f} - \left(\frac{L_d}{f}\right)^2 & 2L_d + \frac{L_d^2}{f} \\ -\frac{L_d}{f^2} & \frac{L_d}{f} + 1 \end{bmatrix}. \quad (3.21)$$

Note that in the first display of the matrix elements in Eq. (3.21), we indicate their general meaning as the first partial derivatives of the final conditions with respect to the initial conditions. The partial derivative form of the matrix shows explicitly that it can also be interpreted as a generalized linear transformation of coordinates in trace space. The determinant of this matrix is known, in the context of coordinate transformations, as the **Jacobian** of the transformation; it is the ratio of differential area elements of the final and original coordinates, $\det \mathbf{M_T} = dx\, dx'/dx_i\, dx_i'$. The fact that this determinant is unity in all linear trace space transformations indicates that they are **area preserving**, as anticipated by application of Liouville's theorem to trace space.

A few salient aspects of the transformation can be deduced by inspection of Eq. (3.21). The matrix element M_{T11} is the spatial magnification ($\partial x/\partial x_i$), and it can be seen in this case to be strictly less than one in our example. Further, if this matrix element is zero, this indicates a parallel-to-point transformation of the trajectory. The "focusing" matrix element M_{T21} ($\partial x'/\partial x_i$) can be compared to that found in Eq. (3.19), to deduce an equivalent thin-lens $f_{thin} = f^2/L_d$ that is always positive (focusing). The focusing indicated by M_{T21} is of second-order in the lens strength f because the two first-order contributions, due to the focusing and defocusing lenses, respectively, cancel. The second-order focusing naturally disappears when $L_d \to 0$, as it must when two thin-lenses of equal and opposite strength are directly joined. If the two lenses were of unequal strength, a first-order effect would appear, but it would be focusing in x and defocusing in y, or vice versa. Thus, in order that there be equal net focusing in both transverse dimensions x and y, the focusing and defocusing lens strengths in a two-lens system should be equal. When the matrix element M_{T12} vanishes, this indicates the point-to-point imaging condition, where the final offset is independent of the initial angle. Finally, M_{T22} is the magnification in x'; when it vanishes, this indicates a point-to-parallel transformation.

Thus far, we have not explicitly required that the matrix transformations discussed correspond to periodic systems. In many scenarios in charged particle optics (e.g. circular accelerators and storage rings) and in light optics (e.g. laser resonators, cf. Section 8.3), the particles are stored in a periodic focusing system like that described by Eq. (3.20) for many traversals of the system. In a large colliding beam storage ring, this may mean billions of turns around the machine. It is, therefore, of primary interest to find out what the effect of negotiating this system many times is on the particle motion—in particular, whether or not the motion is linearly stable. Assurance of the stability of particle motion under forces that are linear in displacement from the design orbit (as in Eq. (3.10)) is a necessary, but not sufficient, condition for absolutely stable motion. Nonlinear forces may also cause unstable orbits to appear at large amplitude in an otherwise linearly stable system.

In order to begin this analysis, we first note that the transformation corresponding to n passes through the system can be written as

$$\vec{x}(NL_p + z_0) = \mathbf{M}_T^n \cdot \vec{x}(z_0).\tag{3.22}$$

This expression is simplified if we note that the transformation of the vector can be written in terms of eigenvectors (akin to the familiar normal mode vectors of coupled oscillator systems) as

$$\vec{x}(L_p + z_0) = \mathbf{M}_T \cdot \vec{x}(z_0) = a_1\lambda_1\vec{d}_1 + a_2\lambda_2\vec{d}_2.\tag{3.23}$$

In Eq. (3.23), the eigenvectors $\vec{d}_j$ have the defining property that the matrix transformation only changes them by a constant factor, $\mathbf{M}_T \cdot \vec{d}_j = \lambda_j\vec{d}_j$. The coefficients defined in Eq. (3.23) are the projections $a_j = \vec{x}(z_0) \cdot \vec{d}_j$ of the initial conditions. Using the eigenvectors, we can recast Eq. (3.22) as

$$\vec{x}(NL_p + z_0) = \mathbf{M}_T^n \cdot \vec{x}(z_0) = a_1\lambda_1^n\vec{d}_1 + a_2\lambda_2^n\vec{d}_2.\tag{3.24}$$

We will see below the physical meaning of these eigenvectors. Before discussing them, we first note, from Eq. (3.24), that the eigenvalues of the transformation must be complex numbers of unit magnitude, or the motion will be exponential—meaning either unbounded (positive exponent), or decaying (negative exponent). In either of these cases, the motion is termed unstable. To find the eigenvalues, we write the transformation of an eigenvector through one period as

$$(\mathbf{M}_T - \lambda_j\mathbf{I}) \cdot \vec{d}_j = 0,\tag{3.25}$$

where $\mathbf{I}$ is the identity matrix. Requiring the determinant of the matrix operating on the eigenvector vanish, we have

$$\lambda_j^2 - (\mathbf{M}_{T11} + \mathbf{M}_{T22})\lambda_j + (\mathbf{M}_{T11}\mathbf{M}_{T22} - \mathbf{M}_{T12}\mathbf{M}_{T21}) = 0,\tag{3.26}$$

or, using the fact that $\det(\mathbf{M}_T) = 1$,

$$\lambda_j^2 - (\mathbf{M}_{T11} + \mathbf{M}_{T22})\lambda_j + 1 = 0. \tag{3.27}$$

Recognizing that the eigenvalue is of unit magnitude, $|\lambda_j| = 1$, when the motion is stable, we now choose to write it as $\lambda_j = \exp(\pm i\mu)$. Here μ, when it is real, is referred to as the **phase advance** per period. With this choice, the solution to Eq. (3.27) becomes

$$\exp(\pm i\mu) = \cos(\mu) \pm i\sin(\mu) = \frac{\mathrm{Tr}(\mathbf{M}_T)}{2} \pm i\sqrt{1 - \left(\frac{\mathrm{Tr}(\mathbf{M}_T)}{2}\right)^2}, \tag{3.28}$$

and we have employed in Eq. (3.28) the definition of the trace of the transformation matrix, $\mathrm{Tr}(\mathbf{M}_T) \equiv M_{T11} + M_{T22}$. It is clear that the absolute value $|\exp(\pm i\mu)| = 1$ if $|\mathrm{Tr}(\mathbf{M}_T)| \leq 2$, and, furthermore, the value of μ is real under the same condition. Thus, the condition for stable motion, in short, is

$$|\mathrm{Tr}(\mathbf{M}_T)| = |\lambda_1 + \lambda_2| \leq 2. \tag{3.29}$$

For the case of the focus–drift–defocus–drift, or FODO, array (in the context of periodic systems, the term FODO **lattice** is applied) discussed above, this stability criterion becomes

$$\frac{L_d}{f} \leq 2, \tag{3.30}$$

which provides an upper limit on the defined phase advance. This limit is given in the FODO system by

$$\cos(\mu) = \frac{\mathrm{Tr}(\mathbf{M}_T)}{2} = 1 - \frac{1}{2}\left(\frac{L_d}{f}\right)^2. \tag{3.31}$$

Note that the maximum stable phase advance per period is $\mu = \pi$. Also, it may be remarked that since the eigenvalues of stable motion are complex, the eigenvectors are generally complex.

The meaning of the phase advance per period and the associated eigenvectors can be illustrated with a few examples of ray tracing, as in Figs 3.3 and 3.4. In these ray plots, the conventional representation of focusing and defocusing lenses as thin convex and concave forms, respectively, is used. The first example, shown in Fig. 3.3, is the case $\mu = \pi/2$ ($\lambda_j = \pm i$), where (the real component of) an eigenvector trajectory (1) with pure offset initial conditions (midway through

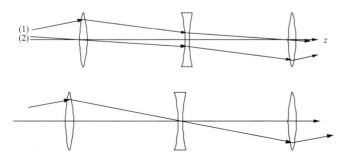

Fig. 3.3 The real components of the eigenvector rays for the case $\mu = \pi/2$ ($L/f = \sqrt{2}$) in a FODO periodic array. Ray (1) is the cosine-like trajectory and ray (2) is the sine-like trajectory (beginning propagation in the center of the first lens).

Fig. 3.4 The real component of the eigenvector ray for the case $\mu = \pi$ ($L/f = 2$) in a FODO periodic array, showing limits of stability. The eigenvalues are degenerate (both equal to -1) in this case.

the first focusing lens), $(x, x') = (x_i, 0)$, is converted to a trajectory with pure angle and no offset in final (conditions midway through the final focusing lens, $(x, x') = (0, x'_f)$). This behavior is reminiscent of the cosine function over the first $90°$ of phase advance. Note that trajectory (2) is the converse of trajectory (1) since it begins with only an angular deviation $(x, x') = (0, x'_i)$, and ends with only an offset $(x, x') = (x_f, 0)$, just as a sine function does over its initial $90°$. This example illustrates why the initial conditions of type (1) are said to generate the cosine-like solution to the motion and the initial conditions of type (2) give rise to the sine-like solution.

This point of view suggests yet another alternative notation for the transformation matrix, indicating the matrix elements as the coefficients of the cosine- and sine-like orbits (C and S), and their derivatives (C' and S'),

$$\mathbf{M}_T = \begin{bmatrix} C & S \\ C' & S' \end{bmatrix}. \tag{3.32}$$

The cosine-like component begins in Fig. 3.3 midway through the first focusing lens with a parallel, unit-offset ray, and the sine-like component begins at this point with an on-axis ray having unit angle(!). These are just mathematical definitions; one need not be concerned with a unit angle being a violation of the paraxial ray approximation, as, in practice, the coefficient (initial angle) that is multiplied by S will always be small compared to unity.

The case of $\mu = \pi$, where $\mathrm{Tr}(\mathbf{M}_T) = -2$ and is therefore directly at the limit of stability, is shown in Fig. 3.4. It is revealed that there is only one non-trivial eigenvector corresponding to the degenerate eigenvalue $\lambda = -1$. With only one eigenvector, there can only be one type of stable trajectory, that which enters and exits the focusing lens with equal and opposite angles, and passes through the axis at the defocusing lens position. One can see that, as this situation is approached, $\mu \to \pi$, and the particles tend to undergo very large excursions at the focusing lens position, while having very small offsets near the defocusing lens. In a beam composed of many particles, this implies that the beam will be much larger near the focusing lens than it is near the defocusing lens.

It should be noted that the trace of the matrix describing a period of the motion is independent of the choice of initial position z_0—the proof of this assertion is left for Exercise 3.5. Thus, the eigenvalues and phase advance μ are independent of this choice. The eigenvectors, in contrast, are dependent on choice of z_0.

3.3   Visualization of motion in periodic focusing systems

In a periodic focusing system such as the FODO lattice introduced in the previous section, the motion in x, when plotted continuously at every point in z (see Fig. 3.5) shows a simple harmonic oscillator-like behavior, with some notable local "errors" in the trajectory. In fact, these errors can be seen to be due to a fast, yet small-amplitude, oscillation about a slower, secular simple harmonic motion. The period of the fast oscillatory motion is clearly identical to that of the FODO lattice, due to the alternating sign of the focusing and defocusing forces in a FODO period, while the secular oscillation period is much longer.

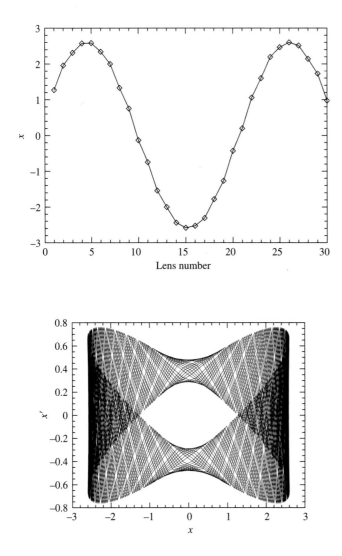

Fig. 3.5 Motion of a particle in a FODO channel with $\mu = 33°$. Lenses are at positions marked with diamond symbols. Note the deviation from simple harmonic motion occurring with the FODO period.

Fig. 3.6 Motion of a particle in a FODO channel of $\mu = 33°$, plotted in trace space. The fast deviations from simple harmonic motion occurring with the FODO period have a large angular spread.

As we shall see below, the fast motion is due to first-order (in the applied field amplitude) forces, while the slower motion may be ascribed to second- (or higher-order) forces that become apparent when averaging over a period of the fast oscillation. The fast motion, despite its small spatial amplitude, will also be seen to have relatively large angles associated with it.

The continuous plotting of the motion in a periodic focusing system is especially troublesome if one follows the motion in trace space, as shown in Fig. 3.6. In this plot, the fast errors in the trajectory have large angular oscillations, and the trace space plot fills in a distorted annular region, yielding unclear information about the nature of the trajectory. On the other hand, if one only plots the trace space point of a trajectory once per FODO period, a method of graphical representation known as a **Poincaré plot**, then the motion is regular. It is in fact an ellipse in trace space, as illustrated by Fig. 3.7, and is thus reassuringly reminiscent of a simple harmonic oscillator. This "stroboscopic" method of plotting allows the secular motion to be displayed without the interference of the fast oscillations so dramatically illustrated in Fig. 3.6.

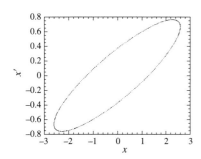

Fig. 3.7 Poincaré plot of the motion of a particle in a FODO channel of $\mu = 33°$, shown previously in Fig. 3.6, but here plotted only at the end of every FODO array.

By examining the eigenvalues and eigenvectors of the matrices for a period of the system, it is clear why the secular motion of a particle in a periodically focusing system is very nearly simple harmonic. The eigenvalues are $\lambda_j = \exp(\pm i\mu)$, and so both indicate motion with a single spatial frequency $k_{sec} = \mu/L_p$, where L_p is the length of one period. This motion is therefore simple harmonic about the real components of the eigenvector directions, which are the major and minor axes of the ellipse shown in Fig. 3.7. Since the real components of the eigenvectors are not in general parallel to the (x, x') axes, the simple harmonic motion does not necessarily yield an ellipse aligned to these axes, but one that is aligned to the eigenvector axes.

This discussion brings up the question of which quantities are independent of where in the period one chooses to interrogate the motion while performing a matrix analysis, or generating a Poincaré plot. As discussed above, and in Exercise 3.5, the trace of a matrix is independent of where in the period z_0 one chooses to begin the transport description. Thus the eigenvalues of the motion and the phase advance per period are independent of the choice of z_0. This is intuitively obvious from Fig. 3.5—the average, secular motion frequency cannot depend on which point in the fast periodic motion one begins the analysis. On the other hand, the eigenvectors depend on the choice of z_0, and so the orientation of the ellipse typified by that shown in Fig. 3.7 is also dependent on choice of z_0. The area inside of the ellipse, being related to the **action**, and therefore the energy of the slow, secular oscillation (see Chapter 1), is independent of z_0. This area and its physical meaning will be discussed further in this chapter, as well as in Chapter 5.

If one is concerned primarily with the approximate secular focusing effects of a periodic lattice, they can be taken into account by use of the **smooth approximation**, in which only the average focusing effect is used in the equation of motion,

$$x'' + k_{sec}^2 x = 0. \tag{3.33}$$

This approximation will be used in several upcoming analyses of oscillatory forces in beam physics phenomena. A direct analytical method for deriving the smooth (secular), focusing strength k_{sec}^2 will be presented in Section 3.4. For now, we restate that the average focusing strength employed in Eq. (3.33) can be simply deduced from $k_{sec} = \mu/L$. In the case of an alternating, focus–defocus (no drift) lattice (see Ex. 3.6), the phase advance is given exactly by

$$\cos(\mu) = \cos\left(\frac{\kappa_0 L_p}{2}\right) \cosh\left(\frac{\kappa_0 L_p}{2}\right). \tag{3.34}$$

For small phase advance per period, $\mu \ll 1$, Eq. (3.34) can be approximated as

$$1 - \frac{\mu^2}{2} \cong \left(1 - \left(\frac{\kappa_0 L_p}{2}\right)^2\right)\left(1 + \left(\frac{\kappa_0 L_p}{2}\right)^2\right) \quad \text{or} \quad \mu \cong \frac{1}{4\sqrt{2}}(\kappa_0 L_p)^2, \tag{3.35}$$

and the average focusing strength is given by

$$k_{sec}^2 \cong \frac{1}{32}k_0^4 L_p^2. \tag{3.36}$$

Assuming magnetic quadrupole focusing, Eq. (3.36) indicates that the focusing strength is proportional to the applied field squared, $k_{sec}^2 \propto B'^2$. This is similar to the situation we have found in solenoid focusing, and it again points to a general result mentioned above—all linear focusing that is equal in both transverse dimensions, and occurs in a macroscopic charge and current-free region, has an average strength that is second-order in the applied field strength. The reasons for this result in the context of quadrupole focusing will be more apparent after the discussion in Section 3.4.

3.4 Second-order (ponderomotive) focusing

As we noted at the beginning of Section 3.3, periodic focusing lattices can be analyzed in a number of ways. In many situations encountered in charged particle optics the forces are piece-wise constant, and the matrix methods of Section 3.3 provide an exact and powerful description of the motion. On the other hand, there are situations where the periodic focusing is not piece-wise constant, and another method is appropriate, one based on a Fourier decomposition of the periodic forcing function $\kappa^2(z)$. The basis of this analysis is an examination of purely harmonic, or sinusoidally varying, focusing. To be specific, we consider an equation of motion of the form

$$x'' + \kappa_0^2 \sin(k_\text{p}z)x = 0, \tag{3.37}$$

where we have chosen the phase of the focusing function for ease of further analysis,[2] and defined the wave number associated with the focusing strength period, $k_\text{p} = 2\pi/L_\text{p}$.

Equation (3.37) is classified in mathematics texts as a form of Hill's equation (an oscillator with periodic "focusing" coefficient), and also more specifically a form of the **Mathieu equation** (an oscillator with a component of sinusoidally periodic focusing). The exact solutions to the Mathieu equation have been studied in detail, but these solutions are not terribly useful or illuminating. For our purposes, therefore, we would like to use approximate **perturbative** methods from which we can learn some lessons about the physics of this system. This type of equation of motion is often found in electrodynamics, most commonly in the case of a charged particle oscillating in a spatially non-uniform electromagnetic wave, e.g. a free-electron near the focus of a powerful laser. We will return to this example later.

The discussion of Section 3.3, where we saw that the secular simple harmonic motion in a periodic lattice has a small, fast "error" trajectory overlaid upon it, motivates our choice of analysis method. The approximation we will employ here assumes that the motion can be broken down into two components, one which contains the small amplitude fast oscillatory motion (the **perturbed** part of the motion), and the other that contains the slowly varying or secular, large amplitude variations in the trajectory. This is written explicitly in one transverse direction as

$$x = x_\text{osc} + x_\text{sec}. \tag{3.38}$$

By this assumption, we expect the case described by Eq. (3.37), to display motion that looks qualitatively similar like the example given in Fig. 3.8.

[2]This form indicates that we will always be choosing the phase of the focusing function so that it is odd with respect to the origin. Since we are interested only in finding the averaged second-order focusing in this analysis, no loss of generality in the discussion is suffered due to this assumption.

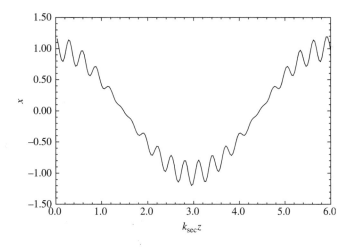

Fig. 3.8 An example of fast, small amplitude oscillatory motion superimposed upon slow, large amplitude, secular oscillation.

The oscillatory component is analyzed by making the approximation that the offset $x = x_{sec} \cong$ constant $(x_{sec}'' \ll x_{osc}'')$ over an oscillation in the second term on the right hand side of Eq. (3.37),

$$x_{osc}'' + \kappa_0^2 \sin(k_p z) x_{sec} = 0, \tag{3.39}$$

where we have used the assumed $|x_{osc}| \ll |x_{sec}|$.

Since we will eventually add the secular solution x_{sec} to the perturbed component x_{osc}, we do not have to examine the full solution to Eq. (3.39) at this point, only the inhomogeneous solution. We can add the homogeneous components along with x_{sec} later, when we apply the initial conditions to the full solution.

The inhomogeneous component of the solution to Eq. (3.39) is

$$x_{osc} = \sin(k_p z) \frac{\kappa_0^2}{k_p^2} x_{sec}. \tag{3.40}$$

This solution for the oscillatory portion of the motion is accurate if the assumptions leading to it are valid, requiring that

$$\frac{|x_{osc}|}{|x_{sec}|} \ll 1 \quad \text{or} \quad \frac{\kappa_0^2}{k_p^2} \ll 1. \tag{3.41}$$

In short, if the fast oscillation amplitude is small compared to the secular amplitude, then we expect the approximate solution to Eq. (3.39) to be accurate. This will be true if $\kappa_0^2 / k_p^2 \ll 1$, implying that the focusing is weak, in the sense that no significant part of a secular oscillation can occur during the focusing strength period $2\pi/k_p$. This condition, of course, also implies that the secular phase advance per period is small, $\mu \ll 1$.

With Eq. (3.40) in hand, it is now possible to substitute it into Eq. (3.39) to obtain

$$x'' = -\kappa_0^2 \sin(k_p z) x \cong -\kappa_0^2 \sin(k_p z) \left[1 + \sin(k_p z) \frac{\kappa_0^2}{k_p^2} \right] x_{sec}. \tag{3.42}$$

The next step in the analysis is to convert Eq. (3.42) into an averaged expression that will give us the behavior of the secular component of the motion, $x_{sec} \equiv \langle x \rangle$, where the indicated average is over a period L_p. We now obtain, by averaging Eq. (3.42) over one period of the fast oscillation,

$$x''_{sec} \equiv \langle x'' \rangle = -\langle \kappa^2 \rangle x_{sec} \cong -\frac{\kappa_0^2}{2} \frac{\kappa_0^2}{k_p^2} x_{sec} \qquad (3.43)$$

or, in standard simple harmonic oscillator form,

$$x''_{sec} + \frac{\kappa_0^4}{2k_p^2} x_{sec} = 0. \qquad (3.44)$$

Equation (3.44) predicts simple oscillations having a spatial frequency (wave number) $k_{sec} = \sqrt{\langle \kappa^2 \rangle} = \kappa_0^2/\sqrt{2}k_p$. The derivation of these oscillations can be viewed as an algorithm for extracting the smooth approximation picture equivalent to Eq. (3.37). Note that one of the assumptions used in generating Eq. (3.39) was that $x''_{sec} \ll x''_{osc}$, which we now see implies $\kappa_0^4/2k_p^2 \ll k_p^2$—essentially the same criterion as given by Eq. (3.41). Equation (3.44) is derived through an averaging technique that is reminiscent of the stroboscopic visualization analysis given in the Section 3.3; the slow oscillations in an alternating periodic focusing system become easily apparent if one examines the system only once per period, á la Poincaré.

   The full solution to Eq. (3.37) is, assuming that our analysis is valid for the parameters of interest,

$$x = \left[1 + \sin(k_p z)\frac{\kappa_0^2}{k_p^2} \right] \left[A \cos\left(\frac{\kappa_0^2}{\sqrt{2}k_p} z \right) + B \sin\left(\frac{\kappa_0^2}{\sqrt{2}k_p} z \right) \right], \qquad (3.45)$$

where A and B are constants of integration. These constants, as mentioned above, can be used to ensure that the initial conditions are properly taken into account.

   Note that, unlike our discussion in Section 3.3, we have no predictions concerning the values of the focusing amplitude that will produce instability, as Eq. (3.44) always indicates stable, simple harmonic motion. This is because Eq. (3.44) loses validity in precisely the parameter range where the system would be found unstable, $\kappa_0^2 \approx k_p^2$. In this range, one observes that the oscillations behave as in the thin/thick lens alternating focusing system (see Ex. 3.6) in an unstable regime—x may change wildly in amplitude during a single focusing period.

   For cases where our approximations are valid, one can trivially predict the phase advance per period μ. This quantity can be seen to be merely the argument of the sinusoidal functions in the secular component of x,

$$\mu \cong \kappa_0^2 L_p/\sqrt{2}k_p = \sqrt{2}\pi(\kappa_0^2/k_p^2) = \kappa_0^2 L_p^2/\sqrt{8}\pi, \qquad (3.46)$$

which is nearly the same as that given by Eq. (3.36). It should be noted again that this approximation is useful only for small values of μ since the validity of the analysis implies $\kappa_0^2 \ll k_p^2$. In fact, if we examine the treatment of the thin-lens FODO system discussed in Section 3.3, we see that the stability limit is reached when $\mu = \pi$. Thus, the estimate given in Eq. (3.46) is valid when we are far below the stability limit. If one is near the stability limit, then the actual value of μ can be obtained by integrating Eq. (3.37) numerically over a period. One algorithm to follow involves starting from a maximum or minimum in the focusing ($k_p z_0 = 0, \pi$), which gives the correct symmetry in focusing with the initial conditions $(x, x')|_{z=z_0} = (1, 0)$ to yield the cosine-like orbit. The phase advance is then found numerically, through $\mu = \cos^{-1}(x(z_0 + L_p))$.

Once one has developed this approximate analysis of a sinusoidal focusing function, a much more general result follows immediately—the secular oscillatory behavior of an arbitrary periodic focusing function $\kappa^2(z)$ can be predicted. Using of a Fourier series representation, one can write (for odd symmetry functions $\kappa^2(z)$)

$$\kappa_x^2(z) = \kappa_0^2 \sum_{n=0}^{\infty} a_n \sin(n k_p z), \tag{3.47}$$

where the Fourier coefficients

$$a_n = \frac{2}{L_p \kappa_0^2} \int_0^{L_P} \kappa_x^2(z) \sin(n k_p z) \, dz. \tag{3.48}$$

Employing the same derivation methods as those given in Eqs (3.38)–(3.43) yields an averaged expression similar to Eq. (3.44),

$$x_{sec}'' + \frac{\kappa_0^4}{2k_p^2} \left[\sum_{n=1}^{\infty} \frac{a_n^2}{n^2} \right] x_{sec} = 0. \tag{3.49}$$

This expression can be used to more accurately compare to the results of the matrix analysis introduced in Section 3.3. For the case of a thick-lens focusing described in Exercise 3.6, we have

$$\kappa^2(z) = \begin{cases} \kappa_0^2, & 0 \leq z < L_p/2, \\ -\kappa_0^2, & L_p/2 \leq z < L_p, \end{cases} \tag{3.50}$$

and the Fourier coefficients are

$$a_n = \begin{cases} 4/(\pi n), & n \text{ odd}, \\ 0, & n \text{ even}. \end{cases} \tag{3.51}$$

The secular focusing in this case is simply described by

$$x_{sec}'' + \frac{8\kappa_0^4}{\pi^2 k_p^2} \left[\sum_{n=1, n \text{ odd}}^{\infty} \frac{1}{n^4} \right] x_{sec} = 0. \tag{3.52}$$

The sum in Eq. (3.52) is very close to unity, $\sum_{n=1, n \text{ odd}}^{\infty} n^{-4} = 1.015$. Therefore, the focusing is almost exclusively due to the fundamental harmonic $n = 1$, which gives an average focusing strength $\langle \kappa^2 \rangle = 8\kappa_0^4/\pi^2 k_p^2$. The predicted approximate phase advance per period, including all harmonics, is

$$\mu \cong \sqrt{\langle\kappa^2\rangle}L_{\mathrm{p}} = 4\sqrt{2.03}\kappa_0^2/k_{\mathrm{p}}^2 = 5.70 \cdot \kappa_0^2/k_{\mathrm{p}}^2. \tag{3.53}$$

The exact phase advance per period can be found by matrix techniques (Eq. (3.34)) and is

$$\mu = \cos^{-1}(\cos(\kappa_0\pi/k_{\mathrm{p}})\cosh(\kappa_0\pi/k_{\mathrm{p}})) \cong \pi^2\kappa_0^2/\sqrt{3}k_{\mathrm{p}}^2 \cong 5.70 \cdot \kappa_0^2/k_{\mathrm{p}}^2, \tag{3.54}$$

which is in complete agreement with Eq. (3.53). We have a curious ancillary result from this exercise, a fundamental mathematical artifact obtained by comparing Eqs (3.53) and (3.54): the series summation $\sum_{n=1,n\text{ odd}}^{\infty} n^{-4} = \pi^4/96$.

This type of system, with a sinusoidally time-dependent focusing force, or more generally, a sinusoidally time-dependent force having a linear gradient in the direction of the force, is found in many situations in physics. A simple example is that of a spatially localized electromagnetic wave (e.g. a laser beam). In this case, a charged particle oscillates under the influence of the light wave, but the force is slightly smaller during the half-cycle that the particle spends in the weaker field region. By this mechanism, the particle is pushed secularly towards regions of smaller wave intensity or energy density. This effect is known as a **ponderomotive force**, and is always characterized by having a secular force (the focusing strength in this case) that is second-order in the applied field amplitude.

We will also see that the results of this section are directly applicable to analyzing the focusing experienced by charged particles accelerating in radio-frequency linear accelerators. In this case, the transverse electromagnetic forces are oscillatory and quickly varying and give rise to a second-order secular, or ponderomotive, focusing effect. It is also interesting to note that a similar second-order ponderomotive effect occurs in the longitudinal dynamics of the electrons in a free-electron laser.

3.5 Matrix description of motion in bending systems

The discussions in Sections 3.3 and 3.4, in which the focusing optics are assumed to be periodic functions of the independent variable, have clearly been motivated by desire to understand circular accelerators, where the periodicity is enforced every turn around the device. Most of the tools needed to understand the motion in circular systems, which of necessity contain magnets that bend the particle trajectories, have been introduced by now, but a few remain. Here we discuss two of these tools: the method of analyzing entrance and exit effects in magnets, and a more detailed discussion of the dispersion function introduced in Chapter 2. Note that we change our notation convention slightly in this section, with the independent variable indicated by s and not z, as we have already done once before for our weak focusing analysis.

We must first examine the coordinate conventions that we are to use before we develop the description of the motion in bends. When one encounters a bend magnet, the most common convention dictates that positive x is in the direction away from the center of curvature. If one has started the calculation of the optics with the opposite convention, it is necessary to flip the coordinates by use of the negative identity matrix, $\vec{x} = -\mathbf{I} \cdot \vec{x} = -\vec{x}$ before proceeding.[3] Then the matrix of a finite-length magnet can be broken up into three separate matrices

$$\mathbf{M}_{\text{mag}} = \mathbf{M}_{\text{exit}}\mathbf{M}_{\text{bend}}\mathbf{M}_{\text{entrance}}, \tag{3.55}$$

where the bend matrix in a combined-function magnet can be deduced simply from Eqs (3.3) and (3.4) to be

$$\mathbf{M}_{\text{bend}} = \begin{bmatrix} \cos(\kappa_{\text{b}}l) & \frac{1}{\kappa_{\text{b}}}\sin(\kappa_{\text{b}}l) \\ -\kappa_{\text{b}}\sin(\kappa_{\text{b}}l) & \cos(\kappa_{\text{b}}l) \end{bmatrix}$$

$$= \begin{bmatrix} \cos(\sqrt{1-n}\,\theta_{\text{b}}) & \frac{R}{\sqrt{1-n}}\sin(\sqrt{1-n}\,\theta_{\text{b}}) \\ -\frac{\sqrt{1-n}}{R}\sin(\sqrt{1-n}\,\theta_{\text{b}}) & \cos(\sqrt{1-n}\,\theta_{\text{b}}) \end{bmatrix}. \tag{3.56}$$

Here the focusing wave number is given by $\kappa_x = \kappa_{\text{b}} = \sqrt{1-n}/R$, and the bend angle is $\theta_{\text{b}} = l/R$.

The edge matrices can be constructed by careful consideration of fringe field and differential path length effects. These effects will be treated here in the thin-lens approximation, meaning that the momentum kick imparted to the particle occurs in such a short distance that the particle does not change its transverse position appreciably during the transit of the edge regions. The horizontal kick can be understood purely in terms of differential path length, as illustrated in Fig. 3.9, which gives a more detailed picture of the magnet entrance edge region shown in Fig. 3.10. The offset trajectory integrates a different total magnetic force, which is linearly dependent on the amount of horizontal offset x, than the design trajectory. This integrated kick is written, following the discussion leading to Eq. (2.13), as the amount of deflecting momentum impulse (force integral) encountered due to the differing path length inside of the magnet edge at a given offset x,

$$\Delta p_{x,\text{edge}} = -q\int_{\text{edge}} B_0\left(1 + \frac{B'x}{B_0R}\right)\,\mathrm{d}s \cong qB_0\tan(\theta_{\text{E}})x. \tag{3.57}$$

In Eq. (3.57), we have only included terms linear in x (due to the field index), dropping quadratic terms in this small quantity. The additional path length for the offset particle is seen to be $(\theta_{\text{E}})x$, where the angle θ_{E}, defined as the angle from the (outward) edge normal to the incoming trajectory, is taken to be positive when it points toward the center of curvature in the magnet. A positive θ_{E} indicates that, for a positive offset x, there is a "missing" field region, and the effect of the edge is defocusing. For a negative θ_{E}, there is "additional" field encountered, and the edge has a focusing effect. For the exit angles, the same convention apply, and the angular kick encountered at an edge is given by

$$\Delta x'_{\text{edge}} = \frac{\tan(\theta_{\text{E}})}{R}x. \tag{3.58}$$

The thin-lens matrix associated with an edge is thus found to be

$$\mathbf{M}_{\text{edge}} = \begin{bmatrix} 1 & 0 \\ \frac{\tan(\theta_{\text{E}})}{R} & 1 \end{bmatrix}. \tag{3.59}$$

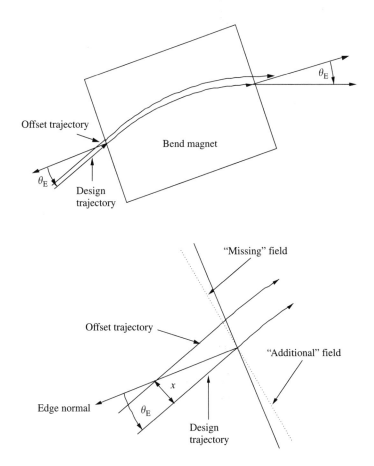

Fig. 3.9 Geometry for defining edge angles and considering the horizontal focusing effects of entrance and exit angles.

Fig. 3.10 Close-up picture of magnet entrance region showing differential path length encountered by offset trajectory. In this case, the offset trajectory sees less field region, and is defocused by the edge.

The focal length of the edge is $f = -R/\tan(\theta_E)$ and is, by convention, positive in the focusing case, $\theta_E < 0$.

As an example of this focusing, the magnet shown in Fig. 3.9 is a **rectangular wedge** magnet. For a field index equal to zero (**flat-field magnet**), it has the following total matrix transformation in x:

$$\mathbf{M}_{T,x} = \mathbf{M}_{edge}\mathbf{M}_{bend}\mathbf{M}_{edge}$$

$$= \begin{bmatrix} 1 & 0 \\ \frac{\tan(\theta_E)}{R} & 1 \end{bmatrix} \begin{bmatrix} \cos(\theta_b) & R\sin(\theta_b) \\ -\frac{1}{R}\sin(\theta_b) & \cos(\theta_b) \end{bmatrix} \begin{bmatrix} 1 & 0 \\ \frac{\tan(\theta_E)}{R} & 1 \end{bmatrix}. \quad (3.60)$$

For the symmetric trajectory shown in Fig. 3.9, the edge angle $\theta_E = \theta_b/2$, and the transformation is written as

$$\mathbf{M}_{T,x} = \begin{bmatrix} 1 & R\sin(\theta_b) \\ 0 & 1 \end{bmatrix}. \quad (3.61)$$

Note that this matrix takes the form of a simple drift, and there is no focusing or magnification associated with it.

In the vertical dimension, the edge transformation can be deduced by noting that, in the fringe field, the vertical component of the field goes from zero to the maximum value as one traces along the edge normal (see Fig. 3.10). Terming this coordinate ξ, we have $\partial B_y/\partial \xi = \partial B_\xi/\partial y \neq 0$ in the fringe region, and thus the field lines have a component in the normal direction, $B_\xi = \int_0^y (\partial B_\xi/\partial y)\, dy \cong (\partial B_y/\partial \xi)_{y=0}\, y$. If the particle enters the fringe field at a non-zero edge angle, θ_E, there is a component of B_ξ normal to the trajectory, and we have a vertical force, $F_y = qv_0 \sin(\theta_E)B_\xi$. This force is gives rise to an integrated angular kick,

$$\Delta y'_{\text{edge}} = -\frac{q}{p_0} \int_{\text{edge}} [B_\xi \sin(\theta_E)]\, ds = -\frac{q}{p_0} \int_{\text{edge}} \left(\frac{\partial B_y}{\partial \xi}\right)_{y=0} y \frac{\sin(\theta_E)}{\cos(\theta_E)}\, dz$$

$$= -\frac{qB_0}{p_0} \tan(\theta_E) y \cong -\frac{\tan(\theta_E)}{R} y. \tag{3.62}$$

The vertical kick is equal in magnitude and opposite in sign to the horizontal edge kick.

For the rectangular, flat-field wedge magnet, the focusing lies entirely in the vertical dimension. The total vertical transformation matrix is written as

$$\mathbf{M}_{T,y} = \begin{bmatrix} 1 & 0 \\ -\frac{\tan(\theta_E)}{R} & 1 \end{bmatrix} \begin{bmatrix} 1 & R\theta_b \\ 0 & 1 \end{bmatrix} \begin{bmatrix} 1 & 0 \\ -\frac{\tan(\theta_E)}{R} & 1 \end{bmatrix}$$

$$= \begin{bmatrix} 1 - \theta_b \tan\left(\frac{\theta_b}{2}\right) & R\theta_b \\ -\frac{\tan\left(\frac{\theta_b}{2}\right)}{R}\left[2 - \theta_b \tan\left(\frac{\theta_b}{2}\right)\right] & 1 - \theta_b \tan\left(\frac{\theta_b}{2}\right) \end{bmatrix}. \tag{3.63}$$

This magnet is focusing and has an effective focal length given by

$$\frac{1}{f_y} = \tan\frac{\left(\frac{\theta_b}{2}\right)}{R}\left[2 - \theta_b \tan\left(\frac{\theta_b}{2}\right)\right]. \tag{3.64}$$

The rectangular wedge magnet displays the opposite characteristics of the zero-edge angle case, which is termed a **sector magnet**. In this case, the edge matrices are ignorable, and $\mathbf{M}_{T,x} = \mathbf{M}_{\text{bend}}$. The effective horizontal focal length is $f_x = R/\sin(\theta_b)$, which for small angle $\theta_b \ll 1$ gives the same value as Eq. (3.64). The vertical focusing is a small, but perhaps non-negligible effect in this case. It arises when the charged particle deflects noticeably in the fringe-field region of the magnet (i.e. when the magnet gap is large). The calculation of this effect is left as an exercise to the reader.

3.6 Evolution of the momentum dispersion function

Now that we have essentially concluded our introductory discussion of betatron motion, where we have been concerned with trajectories of particles possessing the design momentum, but having **offset** and **angular** errors, we can proceed to the discussion of the effects of **momentum errors**. To do this, we must re-examine the momentum dispersion function, which was introduced in Section 2.2 and defined by the differential relation $\eta_x = \partial x/\partial[\delta p/p_0]$. This function allows us to write the horizontal (transverse, in bend-plane) first-order (in betatron and momentum error amplitude) offset as (see Eq. (2.14)) $x = x_\beta + \eta_x(\delta p/p_0)$, with the betatron offset x_β being governed by the analyses of Sections 3.1–3.5. If one needs to analyze a beam optics system with bends that do not lie all in one plane, it is also necessary to introduce a vertical momentum dispersion function, so that $y = y_\beta + \eta_y(\delta p/p_0)$.

The momentum dispersion function, when multiplied by the relative momentum error $\delta p/p_0$, is most usefully thought of as the trajectory of a particle with a (unit) relative momentum error alone, and no betatron error. Therefore, the dispersion is a normalized trajectory and, to the extent that the momentum error is small, will behave nearly as a paraxial betatron trajectory when subjected only to focusing forces—it will approximately obey Eq. (3.10) in such a case. Introducing a bending force on the design orbit produces a qualitatively different effect, however, because the radii of curvature associated with the differing momenta are different. Since the differential acceleration of these orbits is constant, the horizontal dispersion function is also driven by curvature term $x''_{\delta p/p_0} = \eta''_x \cdot (\delta p/p_0) = R(p_0)^{-1} - R(p_0 + \delta p/p_0)^{-1}$, giving a new term in the dispersion evolution equation, $\eta''_x \cong R(p_0)^{-1} \equiv R_0^{-1}$. With the quadrupole and path length focusing terms included as well, we have a differential equation governing the evolution of the horizontal dispersion,

$$\eta''_x + \left(\frac{1}{R_0^2} + \frac{qB'}{p_0} \right) \eta_x = \frac{1}{R_0},$$

or more simply

$$\eta''_x + \kappa_b^2 \eta_x = \frac{1}{R_0}. \tag{3.65}$$

Here, $R_0 = R(p_0)$ is the design radius of curvature and $\kappa_x = \kappa_b = \sqrt{(1-n)}/R_0$ in a bend. The left-hand side of the equation (the homogeneous portion of Eq. (3.65)) has the same form of solutions as given by Eqs (3.11) and (3.15). Equation (3.65) has an inhomogeneous component in a bend and, therefore, has a particular solution in a bend magnet, $\eta_{x,\text{part}} = 1/\kappa_b^2 R_0$. The full solution is, assuming a (possibly combined-function) magnet with net interior horizontal focusing ($\kappa_b^2 > 0$, or $n < 1$), formally

$$\eta_x = A\cos(\kappa_b s) + B\sin(\kappa_b s) + \frac{1}{\kappa_b^2 R_0}, \tag{3.66}$$

where the magnet entrance is taken to be $s = 0$. The full solution can be constructed by matching boundary conditions at the entrance of the bend magnet. Continuity of the horizontal dispersion and its derivative at the entrance to the magnet requires

$$\eta_x(s) = \frac{1}{\kappa_b^2 R_0} + \left[\eta_x(0) - \frac{1}{\kappa_b^2 R_0} \right] \cos(\kappa_b s) + \frac{\eta'_x(0)}{\kappa_b} \sin(\kappa_b s), \tag{3.67}$$

with

$$\eta'_x(s) = \left[\frac{1}{\kappa_b R_0} - \kappa_b \eta_x(0) \right] \sin(\kappa_b s) + \eta'_x(0)\cos(\kappa_b s). \tag{3.68}$$

This linear transformation is written in matrix form by using a 3×3 matrix based on the betatron matrix operating on a dispersion state vector with a dummy third entry, that is,

$$
\begin{pmatrix} \eta_x(s) \\ \eta'_x(s) \\ 1 \end{pmatrix} = \begin{bmatrix} \cos(\kappa_b s) & \frac{1}{\kappa_b}\sin(\kappa_b s) & \frac{1-\cos(\kappa_b s)}{\kappa_b^2 R_0} \\ -\kappa_b\sin(\kappa_b s) & \cos(\kappa_b s) & \frac{\sin(\kappa_b s)}{\kappa_b R_0} \\ 0 & 0 & 1 \end{bmatrix}
$$
$$
\times \begin{pmatrix} \eta_x(0) \\ \eta'_x(0) \\ 1 \end{pmatrix}. \tag{3.69}
$$

For defocusing systems with $\kappa_b^2 < 0$, it is straightforward to show that

$$
\begin{pmatrix} \eta_x(s) \\ \eta'_x(s) \\ 1 \end{pmatrix} = \begin{bmatrix} \cosh(|\kappa_b|s) & \frac{1}{\kappa_b}\sinh(|\kappa_b|s) & -\frac{(1-\cosh(|\kappa_b|s)}{|\kappa_b^2|R_0} \\ \sinh(|\kappa_b|sl) & \cosh(|\kappa_b|s) & \frac{\sinh(|\kappa_b|s)}{|\kappa_b|R_0} \\ 0 & 0 & 1 \end{bmatrix}
$$
$$
\times \begin{pmatrix} \eta_x(0) \\ \eta'_x(0) \\ 1 \end{pmatrix}. \tag{3.70}
$$

The upper left hand 2×2 block in the transformation matrices of Eqs (3.69) and (3.70) are simply the betatron transformation matrices, **M**.

When the beam is in a straight section, $R_0 \to \infty$, and the upper two elements of the third column vanish of the matrix—the third column and third row do not affect the horizontal dispersion or its derivative, as expected. It should be noted in this regard that, since edge matrices are of negligible length, there is no bending of the design orbit and the dispersion transformation matrices are constructed just as in a straight section. That is, the upper left hand 2×2 block in the transformation matrix is the thin-lens matrix, and the final column is $(0,0,1)$.

As with betatron motion, convention dictates that positive η_x be defined in the direction away from the center of curvature. If one has started the calculation of the optics with the opposite convention (for instance, when the direction of a bend changes from one magnet to the next), it is again necessary to flip the sign of η_x and η'_x before proceeding with the matrix calculation of the dispersion and its derivative. In matrix form, this procedure is written

$$
\begin{pmatrix} \eta_x \\ \eta'_x \\ 1 \end{pmatrix}_{new} = \begin{bmatrix} -1 & 0 & 0 \\ 0 & -1 & 0 \\ 0 & 0 & 1 \end{bmatrix} \begin{pmatrix} \eta_x \\ \eta'_x \\ 1 \end{pmatrix}_{old}. \tag{3.71}
$$

For cases where the bending field changes continuously in space, it is not possible to utilize the matrix-based approach to the solution of the dispersion evolution, and one must solve Eq. (3.65) in another manner. As an example of this type of scenario, we recall the case of the magnetic undulator introduced in Section 2.8. In this device, the undulating component of the motion was seen to be

$$
x = \frac{qB_0}{p_0 k_u^2}\sin(k_u z), \tag{3.72}
$$

from which we have the relation

$$
\frac{\partial x}{\partial p_0} = -\frac{qB_0}{p_0^2 k_u^2}\sin(k_u z), \tag{3.73}
$$

and

$$\eta_x = p_0 \frac{\partial x}{\partial p_0} = -\frac{qB_0}{p_0^2 k_u^2} \sin(k_u z) \equiv -\frac{a_u}{\beta_0 \gamma_0 k_u} \sin(k_u z). \qquad (3.74)$$

Here, we introduce the **undulator parameter** $a_u \equiv qB_0/k_u m_0 c$, which we will see (in Chapter 8) plays a central role in the operation of the free-electron laser.

3.7   Longitudinal motion and momentum compaction

While this chapter is nominally only concerned with first-order transverse motion, in this section we open the door to considering the effects of momentum errors on longitudinal motion by further discussing the effects of momentum dispersion. As we shall see below, momentum dispersion plays a critical role in the first-order theory of longitudinal motion and, thus, in the interest of completeness, we include a discussion of this dispersion-based effect. In particle transport systems, longitudinal motion in accelerators can be analyzed approximately using a Taylor series expansion of the motion in the relative longitudinal momentum error. The point of the present analysis is to generate a longitudinal equation analogous to the paraxial betatron equations of motion found in our treatment of transverse motion, for example, Eqs (3.5) and (3.6). In non-bending systems, this analogy is particularly close, because the paraxial betatron equations are also based on first-order expansion of the motion in transverse momentum errors. With bending systems, because of dispersion, the equations of motion are slightly more complicated.

Since the independent variable has been taken to be distance along the design trajectory (s or z), the canonical dependent coordinate in the longitudinal direction is time t, which we then measure in Hamiltonian analyses relative to the time of arrival of the design particle, $\tau = t - t_0$. The variable that plays the role of the "momentum," which is canonical to time "coordinate," is the particle's mechanical energy. However, since we will begin our discussion of longitudinal motion without employing Hamiltonians, and have invested much effort in the previous analyses based on momentum differences, we will continue in this somewhat less rigorous way. To make the present analysis connect more easily to Hamiltonian approaches, we will find it useful to introduce a parameterization of the time through a spatial variable, $\zeta = -v_0 \tau$. This is the distance that must be traveled at the design velocity by the design particle, to reach the position of the temporally advanced (or delayed) particle.

The time of flight of an off-momentum particle to a point along the design orbit is given by

$$\tau(p) = \frac{L(p)}{v(p)}, \qquad (3.75)$$

where $L(p)$ is the distance traveled by the particle as a function of its momentum. The first-order logarithmic expansion of this expression is, assuming the paraxial approximation,

$$\frac{\delta\tau}{t_0} = \frac{\delta L}{L_0} - \frac{\delta v_z}{v_0} \cong \left[\alpha_c - \frac{1}{\gamma_0^2} \right] \frac{\delta p}{p_0}. \qquad (3.76)$$

Here we define the path length parameter $\alpha_c \equiv (\delta L/L_0)/(\delta p/p_0)$ and use $\delta v_z/v_0 \cong (1/\gamma_0^2)(\delta p/p_0)$. Equation (3.76) is used to define the momentum compaction, sometimes termed the time dispersion,

$$\eta_\tau \equiv \frac{\partial(\delta\tau/t_0)}{\partial(\delta p/p_0)} = \alpha_c - \frac{1}{\gamma_0^2}, \tag{3.77}$$

which is analogous to the horizontal dispersion discussed in Section 3.6. For a given section of particle transport, e.g. a complete revolution around a circular accelerator, there is a certain energy at which the time dispersion vanishes, and all particles pass through the system in the same amount of time. This energy, given by

$$\gamma_t = \alpha_c^{-1/2}, \tag{3.78}$$

is termed the **transition energy**. Below transition, particles of higher momentum pass through the system more quickly, which is the natural state of affairs in linear systems. Above transition energy they take—somewhat anti-intuitively—more time to pass through the system, since the added path length of a higher-momentum trajectory outweighs the added advantage in velocity, which becomes progressively smaller as particles become more relativisitic.

To lowest order in momentum error, the path length parameter is

$$\alpha_c = \frac{1}{s - s_0} \int_{s_0}^{s} \frac{\eta_x(\tilde{s})}{R(\tilde{s})} \, d\tilde{s}. \tag{3.79}$$

Note that contributions to the integral in Eq. (3.79) vanish in straight sections, where $R \to \infty$. The path length parameter changes only in bend regions, where the linear dependence of the radius of curvature on the momentum produces a first-order difference in path length. Since the dispersion is "naturally" positive (see Ex. 3.13), one must work at making the dispersion negative and thus usually $\alpha_c > 0$. It is possible under some circumstances to make α_c vanish or be negative, in which case the transport is always "above transition," regardless of energy.

One particular magnetic system, the magnetic undulator, lends itself easily to path length parameter analysis. From Eq. (3.74), we can derive the paraxial (small bend limit) expression

$$\alpha_c = -\frac{1}{s - s_0} \int_{s_0}^{s} \frac{a_u \sin(k_u \tilde{z})}{k_u \beta_0 \gamma_0 R(\tilde{s})} \, d\tilde{s} \cong -\frac{1}{s - s_0} \int_{s_0}^{s} \frac{a_u^2 \sin^2(k_u \tilde{s})}{\beta_0^2 \gamma_0^2} \, d\tilde{s} = -\frac{a_u^2}{2\beta_0^2 \gamma_0^2}. \tag{3.80}$$

and

$$\eta_\tau = -\frac{1}{\gamma_0^2}\left[1 + \frac{a_u^2}{2\beta_0^2}\right]. \tag{3.81}$$

This result should be compared with the one that can be directly deduced from the longitudinal momentum calculation of the paraxial undulator given by

Eq. (2.75). In this limit we can write, with $v_z = p_z c^2 / \sqrt{p_0^2 c^2 + (m_0 c^2)^2}$,

$$\eta_\tau = -\frac{\partial v_z}{\partial p_z}\frac{p_0}{v_0} = -\left[\frac{1}{m\gamma_0^3} + \left(\frac{qB_0}{2k_u p_0}\right)^2 \left(\frac{1+\beta_0^2}{\beta_0^2(\gamma_0 mc)^3}\right)\right] m\gamma_0$$

$$\cong -\frac{1}{\gamma_0^2}\left[1 + \frac{a_u^2}{2}\right], \tag{3.82}$$

in good agreement with Eq. (3.81).

3.8 Linear transformations in six-dimensional phase space*

Previously in this chapter, we have introduced analyses of the motion based on linear dependences of the dynamics on errors in the momenta in one phase plane (actually, one trace space) at a time. In practice, as one needs to keep track of the motion in the entire six-dimensional phase space at once when designing an actual accelerator, the 2×2 matrices that allow transformation of one two-entry trace space vector at a time is replaced by a 6×6 matrix that contains all three $((x, x')(y, y'), (\zeta, \zeta'))$ trace space planes, plus additional information, the possible coupling between trace spaces. This generalized transport matrix transforms the six-entry phase space vector $\vec{\Phi} \equiv (x, x', y, y', \zeta, \zeta')$, where for symmetry we have changed our notation somewhat in writing the normalized longitudinal momentum error, $\zeta' = \delta p_z/p_0 \cong \delta p_0/p_0$. The standard form of this transformation is

$$\vec{\Phi}(s) \equiv \mathbf{R}(s, s_0) \cdot \vec{\Phi}(s_0), \tag{3.83}$$

where the 6×6 $\mathbf{R}(s, s_0)$ matrix is specified to transform the $\vec{\Phi}$-vector from one position in the beamline s_0 to another s. The matrix $\mathbf{R}(s, s_0)$ is formally written

$$\mathbf{R}(s, s_0) = \begin{bmatrix} \frac{\partial x_f}{\partial x_i} & \frac{\partial x_f}{\partial x_i'} & \frac{\partial x_f}{\partial y_i} & \frac{\partial x_f}{\partial y_i'} & \frac{\partial x_f}{\partial \zeta_i} & \frac{\partial x_f}{\partial \zeta_i'} \\ \frac{\partial x_f'}{\partial x_i} & \frac{\partial x_f'}{\partial x_i'} & \frac{\partial x_f'}{\partial y_i} & \frac{\partial x_f'}{\partial y_i'} & \frac{\partial x_f'}{\partial \zeta_i} & \frac{\partial x_f'}{\partial \zeta_i'} \\ \frac{\partial y_f}{\partial x_i} & \frac{\partial y_f}{\partial x_i'} & \frac{\partial y_f}{\partial y_i} & \frac{\partial y_f}{\partial y_i'} & \frac{\partial y_f}{\partial \zeta_i} & \frac{\partial y_f}{\partial \zeta_{i_0}'} \\ \frac{\partial y_f'}{\partial x_i} & \frac{\partial y_f'}{\partial x_i'} & \frac{\partial y_f'}{\partial y_i} & \frac{\partial y_f'}{\partial y_i'} & \frac{\partial y_f'}{\partial \zeta_i} & \frac{\partial y_f'}{\partial \zeta_i'} \\ \frac{\partial \zeta_f}{\partial x_i} & \frac{\partial \zeta_f}{\partial x_i'} & \frac{\partial \zeta_f}{\partial y_i} & \frac{\partial \zeta_f}{\partial y_i'} & \frac{\partial \zeta_f}{\partial \zeta_i} & \frac{\partial \zeta_f}{\partial \zeta_i'} \\ \frac{\partial \zeta_f'}{\partial x_i} & \frac{\partial \zeta_f'}{\partial x_i'} & \frac{\partial \zeta_f'}{\partial y_i} & \frac{\partial \zeta_f'}{\partial y_i'} & \frac{\partial \zeta_f'}{\partial \zeta_i} & \frac{\partial \zeta_f'}{\partial \zeta_i'} \end{bmatrix}, \tag{3.84}$$

where the subscript i indicates the initial value of the vector element at s_0, and the subscript f indicates its final value at s. Equations (3.83) and (3.84) explicitly give the final state phase space vector components in terms of their first-order dependences on the initial phase space vector components, e.g.

$$x_f = \frac{\partial x_f}{\partial x_i}x_i + \frac{\partial x_f}{\partial x_i'}x_i' + \frac{\partial x_f}{\partial y_i}y_i + \frac{\partial x_f}{\partial y_i'}y_i' + \frac{\partial x_f}{\partial \zeta_i}\zeta_i + \frac{\partial x_f}{\partial \zeta_i'}\zeta_i'. \tag{3.85}$$

With this form in mind, it is possible to identify the components of $\mathbf{R}(s, s_0)$ in terms of familiar quantities. The upper left diagonal 2×2 block in $\mathbf{R}(s, s_0)$

is the horizontal betatron transformation matrix; the middle diagonal 2×2 block is the vertical betatron transformation matrix; the lower right diagonal 2×2 block is the longitudinal linear transformation matrix (which will be more familiar when we study acceleration in Chapter 4). Other matrix elements outside of these blocks have also been previously discussed—for example R_{16} and R_{26} are η_x and η_x', respectively, and R_{56} is $-\eta_\tau \Delta s$. Matrix elements linking the initial and final x and y phase planes would be non-zero if the planes are coupled, as happens in the solenoid, where the Larmor rotation completely mixes the upper left 4×4 diagonal block (cf. Ex. 3.20).

The 6×6 **transport matrix** $\mathbf{R}(s, s_0)$ has much in common with its 2×2 diagonal blocks. For instance, in the absence of acceleration, it has unit determinant, which is a manifestation of Liouville's theorem concerning phase space density. The 6×6 transport matrix transformation of the $\vec{\Phi}$-vector is used in most of the computer codes employed for accelerator beam dynamics calculations.

3.9   Summary and suggested reading

This chapter began with a discussion that deepened the introductory remarks made on betatron oscillations in Chapter 2, introducing the notion of weak focusing in circular accelerators. In such a scheme, one can have simultaneous stability in both vertical and horizontal transverse dimensions.

In order to circumvent the natural scaling of weak focusing, strong focusing based on periodic arrays of focusing and defocusing (alternating gradient) quadrupoles has been introduced. The piece-wise periodic focusing arising from these arrays is analyzed by powerful matrix methods. These methods, along with the stroboscopic Poincaré trace space-mapping visualization tool, have been explored in detail in this chapter. An alternative approach to understanding periodic focusing, which employs a perturbative analytical technique, has allowed us to identify alternating gradient focusing as a type of ponderomotive focusing.

Periodic focusing naturally arises in the context of circular machines, so we, of necessity, examined strong focusing effects, using matrix techniques, in bending systems. The bending of the design trajectory also introduces a new class of paraxial trajectory error—momentum dispersion. This phenomenon was also treated in this chapter by use of matrix methods.

Momentum dispersion couples the longitudinal phase plane with the transverse phase plane, by causing trajectory errors in the bend plane. Likewise, these trajectory errors can also affect the longitudinal motion of a particle, by changing its path length through a section of beamline. We have analyzed the competition between this effect and the change in the velocity of the particle on the time-of-flight through the system. The results we have obtained will be needed in Chapters 4 and 5 when we discuss longitudinal motion in circular accelerators.

Trace (or phase) planes are, in general, coupled by effects like dispersion and rotation (e.g. in solenoids). Thus, we ended this chapter by introducing the general six-dimensional square matrix description of the six-dimensional phase space dynamics of a charged particle.

The material in this chapter contains the core concepts of charged particle transverse optics. As such, many other texts in accelerator physics also treat the subjects we have introduced here. There is a wide variety of approaches to discussing linear transverse motion and many levels at which the discussion is given in other books. The following texts may be recommended as a supplement to this chapter:

1. P. Dahl, *Introduction to Electron and Ion Optics* (Academic Press, 1973). A primer on optics.
2. D.C. Carey, *The Optics of Charged Particle Beams* (Harwood Academic Publishers, 1987). A complete treatment of transverse charged particle motion, including an excellent treatment of higher-order (nonlinear) forces.
3. D. Edwards and M. Syphers, *An Introduction to the Physics of High Energy Accelerators* (Wiley, 1993).
4. H. Wiedemann, *Particle Accelerator Physics I: Basic Principles and Linear Beam Dynamics* (Springer-Verlag, 1993). The rigor of the presentation in this text should be helpful in clarifying issues arising from our less formal presentation.
5. S.Y. Lee, *Accelerator Physics* (World Scientific, 1996).

A number of texts may be used as an introduction to advanced topics in transverse beam dynamics:

6. M. Reiser, *Theory and Design of Charged Particle Beams*. This is an encyclopedic reference on transverse motion in beams, with and without collective effects.
7. M. Berz, *Modern Map Methods in Particle Beam Physics*. Rigorous treatment of modern analytical methods for study of linear and nonlinear beam dynamics. Written at the graduate-to-professional level.
8. H. Wiedemann, *Particle Accelerator Physics II: Nonlinear and Higher-order Beam Dynamics* (Springer-Verlag, 1999). Advanced topics in nonlinear dynamics, using the first volume in the series as a basis. Written at the graduate-to-professional level.

Exercises

(3.1) As will be discussed in detail in a following chapter, the surfaces of the iron in a ferromagnet are roughly magnetic equipotentials. Given this fact, what mathematical form should a combined-function magnet surface like that shown in Fig. 3.1 have to support a combination of dipole and quadrupole fields?

(3.2) Assuming a mechanism for equipartion of energy, and thus temperature equilibrium, between phase planes, so that the root-mean-square (rms) angle $x'_{rms} = \sqrt{\langle x'^2 \rangle} = \sqrt{kT/m_0c^2}/\beta\gamma = y'_{rms}$, what field index n should one use to guarantee that the beam sizes in a betatron obey $x_{rms} = 2y_{rms}$? You can use $x_{rms} = Rx'_{rms}/\nu_x$, $y_{rms} = Ry'_{rms}/\nu_y$.

(3.3) Explain, in terms of the transverse velocity components of the motion, why the oscillator strength associated with betatron oscillations in solenoids is of second-order, even though the force on the particle is first order in the field amplitude.

(3.4) Consider a thin-lens system in which there is a repetitive application of a focusing lens with focal length f, each separated by a drift of length L.

    (a) What is the total transformation matrix corresponding to one period of the system in this case?

    (b) What is the relationship between f and L that guarantees linear stability of the transformation?

(3.5) To prove that the trace of a matrix $\mathbf{M}_1$ is the same as that of another matrix $\mathbf{M}_2$, it is sufficient to show that the two matrices are related by a similarity transformation,

$$\mathbf{M}_1 = \mathbf{A}^{-1}\mathbf{M}_2\mathbf{A},$$

where $\mathbf{A}$ is a matrix of unit determinant. Show that any two matrices representing a periodic focusing system $\mathbf{M}_1$ and $\mathbf{M}_2$, corresponding to two different choices of $z_0(z_{0,1}$ and $z_{0,2})$, are related by such a transformation. Hint: examine matrix $\mathbf{A}$, which is the transformation matrix ($\mathbf{M}$) from $z_{0,1}$ to $z_{0,2}$.

(3.6) Consider a charged particle transport system, created by two thick lenses of length l, with opposite focusing strengths κ_0^2 and $-\kappa_0^2$, so that the matrix mapping for the focusing and defocusing lenses are

$$\mathbf{M}_F = \begin{bmatrix} \cos(\kappa_0 l) & (1/\kappa_0)\sin(\kappa_0 l) \\ -\kappa\sin(\kappa_0 l) & \cos(\kappa_0 l) \end{bmatrix}$$

and

$$\mathbf{M}_D = \begin{bmatrix} \cosh(\kappa_0 l) & (1/\kappa_0)\sinh(\kappa_0 l) \\ \kappa_0\sinh(\kappa_0 l) & \cosh(\kappa_0 l) \end{bmatrix},$$

respectively. These two lenses are applied repetitively so that the transformation of the coordinate vector at the nth step $\vec{x} = \begin{pmatrix} x \\ x' \end{pmatrix}$ is $\vec{x}_{n+1} = \mathbf{M}_F \cdot \mathbf{M}_D \cdot \vec{x}_n$.

(a) What is the phase advance per period of the oscillation as a function of $\kappa_0 l$?
(b) Like the case of a thin-lens-based system, when κ_0^2 becomes large enough the transformation is unstable. Unlike the thin-lens case, other regions of stability are encountered when κ_0^2 is raised even. Plot a trajectory from the second stability region (the one encountered after the first unstable region). Hint: It is easy to construct this plot from the piece-wise solutions to the motion if you start with conditions $\vec{x} = \begin{pmatrix} x_i \\ 0 \end{pmatrix}$ at the middle of one of the lenses. Can you tell from this trajectory why the second stability region exists for the alternating thick lens, but not the alternating thin-lens system?

(3.7) Consider a thin-lens system in which there is a repetitive application of a focusing lens with focal length f_1 and defocusing lens with focal length $-f_2$, each separated by a drift of length L. Let us examine what happens if $f_1 \neq f_2$:

(a) What is the total transformation matrix corresponding to one period of the system?
(b) Sketch out the region of stability in the parameters f_1 and f_2. This is best accomplished by drawing the borders of stability on a two-dimensional plot where the axes are f_1/L and f_2/L.

(3.8) The **Wronskian determinant** (or simply, the Wronskian) of a linear second-order differential equation

$$x'' + v(z)x' + w(z)x = 0$$

can be formed as the product of the two linearly independent solutions of the equation x_1 and x_2, and their first derivatives,

$$W(z) \equiv x_1(z)x_2'(z) - x_1'(z)x_2(z).$$

(a) Derive the differential equation governing the Wronskian,

$$W'(z) + v(z)W(z) = 0.$$

For $v(z) = 0$, this clearly implies that the Wronskian is constant. This constant is determined in the (simple harmonic oscillator) example of constant $w(z)$, in which case one has $W = 1$.

(b) Show that the solution to the Wronskian differential equation is

$$W(z) = W(0)\exp\left[-\int_0^z v(\tilde{z})\,d\tilde{z}\right].$$

For the damped oscillator equation describing betatron oscillations during acceleration in a solenoid (Eq. (2.56)), show that the Wronskian damps along with the trace space area, that is, $W(z) = \beta\gamma(0)/\beta\gamma(z)$.

(3.9) Construct a Poincaré plot of a FODO lattice with a phase advance per period equal to 50°. Verify that the plot produces an ellipse aligned to the (x, x') axes if one begins the matrix construction in the middle of either the focusing or defocusing lenses. This requires that the thin-lens matrix of one of the lenses be split into two equal thin-lens matrices with twice the focal length.

(3.10) Consider the smooth approximation applied to a FODO lattice. Find the value of k_{sec}^2 in the limit that $\mu \ll 1$. How does this result compare to that of Eq. (3.36)?

(3.11) Consider a charged particle transport system created by two repetitively applied thin lenses with opposite focal lengths f and $-f$, separated by a distance L_d.

(a) Using matrix analysis, obtain the phase advance per period μ. In order to compare this result to part (b), expand Eq. (3.31) for small μ.
(b) Now using the harmonic analysis of Section 3.4, find the value of μ predicted for this system. In order to do this, you should take the periodic focusing to be given by

$$\kappa^2(z) = -\frac{1}{f}\delta\left(z - \frac{L_p}{4}\right) + \frac{1}{f}\delta\left(z - \frac{3L_p}{4}\right),$$

where $L_p = 2L_d$ is the period of the system. Compare to the result of part (a).

3.12) Consider a so-called FOFO particle focusing system created by periodic application of a thick lens of length l with focusing strength κ_0^2, followed by a drift of length l.

    (a) Using matrix methods, find the phase advance per period μ as a function of κ_0 and l.

    (b) If $\kappa_0 l \ll 1$, find from the matrix result an approximate expression for the smooth approximation focusing strength k_{sec}^2 in this case. Hint: the approximate extraction of $\mu^2 \cong 4k_{\text{sec}}^2 l^2$ from the matrix calculation must include all terms up to fourth order in κ_0 and second-order in μ^2.

    (c) Now consider the analysis of this system as the superposition of harmonic (sine) components in focusing strength. Write the Fourier decomposition of the focusing strength.

    (d) What is the average secular focusing strength of this system? Hint: this system is a superposition of a uniform focusing system of strength $\kappa_0^2/2$, and the FD system analyzed at the end of Section 3.4, also with half-strength $\pm\kappa_0^2/2$.

    (e) Compare the results of parts (b) and (d). (You should obtain near agreement by using the expansion of $\langle \kappa^2 \rangle$ found in Eq. (3.49) up to fourth order in κ_0 and exact agreement if you sum the entire series in Eq. (3.49).)

3.13) An ultra-relativistic charged particle accelerates uniformly under the influence of a longitudinal electric field $\vec{E} = E_0 \hat{z}$. It is also confined by a sinusoidally varying gradient focusing quadrupole channel, mathematically stated as $B'(z) = B_0' \sin(k_p z)$.

    (a) Using the approximation $\beta \cong 1$ and assuming that the energy $\gamma m_0 c^2$ does not appreciably change over a period of the focusing, find an expression for the secular focusing strength.

    (b) Write the transverse equation of motion á la Eq. (2.57) and solve.

3.14) A common type of spectrometer magnet is shown in Fig. 3.11. The particles emitted from the point source are dispersed in momentum by the differences in radius of curvature. The momenta are well determined at the horizontal focus after the final drift B.

    (a) What is the final edge angle as a function of the bend angle θ_b? Hint: refer to Fig. 2.6 to see the relation between θ_b and the pole edge orientation.

    (b) Write the horizontal matrix transformation including the drifts A and B.

    (c) Determine the length B that yields a point-to-point horizontal focus as a function of R, θ_b, and A. As one changes R, what curve do these focal points describe?

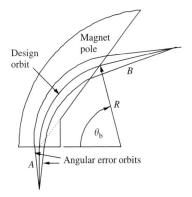

Fig. 3.11 Flat field spectrometer magnet, showing outline of pole, and focal properties.

(3.15) In the smooth approximation, one may assign the focusing in an entire strong focusing circular accelerator to be proportional to the tune, $k_{\text{sec},x} = \nu_x/R_0$. In this case, the dispersion may also have an average value—deduced from the particular solution to Eq. (3.65). Find this value for the Tevatron at Fermilab, in which $\nu_x = 19.4$, and the average radius of curvature is $R_0 = 1$ km.

(3.16) The flat-field spectrometer magnet discussed in Exercise 3.14 has interesting dispersive properties that are displayed schematically in Fig. 3.12.

    (a) Show that, in the region after the magnet, the momentum dispersion is constant, $\eta_x' = 0$.

    (b) Find the value of η_x in this region as function of θ_b and R_0.

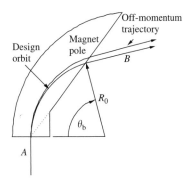

Fig. 3.12 Flat field spectrometer magnet of Fig. 3.11, showing dispersive properties.

(3.17) Evaluate α_c for the spectrometer magnet in Exercise 3.14.

(3.18) A transport line that translates the beam to the side while allowing the dispersion and its derivative to vanish at the second bend exit can be constructed by the deployment of magnets shown in Fig. 3.13. In this case it can be seen that the dispersion vanishes at the mid-point between the bend magnets.

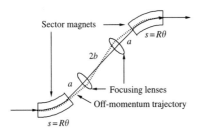

Fig. 3.13 Schematic of side-translating beamline with no residual dispersion at exit.

(a) Derive the focal length associated with the focusing lenses that gives this behavior of the dispersion as a function of a, b, R, and θ. Hint: it is best to normalize all lengths in the problem to the radius of curvature in the dipoles, R.

(b) Evaluate α_c for this transport system. Remember that even though the dispersion function changes sign in between the bend magnets, the convention of the sign of the dispersion in the second bend, with opposing sign radius of curvature, must also change. If you do not take this into account, α_c will vanish, which is clearly not true, as indicated by the above picture.

(3.19) A common magnet configuration used to rearrange particles longitudinally is the so-called chicane, as shown below in Fig. 3.14. It has flat-field dipoles of magnetic field into and out of the viewed plane, as indicated. Note that it is similar to one period of an undulator magnet, but with a large bend angle (not small compared to unity).

The chicane consists of four bend magnets of equal strength and size so that, for the design energy, R and θ are the same in each. The first bends out an angle θ, the second and third bend in by $-\theta$, and the fourth bends out again by θ.

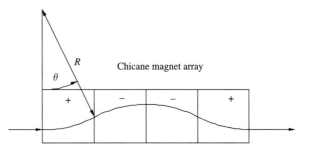

Fig. 3.14 Chicane magnet array, with design trajectory.

(a) Find the total matrix transformation in x and y. This exercise should illustrate that this array of magnets is equivalent to a drift in x and is focusing in y.

(b) Find the dispersion everywhere inside of the magnets. The quantities η_x and η_x' should disappear at the end of the fourth magnet.

(c) Find the momentum compaction $\alpha_c = 1/s - s_0 \times \int_{s_0}^{s} \eta_x(\tilde{s})/R(\tilde{s}) \, d\tilde{s}$ in this device.

(d) For an electron initially trailing the 20 MeV design electron by 1 psec, chicane magnets with $R = 0.5$ m and $\theta = 30°$, what must its momentum offset δp be for it to catch up to the design particle at the chicane exit?

(3.20) Write the $\mathbf{R}(s, s_0)$ matrix describing full passage of a paraxial particle through a solenoid magnet. Hint: write the matrix as the product of three matrices—an entrance matrix that projects the initial conditions into the Larmor frame, a simple harmonic Larmor oscillation matrix (decoupled in x_L and y_L phase planes), and an exit matrix that projects the final Larmor (rotated by the Larmor angle in the x–y plane) conditions back into the x and y phase planes.

Acceleration and longitudinal motion

Now that we have surveyed the basic concepts of linear transverse motion in charged particle optics systems, it is time to turn our attention to the problem of acceleration. This chapter begins with an introduction to acceleration due to confined electromagnetic waves, as a way of introducing the physics of radio-frequency linear accelerators (rf linacs). Within this context, it is possible to study both the strong acceleration typical of electron linear accelerators, and the comparatively gentle acceleration found in ion linacs. Once the fundamental ways of analyzing linear acceleration and related longitudinal (along the direction of beam propagation) dynamics processes are discussed, these methods are extended to allow an understanding of longitudinal motion in the circular accelerator based on rf acceleration, the synchrotron. We end this chapter on a complementary note to the last sections of Chapter 3—we examine the possible effects of the acceleration process on transverse motion.

4.1 Acceleration in periodic electromagnetic structures

As will be discussed further in Chapter 7, in free space, the solutions to the electromagnetic wave equation are transversely polarized waves (the electric field is transverse to the propagation vector) that have phase velocity c, the speed of light. These properties are problematic from the viewpoint of charged particle acceleration, because in order for a charged particle to absorb energy from an applied electric force, the motion of the charged particle must have a component parallel to the electric field. This statement is quantified by the expression that gives the time-rate-of-change of the particle energy,

$$\frac{dU}{dt} = q(\vec{v} \cdot \vec{E}). \tag{4.1}$$

In all of the cases we consider in this chapter, the motion of the particle in an accelerating wave will be rectilinear in the z-direction, and thus the electric field must be rotated to have a longitudinal component in order for acceleration to occur. This can be accomplished by using a smooth-walled waveguide, in which case we note the existence of the familiar transverse magnetic (TM) modes.[1] These modes have a longitudinal electric field, but have phase velocity larger than c, and thus cannot remain phase synchronous with a charged particle whose velocity must always be less than c. This means that in order to allow a traveling wave to stay in a nearly constant phase relationship with an accelerating particle,

[1] If TEM modes are not familiar, please see the discussion of electromagnetic modes in waveguides given in Section 7.3.

Fig. 4.1 Bisected view of a cylindrically symmetric, standing wave linear accelerator structure. The hatched portion is a conducting wall, which is typically made of copper, or a superconducting material. The electric field lines indicate the structure is operating in the π-mode, in which the longitudinal electric field changes sign every structure period.

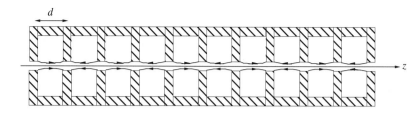

[2]Because the electromagnetic waves used in most linear accelerator structures are in the radio band, they are often termed radio-frequency linear accelerator structures, or more compactly, rf linacs.

the phase velocity of the wave must be slowed down. This is accomplished by loading the wave-guide with obstructions. An example of this loading i displayed in Fig. 4.1. Here, disks with irises cut into them about the axis o symmetry form the obstructions. The irises have two purposes: (1) to allow flow of electromagnetic energy along the z-axis, and (2) to allow the unimpede passage of paraxial beam particles.

The linear accelerator structure[2] shown in Fig. 4.1 does not in fact displa a traveling wave, but a standing wave, as can be seen from examining th longitudinal electric field pattern. The actual dependence of the solutions to th electromagnetic wave equation in such a structure is examined in detail below with the present discussion limited to an idealized, on-axis representation of th pure harmonic (at the *rf frequency*) standing wave longitudinal electric field,

$$E_z(z,t) = 2E_0 \sin(k_z z) \cos(\omega t) = E_0[\sin(k_z z - \omega t) + \sin(k_z z + \omega t)]. \quad (4.2)$$

It can be seen from this expression that the standing wave can be written a the superposition of two traveling waves. The **forward wave** component of th standing wave has argument, $k_z z - \omega t = k_z(z - v_\varphi t)$ where $v_\varphi \equiv \omega/k_z$. Th **backward wave** component has argument $k_z z + \omega t$, and thus has equal an opposite phase velocity $v_\varphi = -\omega/k_z$. For a standing wave, both forward an backward wave components have equal amplitude.

It should be noted that the structure shown in Fig. 4.1 is periodic, with perio length d. The two traveling wave components listed in Eq. (4.2) are a subset o all possible solutions of the wave equation with periodic boundary conditions In fact, a theorem due to Floquet states that the **spatial component** of th solutions to the Helmholtz equation (the simplified wave equation obtaine after substitution of a harmonic time dependence $\exp(-i\omega t)$),

$$\left[\vec{\nabla}^2 + \frac{\omega^2}{c^2} \right] \left\{ \begin{matrix} \vec{E} \\ \vec{B} \end{matrix} \right\} = 0 \quad (4.3)$$

with spatial periodicity enforced by boundary conditions, can always be writte in the form $E_i(z + d) = E_i(z) \exp(i\psi)$. Thus, a given solution is characterize by a **phase shift per period** ψ.

Simple Fourier decomposition of the on-axis solution then gives th useful form

$$E_z(z) = E_0 \operatorname{Im} \sum_{n=-\infty}^{\infty} a_n \exp\left[i \frac{(2\pi n + \psi)}{d} z \right]. \quad (4.4)$$

With this general form of the solution, the field can be viewed as the sum o many wave components, which are termed **spatial harmonics**, having differen longitudinal wave numbers $k_{z,n} = (2\pi n + \psi)/d$, and thus different phas velocities $v_{\varphi,n} = \omega/k_{z,n}$. A pure traveling wave solution has only one non-vanishing amplitude coefficient a_n, whereas a pure standing wave solution has two non-vanishing a_n. In general, in order to have the full solution obey the conducting boundary conditions, all of the components of the Floquet expansion in Eq. (4.4) must be considered. To settle on a normalization convention, will take $a_0 = 1$ in this text, so that the average accelerating field at the optimal phase in the wave will always be E_0.

The structure shown in Fig. 4.1 illustrates the concept of the phase shift per period well, as it clearly shows a field reversal every period, indicating $= \pi$. This is the so π-called-mode, which is a common field configuration for standing wave accelerators. For a pure harmonic standing wave field, we have $a_{-1} = -1$ and $a_0 = 1$, with all other components vanishing.

If we now consider a charged particle traveling on-axis through an accelerator structure at approximately constant velocity $v_z \cong v_{\varphi,0} = \omega/k_{z,0}$, and integrate Eq. (4.1) through an integer number of periods M, we obtain an energy gain due to each spatial harmonic of the field,

$$\Delta U = \begin{cases} qE_0Md\sin(\varphi), & n = 0 \\ 0, & n \neq 0. \end{cases} \qquad (4.5)$$

In Eq. (4.5) we have introduced φ, the phase of the particle with respect to the $n = 0$ wave crest. The $n = 0$ component is termed the **fundamental spatial harmonic**. Because it travels in this case at approximately the same velocity as the particle (we have ignored the fact that the velocity may change slightly during the acceleration over the integration interval), it is also termed the **synchronous wave**. All other non-synchronous components do not contribute in this constant-velocity approximation to the **secular** (averaged over a period) acceleration occurring over lengths greater than a period of the structure.

What this discussion illustrates is that, for cases where the velocity does not change appreciably during a period of the structure, in calculating energy gain, all components of the electric field may be neglected except the synchronous wave. In fact, even in cases where the particle changes velocity dramatically during passage through a structure, one finds that consideration of the dominant synchronous component is enough to describe the long-term secular acceleration of the particle. Any notable effects that are due to backward or other non-synchronous components are therefore typically localized to a region smaller than a structure period.

4.2 Linear acceleration in traveling waves

We begin our analysis of acceleration in traveling wave structures by adopting a Hamiltonian approach. To construct the Hamiltonian, we note that the longitudinal electric field associated with a single traveling wave in an accelerating structure can be derived from a vector potential with only a longitudinal component,[3]

$$A_z(z - v_\phi t) = -\frac{E_0}{k_z v_\phi}\cos[k_z(z - v_\phi t)], \qquad (4.6)$$

as

$$E_z(z - v_\phi t) = -\frac{\partial A_z}{\partial t} = E_0\sin[k_z(z - v_\phi t)]. \qquad (4.7)$$

[3] We have at this point decided on a phase convention in our description of the sinusoidal traveling wave. There are of course other possible conventions which are followed in other analyses (we could have used, e.g. a cosine), but since we need to discuss such a wide array of physical scenarios, and since stable motion will occur at different phases depending on which scenario is discussed, we will stay with this convention throughout our analysis.

The Hamiltonian associated with this vector potential can be written as (se Eq. 1.63)

$$H = \sqrt{\left(p_{z,c} + \frac{qE_0}{k_z v_\phi}\cos[k_z(z - v_\phi t)]\right)^2 c^2 + (m_0 c^2)^2}, \qquad (4.8$$

where we are, consistent with a paraxial ray approximation, only consid ering longitudinal motion. This Hamiltonian generates the correct canonica equations of motion,

$$\frac{dz}{dt} = \frac{\partial H}{\partial p_{z,c}} = \frac{p_z c^2}{\sqrt{p_z^2 c^2 + (m_0 c^2)^2}} = v_z \qquad (4.9$$

and

$$\frac{dp_{z,c}}{dt} = -\frac{\partial H}{\partial z} = \frac{p_z c^2 (qE_0/v_\phi)\sin[k_z(z - v_\varphi t)]}{\sqrt{p_z^2 c^2 + (m_0 c^2)^2}} = \frac{qE_0 v_z}{v_\varphi}\sin[k_z(z - v_\varphi t)],$$

$$(4.10$$

where we have used the relationship between the mechanical and canonica momentum, $p_z = p_{z,c} - qA_z$. The equation of motion for the mechanica momentum is recovered from Eqs (4.6) and (4.10),

$$\frac{dp_z}{dt} = \frac{dp_{z,c}}{dt} - q\frac{dA_z}{dt} = \frac{dp_{z,c}}{dt} - q\left[\frac{\partial A_z}{\partial t} + v_z\frac{\partial A_z}{\partial z}\right] = qE_0\sin[k_z(z - v_\varphi t)],$$

$$(4.11$$

where we have evaluated the total time derivative at the particle position usin the sum of the partial and the convective derivatives, $d/dt = \partial/\partial t + v_z \partial/\partial z$.

The main problem with the form of the Hamiltonian given in Eq. (4.8) is tha it is not a constant of the motion, as its partial time derivative does not vanish In order to make phase plane plots of the longitudinal motion, we must conver the form of the Hamiltonian to one in which it is constant in time. This is don by use of a canonical transformation (a Galilean, not Lorentz transformation as in Section 1.3) of coordinate[4]

$$\zeta = z - v_\varphi t. \qquad (4.12$$

With this choice of new coordinate, the new momentum is set equal to the ol $p_\zeta = p_z$, and the new Hamiltonian is obtained from the old Hamiltonian as

$$\tilde{H}(\zeta, p_{\zeta,c}) = H(\zeta, p_{\zeta,c}) - v_\phi p_{\zeta,c}$$

$$= \sqrt{\left(p_{\zeta,c} + \frac{qE_0}{k_z v_\varphi}\cos[k_z\zeta]\right)^2 c^2 + (m_0 c^2)^2} - v_\varphi p_{\zeta,c}. \qquad (4.13$$

It is clear that the new Hamiltonian is in fact a constant of the motion, and ca be used as such. With this choice of coordinate, the equations of motion derive

[4]This is a Galilean transformation to the wave frame. It is only a mathematical transformation, with a precise interpretation in the context of classical mechanics theory. It is not to be confused with a Lorentz transformation— it is not a physical, only a mathematical, change of variable descriptions.

om the new Hamiltonian are thus

$$\frac{d\zeta}{dt} = \frac{\partial \tilde{H}}{\partial p_{\zeta,c}} = \frac{p_{\zeta,c} + (qE_0/k_z v_\varphi)\cos[k_z\zeta]}{\gamma m_0} - v_\varphi = \frac{p_\zeta}{\gamma m_0} - v_\varphi = v_z - v_\varphi,$$

(4.14)

ıd

$$\frac{dp_{\zeta,c}}{dt} = -\frac{\partial \tilde{H}}{\partial \zeta} = \frac{qE_0 v_z}{v_\varphi} \sin[k_z(\zeta)],$$ (4.15)

r, writing Eq. (4.15) in terms of the mechanical momentum,

$$\frac{dp_\zeta}{dt} = qE_0 \sin[k_z\zeta].$$ (4.16)

As can be seen from this short discussion, use of canonical variables is a
ıt trickier in this case than use of familiar mechanical variables. Because we
ave concluded the formal discussion of the Hamiltonian, however, we can
ow revert to the mechanical description, to write the constant of the motion
ınctionally as

$$\tilde{H}(\zeta, p_\zeta) = \sqrt{p_\zeta^2 c^2 + (m_0 c^2)^2} - v_\varphi p_\zeta + \frac{qE_0}{k_z} \cos[k_z\zeta].$$ (4.17)

his form of the Hamiltonian could even be used (with less than perfect rigor!)
o generate equations of motion, even though it is nominally not written in
erms of the canonical momentum. In fact, the mechanical momentum in this
me-independent case can effectively be treated as if it were canonical, since
ıe form of Eq. (4.17) is identical to that which would be derived from an
lectrostatic potential giving the same acceleration as Eq. (4.16).

Since we have already derived the correct equations of motion in Eqs (4.13)–
4.16), the more important use of Eq. (4.17) is that it can be used to visualize
ıe motion of charged particles in the longitudinal phase plane (ζ, p_ζ). Before
ve move into discussion of specific examples of such motion, let us note that
.q. (4.17) can be written in normalized form as

$$\frac{\tilde{H}}{m_0 c^2} = \sqrt{(\beta_z\gamma)^2 + 1} - \beta_\phi\beta_z\gamma + \alpha_{rf}\cos[k_z\zeta].$$ (4.18)

ı Eq. (4.18), the quantity

$$\alpha_{rf} \equiv \frac{qE_0}{k_z m_0 c^2} = \frac{\gamma'_{max}}{k_z}$$ (4.19)

defined as the ratio of the maximum spatial rate of change of the normalized
article energy (Lorentz factor γ), to the maximum rate spatial rate of change
f the particle's phase in the wave.

It will be seen below that the type of behavior one observes in particle accel-
ration by a traveling wave can be divided into two distinct regimes. The first
ccurs when $\alpha_{rf} \ll 1$, and is typically encountered in heavy particle (proton or
on) linacs, in which the acceleration is very gentle. The second regime occurs
vhen α_{rf} is of order unity, or above. This is the regime of violent acceleration,
vhich occurs in electron linacs. The physics of violent acceleration is discussed
ı the next section.

4.3   Violently accelerating systems

The case of $\alpha_{rf} \geq 1$ corresponds to an accelerating wave in which the particle can gain more than one unit of rest energy by remaining in synchronism with the wave for a radian or less of spatial propagation, $k_z \Delta z \leq 1$. This is a violent acceleration scenario, in the sense that a particle can be picked up from rest and accelerated to relativistic velocities in less than one wave cycle, thus being captured by the wave. As illustrated by the examples given in Exercise 4.2, this type of acceleration in practice can only be achieved by use of the lightest charged particle, the electron (or its antiparticle, the positron). In analyzing this system, we make use of the approximation that the phase velocity of the wave reaches its ultra-relativistic limit, $v_\varphi \to c$. We adopt this approximation precisely because the acceleration in this case is violent, and thus the charged particles are expected to asymptotically (actually, within a few rf wavelengths) attain ultra-relativistic velocities. If the v_φ is chosen to be noticeably less than c, the particles can accelerate past this phase velocity, and eventually outrun the wave to the point where they may enter a decelerating phase.

In the approximation $v_\varphi = c$, we may write the mechanical version of the Hamiltonian relation, Eq. (4.17), simply as

$$\tilde{H}(\zeta, p_\zeta) = m_0 c^2 [\gamma - \beta_z \gamma + \alpha_{rf} \cos[k_z \zeta]] = m_0 c^2 \left[\sqrt{\frac{1 - \beta_z}{1 + \beta_z}} + \alpha_{rf} \cos[k_z \zeta] \right]$$

(4.20)

This form of the Hamiltonian allows us to both perform rudimentary analysis and also to draw modified phase plane plots to illustrate the dynamics of the acceleration process. An example of such a plot is shown in Fig. (4.2), in which the momentum axis is parameterized by

$$\chi = \frac{m_0 c^2}{U - p_z c} = \sqrt{\frac{1 + \beta_z}{1 - \beta_z}},$$

(4.21)

in order to have a positive quantity displayable in a semi-log plot. The vertical axis chosen to be logarithmic in order to show the large changes in momentum as the particle becomes relativistic.

Figure 4.2, which illustrates a number of curves of constant $\tilde{H}$, shows several interesting aspects of the longitudinal motion. The first is that for the

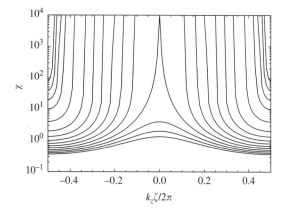

Fig. 4.2 Modified longitudinal phase plane plot, with $\alpha_{rf} = 1$.

case shown ($\alpha_{rf} = 1$), particles initially at rest ($\chi = 1$) can, for certain initial phases in the wave $\varphi_0 = k_z \zeta_0$, be accelerated to very high energy $\chi \gg 1$). In addition, at high energy the phase the particle occupies in the wave is approximately stationary,

$$\tilde{H} = m_0 c^2 [\chi^{-1} + \alpha_{rf} \cos[k_z \zeta]] \Rightarrow \alpha_{rf} m_0 c^2 \cos[k_z \zeta] = \alpha_{rf} m_0 c^2 \cos(\varphi_f).$$
$$(4.22)$$

Since the Hamiltonian is a constant of the motion, as $\chi^{-1} \Rightarrow 0$ the phase of the particle in the wave asymptotically approaches φ_f. There is no notable further slippage of the particle in the wave as it becomes ultra-relativistic, (that is as its velocity approaches the wave phase velocity, $v \Rightarrow c = v_\varphi$), thus giving the stationary final value of φ indicated by Eq. (4.22).

Note that in Fig. 4.2 all motion in φ is towards the left following the constant $\tilde{H}$ curves, as the particles are always traveling slower than the wave. At times this velocity difference is significant, and at other times it is imperceptible, in which case the direction of the curve is near vertical. Since the motion in φ is unidirectional, for positive φ the motion is seen to be accelerating, while for negative φ it is decelerating, as could have easily been deduced from Eq. (4.16).

One may ask what the minimum field amplitude is that allows the phenomenon of capture, where a particle starts from rest and is subsequently accelerated to very high energy. This field value is termed the **trapping threshold**, and gives an idea, for example, of when electrons that are field-emitted from the linac structure walls (so-called **dark current**) can actually be accelerated to high energy. An electron that is barely trapped by the accelerating wave will approach the fixed point at $(\varphi_f, \chi) = (0, \infty)$ (here both the accelerating field and the particle's velocity relative to the wave vanish), so the final value of the Hamiltonian is

$$\tilde{H} = \alpha_{rf} m_0 c^2 \cos(\varphi_f) \Rightarrow \alpha_{rf} m_0 c^2.$$
$$(4.23)$$

The initial state is given by

$$\tilde{H} = m_0 c^2 [1 + \alpha_{rf} \cos(\varphi_0)].$$
$$(4.24)$$

We can examine the minimal conditions for trapping by assuming initial phase $\varphi_0 = \pi$, at the other vanishing phase of the field. Equating the right-hand sides of Eqs (4.23) and (4.24) (with $\varphi_0 = \pi$) gives a value for the field corresponding to $\alpha_{rf} = 0.5$. The barely trapped particle is launched at $\varphi_0 = \pi$ and slips back to at final state at $\varphi_f = 0$ for this value of α_{rf}.

4.4   Gentle accelerating systems

The longitudinal motion of a heavy charged particle (a proton, or heavier ion) in an accelerating wave is not similar to that discussed in the last section for any reasonable values of the electric field amplitude and wavelength (see Ex. 4.2(b)). For heavy particles, one always finds that $\alpha_{rf} \ll 1$, and the resulting gentle accelerating motion is qualitatively different. This situation requires a different approach to the analysis than employed in the previous section. For **gentle acceleration** systems, the energy gain over a wavelength of the accelerator is much less than the rest mass, and so we are not led to assume that the motion is ultra-relativistic, nor that the phase asymptotically approaches a constant value. In fact we will see that the motion in these systems is characterized by simple harmonic motion near the stable fixed point of the system. At this fixed point, the accelerating field vanishes, and the particle has the same velocity as that of the wave. In Sections 4.2 and 4.3, we have assumed that this phase velocity is constant. While we will begin the discussion of gentle acceleration under this assumption, we will see that it is necessary to allow this phase velocity to increase in order for significant, long-range gentle acceleration of particles to occur. This generalization will also cause our view of the phase plane fixed point to change somewhat.

We begin by rewriting the Hamiltonian of Eq. (4.17) by expanding it for small amplitude motion about the design momentum

$$p_0 = \gamma_\varphi m v_\varphi = \frac{m v_\varphi}{\sqrt{1 - (v_\varphi/c)^2}} = \frac{m v_0}{\sqrt{1 - (v_o/c)^2}} = \gamma_0 m v_0. \tag{4.25}$$

which is resonant with the phase velocity of the wave. Keeping terms up to second order in $\delta p = p_\zeta - p_0$, we have the expression

$$\tilde{H}(\zeta, \delta p_\zeta) \cong \gamma_0 m_0 c^2 + v_0 \delta p + \frac{\delta p^2}{2\gamma_0^3 m_0} - v_0(p_0 + \delta p) + \frac{qE_0}{k_z} \cos(k_z \zeta)$$

$$= \frac{m_0 c^2}{\chi_0} + \frac{\delta p^2}{2\gamma_0^3 m_0} + \frac{qE_0}{k_z} \cos(k_z \zeta). \tag{4.26}$$

The addition and subtraction of constants in the Hamiltonian have no effect on the form of the phase plane curves, or on the derived equations of motion. We therefore are free to reformulate Eq. (4.26) in more suggestive form,

$$\tilde{H}(\zeta, \delta p_\zeta) = (m_0 c^2) \left[\frac{\beta_0^2}{2\gamma_0 p_0^2} (\delta p^2) + \alpha_{rf}[\cos(k_z \zeta) + 1] \right]. \tag{4.27}$$

As a check on the derivation of Eq. (4.27), it is instructive to extract the equations of motion from it by differentiation

$$\dot{\zeta} = \frac{\partial \tilde{H}}{\partial(\delta p)} = \frac{m_0(\beta_0 c)^2}{\gamma_0 p_0^2} \delta p = \frac{\delta p}{\gamma_0^3 m_0}, \tag{4.28}$$

$$\delta \dot{p} = -\frac{\partial \tilde{H}}{\partial \zeta} = \alpha_{rf} k_z m_0 c^2 \sin(k_z \zeta) = qE_0 \sin(k_z \zeta). \tag{4.29}$$

The effective longitudinal mass $\gamma_0^3 m_0$ first encountered in Eq. (2.34) is again displayed in Eq. (4.26).

It can be seen from Eq. (4.27) that the maximum fractional momentum change that can be imparted to a particle in this potential is of the order

$$\frac{\delta p_{max}}{p_0} \cong \frac{\sqrt{2\alpha_{rf} \gamma_0}}{\beta_0}.$$

This quantity is much smaller than one by design, however, as we are assuming that the particles are only moderately relativistic (γ_0 is not many orders of

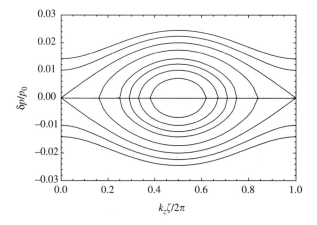

Fig. 4.3 Phase plane trajectories showing the stable region ("bucket") of vibrational motion, bounded by a separatrix, with unbounded phase motion (librational motion) outside. This case is for $\beta_0^2/\gamma_0 = 0.5$, and $\alpha_{\mathrm{rf}} = 10^{-4}$.

magnitude larger than unity), and also that $\alpha_{\mathrm{rf}} \ll 1$. Also, it can be seen that the Hamiltonian in Eq. (4.27) is of the form corresponding to a pendulum, where the stable phase—the minimum "potential energy"—of the pendulum is chosen as $\varphi_{\min} = \pi$ (as above, the particle phase is defined as $\varphi = k_z\zeta$).

The phase plane plot of the pendulum-like trajectories (rigorously, it is a trace space, as we are normalizing the momentum error to the design momentum) is displayed in Fig. 4.3 for a physically realistic case, with $\beta_0^2/\gamma_0 = 0.5$ and $\alpha_{\mathrm{rf}} = 10^{-4}$. The phase plane, which should look familiar to any student of Hamiltonian mechanics, is divided into two regions: one having stable, bounded orbits (vibrational motion) and one displaying unbounded trajectories (librational motion). The boundary between the two regions referred to as the **separatrix**. Note that the existence of an unstable (unbounded) region in phase space is due to the nonlinearity of the applied force, as anticipated by the discussion of Section 3.2.

In the stable region, or **bucket**, the small amplitude motion in the neighborhood of the fixed point at $(\varphi, \delta p) = (\pi, 0)$ is nearly simple harmonic, as is discussed further below. The motion inside of the bucket at larger amplitudes is nonlinear—the oscillations slow down as the potential becomes less well approximated by a parabola, and the longitudinal focusing becomes effectively weaker. As the amplitude approaches that of the separatrix, the period of the motion becomes infinite, since on the separatrix the particles are unable to traverse the unstable fixed points at $(\phi, \delta p) = (0, 0)$, and $(\phi, \delta p) = (2\pi, 0)$. Note that the motion along the constant $\tilde{H}$ curves in this case is towards positive ζ for $\delta p > 0$, and negative ζ for $\delta p < 0$. This is needed for the vibrational orbits to exist, of course, and also points to the fact that the librational orbits always proceed in one direction in ζ above the separatrix, and another direction in the region below.

The equation for the separatrix can be obtained by evaluating the value of the Hamiltonian at an unstable fixed point, for example,

$$\tilde{H}(0,0) = 2\alpha_{\mathrm{rf}}m_0c^2. \tag{4.30}$$

Use of Eqs (4.27) and (4.30) together yields

$$\frac{\delta p_{\mathrm{sep}}}{p_0} = \pm\frac{1}{\beta_0}\sqrt{2\alpha_{\mathrm{rf}}\gamma_0[1 - \cos(k_z\zeta)]} = \pm\sqrt{\frac{4\alpha_{\mathrm{rf}}\gamma_0}{\beta_0^2}}\sin\left(\frac{k_z\zeta}{2}\right). \tag{4.31}$$

Thus the peak momentum offset encountered in the bucket (at $k_z\zeta = \pi$) is simply

$$\frac{\delta p_{\max}}{p_0} = \pm\sqrt{\frac{4\alpha_{\mathrm{rf}}\gamma_0}{\beta_0^2}}. \tag{4.32}$$

The area of the stable phase plane $(\zeta, \delta p)$ region or **bucket area**, A_b, can be found by integrating the area between the curves of the functions given by Eq. (4.31),

$$A_b = 4p_0\sqrt{\frac{\alpha_{\mathrm{rf}}\gamma_0}{\beta_0^2}}\int_0^{2\pi/k_z}\sin\left(\frac{k_z\zeta}{2}\right)d\zeta = \frac{16p_0}{k_z}\sqrt{\frac{\alpha_{\mathrm{rf}}\gamma_0}{\beta_0^2}}. \tag{4.33}$$

These phase plane dynamics are quite unlike those of the transverse motion, which are stable to all amplitudes under linear transformations of the type discussed in Chapter 3. Because the longitudinal motion is mediated by a sinusoidal force (instead of one linearly proportional to the offset from a stable fixed point), it is inherently **nonlinear**. In fact, since the force is periodic, we observe unstable fixed points one half of a wavelength away from the stable fixed points. The existence of both types of fixed points implies that only a finite region of the phase plane about the stable fixed point has vibrational orbits.

Even though this large amplitude motion (with its nonlinear characteristics) is unfamiliar, the small amplitude motion about the stable fixed point is quite familiar. If we expand the Hamiltonian near this point, we have

$$\tilde{H}(\zeta, \delta p) \cong (m_0 c^2)\left[\frac{\beta_0^2}{2\gamma_0 p_0^2}(\delta p^2) + \frac{\alpha_{\mathrm{rf}}(k_z\delta\zeta)^2}{2}\right], \tag{4.34}$$

where $\delta\zeta = \zeta - \pi/k_z$. This small amplitude Hamiltonian can be used to obtain the equations of motion for $\delta\zeta$ and δp, which can be combined to give a single simple harmonic oscillator equation, that is,

$$\ddot{\delta\zeta} + \frac{\alpha_{\mathrm{rf}}(k_z c)^2}{\gamma_0^3}\delta\zeta = 0. \tag{4.35}$$

Equation 4.35 gives solutions, termed **synchrotron oscillations**, that are harmonic with the **synchrotron frequency**

$$\omega_s = k_z c\sqrt{\frac{\alpha_{\mathrm{rf}}}{\gamma_0^3}} = \sqrt{\frac{\alpha_{\mathrm{rf}}}{\gamma_0^3}}\frac{\omega}{\beta_0}. \tag{4.36}$$

As the ratio inside of the square-root sign on the right-hand side of Eq. (4.36) is much smaller than one, the synchrotron frequency is much smaller than the frequency of the wave ω. Note that we have used time as the independent variable in writing Eq. (4.35), as opposed to the distance along the design trajectory to allow simple comparison between the rf and synchrotron frequencies.

4.5   Adiabatic capture

The phase plane plot displayed in Fig. 4.3 only tells part of the story, as it maps the trajectories of particles in the phase plane given a constant value of $\tilde{H}$

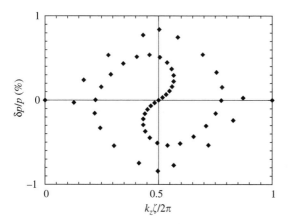

Fig. 4.4 The phase plane distribution after $\omega_s t = 5\pi/2$, starting from initial conditions uniformly populating the $\delta p = 0$ axis, with a constant amplitude wave giving a separatrix value $\delta p_{\max}/p_0 \cong 1\%$.

 says nothing, however, about the time dependence of the traversal of these urves. We know from the above discussion that the motion is simple harmonic or small amplitudes, that is, the particles rotate about the stable fixed point at $\varphi, \delta p) = (\pi, 0)$ with constant frequency. This is not true for arbitrary stable mplitudes, however, as the frequency must approach zero near the separatrix. his fact causes trouble when one attempts the capture of a coasting beam into a unched state inside of the bucket, which would be a relatively straightforward peration if linear forces were employed.

The most naive approach to this capture process would be to take the coasting eam, which we take to be monochromatic ($\delta p = 0$) and uniformly distributed n phase (in terms of a probability distribution function, $f(\varphi) = N(2\pi)^{-1}$, vhere N is the number of particles found in an rf wavelength), and apply a onstant amplitude accelerating wave.

The results of this process are shown in Fig. 4.4, which displays the phase lane distribution after a time $\omega_s t = 5\pi/2$. The variation of the oscillation eriod (note the particles starting nearest $\varphi = 0$ and 2π have hardly moved luring this time) as a function of amplitude has caused phase space filament- tion, which continues until the bucket is fully and nearly smoothly populated vith particles. The effective area in the phase plane occupied by particles has one from zero to essentially the full bucket area[5] A_b, which is obviously not lesirable. In addition, one has not noticeably localized the distribution in time, nerely modulated it. As we shall see below, there is a better way to "bunch" he beam from its initial uniform spatial distribution.

One way of viewing the problem with bunching based on a constant amplitude vave is that a particle, which starts on an oscillation from a given phase offset $\Delta\varphi_0$ from the fixed point, will always turn at a position with the same offset elative to the other side of the fixed point. On the other hand, if one increases he wave strength E_0 (which enters the dynamics description through Eq. 4.29) lowly over time, a particle experiences a growing potential as it passes through he center of oscillation (the stable fixed point), and thus the turning points of he oscillation will move ever closer to this center. The results of this process, nown as *adiabatic capture*, are shown in Fig. 4.5. This picture illustrates he phase plane after the accelerating field has been raised exponentially, yet lowly, from a value five e-foldings below the level illustrated in Fig. 4.4, to he same level as that in Fig. 4.4. By increasing the field slowly, we invoke the

[5]Even after many synchrotron periods, the area in the phase plane occupied by the particles is still zero in the microscopic sense, as the particle distribution is merely a curved line wrapped many times around the origin. In a coarser grained view, however, where one examines any finite area inside of the bucket, one finds particles, and therefore distribution is said to occupy the entire bucket.

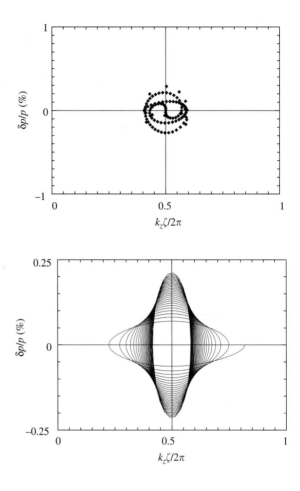

Fig. 4.5 The final phase plane distribution for an adiabatically growing wave potential, with slow exponential increase of the wave amplitude to the same final amplitude as in Fig. 4.4.

Fig. 4.6 The phase plane trajectory for particle with initial conditions $(\varphi, \delta p) = (2 + \pi, 0)$ in the adiabatically growing wave potential associated with the final phase plane distribution shown in Fig. 4.5.

adiabaticity condition, which for an exponential increase in field amplitude $E_0(t) \propto \exp(\nu_a t)$, implies that the relative growth of the field in one radian of longitudinal oscillation should be small compared to unity,

$$\frac{dE_0(t)}{dt} \frac{1}{E_0(t)} = \nu_a \ll \omega_s. \tag{4.37}$$

Because the synchrotron frequency $\omega_s \propto E_0^{1/2}$, the ratio ν_a/ω_s is not held constant during the exponential growth of the field, and in fact the calculation shown in Fig. 4.4 begins with $\nu_a/\omega_s = \frac{1}{8}$, and ends with $\nu_a/\omega_s = \frac{1}{80}$. This points to the fact that one must take care to ensure the adiabaticity condition is not violated at the beginning of the capture process. It should also be noted that the frequency of particle oscillation is smaller for large amplitude particles. In this process, particles are at large amplitude at early times, so extra care must be taken to capture these particles adiabatically.

The physics of adiabatic capture is well illustrated by examining a single particle's motion over the entire capture process, as shown in Fig. 4.6. The particle's motion begins at an arbitrary large amplitude phase offset, $(\varphi, \delta p) = (2 + \pi, 0)$, and is brought ever closer in position to the stable fixed point. On the other hand, the amplitude of maximum momentum offset is enhanced over time in such a way to conserve the area of the motion in the phase plane, the longitudinal action $J_z = \oint \delta p \, d\zeta = \pi \zeta_{max} \delta p_{max}$. The conservation of action, as discussed in Chapters 1 and 2, is a well-known attribute of adiabatically (slowly) varying systems. In fact, we can use it to predict the behavior of the capture process.

The value of J_z is a function of the distance of the initial phase plane coordinates from the stable fixed point, and of the degree of adiabaticity—for a slower initial turn-on of the field, J_z is smaller. When the particle is in the small amplitude region, the Hamiltonian can be written as $\tilde{H} = J_z \omega_s \propto \omega_s^2 \zeta_{max}^2 \propto \delta p_{max}^2$. As the oscillation frequency is raised, $\omega_s \propto E_0^{1/2}$, because of the invariance of Jz the energy in the oscillation increases $\tilde{H} \propto E_0^{1/2}$ (see Ex. 2.12). On the other hand, the curvature of the potential well (in which the particles are oscillating) increases quadratically with ω_s and linearly with field amplitude E_0. Therefore, the maximum offset in position diminishes as $\zeta_{max} \propto \omega_s^{-1/2}$, and its maximum momentum offset grows as $\delta p_{max} \propto \omega_s^{1/2}$.

In practice, a beam always has a finite initial energy spread before the capture process begins, and so particles with initial $\delta p \neq 0$, unlike on-momentum particles, must actually "cross the separatrix" from unbounded motion to trapped vibrational motion about the stable fixed point. The separatrix of course does not actually exist in this time-dependent, growing field amplitude case (if it did, particles would not cross it). Nevertheless, it is worth explaining the process by which particles enter the growing bucket. For the moment consider the field amplitude to be constant. In the capture process, a particle's motion is unbounded when its maximum momentum offset is larger than δp_{max}. When the accelerating potential (and thus the bucket height) grow, the particle's energy falls below the height of the potential well, and the particle is trapped, eventually being attracted towards the stable fixed point.

4.6   The moving bucket

The alert reader at this point may protest that, in our discussion of gentle accelerating systems, we have examined motion (limited to twice the bucket height $\Delta(\delta p) < 2\delta p_{max}$) of the charged particles in stable "buckets" which do not allow significant acceleration. Once particles are trapped inside of the bucket, this situation can be remedied by changing the properties of the wave that creates the bucket. In particular, the design velocity (or synchronous velocity) v_0 can be raised slowly, so that on average all of the particles in the moving bucket can gain energy. This can be accomplished in a traveling or standing wave linear accelerator, where the accelerating wave frequency ω is held constant by changing the spatial periodicity d, so that $k_z(z)$ is a decreasing function of distance and $v_0(z) = \omega/k_z(z)$ increases. On the other hand, as we shall see in the next section, in circular accelerators, where the periodicity in space (set by the circumference of the accelerator) is constant, the synchronous velocity, $v_0(t) = \omega(t)/k_z$, is raised by increasing the frequency of the applied accelerating field in a localized accelerating structure. We now analyze the effects on particle dynamics that the moving bucket causes in a linear accelerator.

We begin by noting that in order for a design particle, which is synchronous with the wave phase velocity, to exist, it must accelerate in the wave to satisfy the synchronous condition, $v_0(z) = \omega/k_z(z)$. With $v_0(z)$ not constant in space, we can write the following differential equation,

$$\frac{dv_0}{dz} = \frac{1}{v_0}\frac{dv_0}{dt} = \frac{F_z(z)}{\gamma_0^3 m_0 v_0} = \frac{qE_0(z)\sin(\varphi_0(z))}{\gamma_0^3 m_0 v_0}, \tag{4.38}$$

which defines the **synchronous phase** φ_0. Note that the spatial rate of change in the synchronous velocity is strictly limited, having its maximum at $\varphi_0 = \pi/2$.

The acceleration of the synchronous particle means that the entire reference frame is accelerated in the forward direction. This yields an effective uniform force in the reverse direction (like the force one feels in an accelerating vehicle), $F_{z,\text{eff}} = -qE_0\sin(\phi_0)$. This force can be accounted for in a new Hamiltonian by inclusion of a uniformly accelerating potential that gives a fixed point at φ_0,

$$\hat{H}(\zeta, \delta p) = (m_0 c^2)\left[\frac{\beta_0^2}{2\gamma_0 p_0^2}(\delta p^2) + \alpha_{\text{rf}}[\cos(k_z\zeta) + k_z\zeta\sin(\varphi_0) + 1]\right]. \tag{4.39}$$

Equation (4.39) can be rewritten in terms of particle phase to aid visualization,

$$\hat{H} = (m_0 c^2)\left[\frac{\beta_0^2}{2\gamma_0 p_0^2}(\delta p^2) + \alpha_{\text{rf}}[\cos(\varphi) + \varphi\sin(\varphi_0) + 1]\right]. \tag{4.40}$$

By this construction, the Hamiltonian is independent of time, which cannot be entirely correct, as it has some time dependence embedded in the quantities β_0, γ_0, etc. This approximation is tolerable because in a gentle accelerating system it is not possible for these quantities to change appreciably in a synchrotron period. Thus any visualization of the motion derived from plotting constant $\tilde{H}$ curves using Eq. (4.39) or (4.40) will be fairly accurate.

The discussion of these curves begins with the stable fixed point and moves outward in the phase plane. The fixed point at φ_0 has stable synchrotron oscillations about it, and a repetition of the analysis leading to Eq. (4.35) describes these oscillations in $\delta\zeta = \zeta - \varphi_0/k_z$ by

$$\ddot{\delta\zeta} - \frac{\alpha_{\text{rf}}(k_z c)^2}{\gamma_0^3}\cos(\varphi_0)\delta\zeta = 0. \tag{4.41}$$

Equation (4.41) indicates a new, smaller value of the synchrotron frequency, for $\pi/2 < \varphi_0 < \pi$,

$$\omega_s = \sqrt{\frac{\alpha_{\text{rf}}}{\gamma_0^3}|\cos(\varphi_0)|}\frac{\omega}{\beta_0}. \tag{4.42}$$

It can be seen that the longitudinal "focusing strength", which is dependent on the local gradient of the applied force about the stable fixed point, is diminished by the factor $\cos(\varphi_0)$. The small amplitude, nearly elliptical orbits are illustrated in the case shown in Fig. 4.7, with $\varphi_0 = 2.85$, or a shift of 17° off from the non-accelerating bucket.

At larger amplitudes, we must be concerned with the limits of the stable motion, and so we examine the characteristics of the separatrix, which is shown in Fig. 4.7 as a "fish-shaped" curve with one unstable fixed point (where the fish

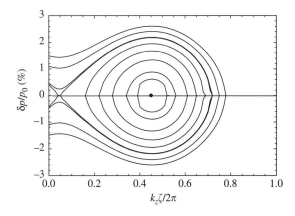

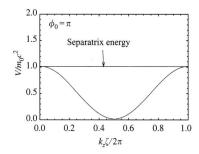

Fig. 4.7 Longitudinal phase plane orbits in the case of a moving bucket with $\varphi_0 = 2.85$, or a shift of $17°$ off of the stationary bucket ($\beta_0^2/\gamma_0 = 0.5$, and $\alpha_{rf} = 10^{-4}$).

ail joins the body), at $(\varphi, \delta p)_{\text{fixed}} = (\pi - \varphi_0, 0)$. The value of the Hamiltonian at the separatrix is thus

$$\hat{H}(\pi - \varphi_0, 0) = \alpha_{rf} m_0 c^2 [1 - \cos(\varphi_0) + (\pi - \varphi_0)\sin(\varphi_0)]. \quad (4.43)$$

The separatrix goes through the $\delta p = 0$ axis at the turning point, where the phase satisfies $\cos(\varphi_{\text{turn}}) + \varphi_{\text{turn}}\sin(\varphi_0) = -\cos(\varphi_0) + (\pi - \varphi_0)\sin(\varphi_0)$. Using Eqs (4.40) and (4.43), we can plot the separatrix, and analytically find the bucket height and its area, as in Section 4.4.

While this effort is left to the exercises, it is instructive to visualize the stable area in terms of the potentials acting on the particles. The trajectory of the barely trapped particle in the case of stationary bucket, which has turning points at the top of the potential maxima, is shown in Fig. 4.8. Since the potential is at a maximum at these points, they are not true turning points, but are unstable fixed points. This trajectory's motion is clearly symmetric about $\varphi = \pi$. Possible positions of stable particles at energies lower than that of the separatrix extend from $\varphi = 0$ to $\varphi = 2\pi$.

This situation changes dramatically for the moving bucket, as is illustrated in Figs 4.9 and 4.10, which display the cases of $\varphi_0 = 2.85$ and $\varphi_0 = 2$, respectively. The unstable fixed point at larger phase disappears, and is replaced by a true turning point. Both the unstable fixed point and the turning point approach φ_0 as $\varphi_0 \to \pi/2$, and the region of stable motion disappears in this limit. In fact, at $\varphi_0 = \pi/2$, the stable and unstable fixed points collide to form a saddle point, which has no oscillatory, stable motion in its neighborhood. As the region of stable motion disappears, the depth of the associate confining potential well obviously tends to zero, as can be seen in Figs 4.8–4.10.

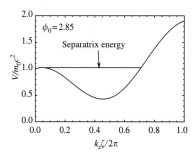

Fig. 4.8 The potential energy diagram associated with a stationary bucket.

Fig. 4.9 The potential energy diagram associated with a moving bucket, $\varphi_0 = 2.85$.

4.7 Acceleration in circular machines: the synchrotron

The idea of using an rf accelerating cavity in a circular accelerator arises from the simple desire to reuse the rf cavity in the acceleration process many times. This is especially important in the case of heavy particles, in which it is difficult to accelerate in a linear structure using moving buckets—the phase space dynamics simply do not allow rapid acceleration with the available wave strengths. Thus,

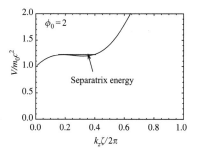

Fig. 4.10 The potential energy diagram associated with a moving bucket, $\varphi_0 = 2$.

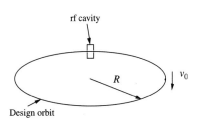

Fig. 4.11 Schematic picture of the synchrotron, showing circular design orbit and accelerating cavity.

we are driven to consider the synchrotron, a device schematically shown in Fig. 4.11, in which a short accelerating cavity that may consist of only one cell, is installed in a ring. The approximately circular design orbit (of average radius R) in this device is bent by dipole magnets, which, unlike the betatron geometry, do not cover a region much larger than the band around the design orbit (having a width of a few times the maximum expected betatron amplitude). Transverse stability in this device is typically provided by alternating gradient, strong focusing quadrupoles that give transverse betatron tunes $\nu_{x,y} \gg 1$.

Longitudinal stability, or phase focusing, is provided in the synchrotron by use of the time varying fields inside of the rf cavity, which in the analysis in this section is considered to be of negligible length. Thus we will be concerned only with the energy change induced during passage through the rf cavity's time-varying integrated field, or voltage

$$\Delta U(t) = -qV_0 \sin(\omega t + \varphi_0) = -qV_0 \sin(h\omega_c t + \varphi_0). \tag{4.44}$$

Here, we have introduced the harmonic number h, which is the integer ratio of the rf frequency to the circulation frequency in the synchrotron, $\omega/\omega_c = h$, implying that the rf goes through an integer number of cycles per revolution of the design particle. For a non-accelerating synchrotron (a storage ring), both of these frequencies are constant. In a ring that accelerates particles, these frequencies increase while the ring circumference remains constant. This contrasts to the case of the linear accelerator, in which the frequency remains constant and the spatial periodicity of the wave changes.

To see how the longitudinal dynamics in the circular accelerator compare as well as contrast to the linear accelerator case, we construct the dynamics "turn-by-turn", looking at the motion at only one position in the ring (as in Poincaré plotting), the rf cavity. We postulate the existence of a particle on the design orbit, with constant design velocity $v_0 = R\omega_c$, and with constant phase (modulo 2π) with respect to the accelerating voltage waveform in the cavity, $\varphi_0 = \pi$, as in the case of a stationary bucket in the linear accelerator. The energy of a particle after its $(n + 1)$th traversal of the rf cavity is related to its energy on the previous turn by

$$\delta U_{n+1}(\tau_n) = \delta U_n - qV_0 \sin(\omega\tau_n + \pi) = \delta U_n + qV_0 \sin(\omega\tau_n), \tag{4.45}$$

where τ is the time of arrival of the particle at the rf cavity with respect to the arrival of the design particle, and $\delta U = U - U_0$ is the difference in particle energy from the design value.

The description of the longitudinal dynamics can be closed by examining the change in time of arrival (as discussed in Section 3.7) turn-by-turn, as a function of energy offset,

$$\tau_{n+1}(\delta U_n) = \tau_n + \frac{2\pi\eta_\tau}{\omega_c}\frac{\delta p_n}{p_0} = \tau_n + \frac{2\pi\eta_\tau}{\beta_0^2\omega_c}\frac{\delta U_n}{U_0}, \tag{4.46}$$

where we have used $\beta_0^2(\delta p/p_0) = \delta U/U_0$. Equations (4.45) and (4.46) are difference equations, and can be understood by viewing them as numerically equivalent to the differential equations

$$\frac{d(\delta U)}{dt} = \lim_{t_0 \to 0} \frac{\Delta U}{t_0} = \frac{q\omega_c V_0}{2\pi} \sin(\omega\tau), \qquad (4.47)$$

and

$$\frac{d\tau}{dt} = \lim_{t_0 \to 0} \frac{\tau}{t_0} = \frac{\eta_\tau}{\beta_0^2 U_0} \delta U, \qquad (4.48)$$

respectively, if the changes in the variables δU and τ are not significant in one turn. The second order differential equation derived from these expressions is

$$\frac{d^2\tau}{dt^2} - \frac{\eta_\tau \omega_c q V_0}{2\pi\beta_0^2 U_0} \sin(\omega\tau) = 0, \qquad (4.49)$$

which, assuming an energy in the device below transition, $\eta_\tau < 0$, can be expanded near $\tau = 0$ to give

$$\frac{d^2\tau}{dt^2} + \frac{|\eta_\tau| h\omega_c^2 q V_0}{2\pi\beta_0^2 U_0}\tau = 0. \qquad (4.50)$$

Thus, the small amplitude frequency of oscillation, or synchrotron frequency, is

$$\omega_s = \sqrt{\frac{|\eta_\tau| hqV_0}{2\pi\beta_0^2 U_0}}\,\omega_c, \qquad (4.51)$$

from which we can define the **synchrotron tune,**

$$\nu_s \equiv \frac{\omega_s}{\omega_c} = \sqrt{\frac{|\eta_\tau| hqV_0}{2\pi\beta_0^2 U_0}}. \qquad (4.52)$$

The assumption that the difference equations (Eqs 4.45 and 4.46) are well approximated by the differential equations (Eqs 4.47 and 4.48) is valid if the synchrotron tune is much smaller than unity (and thus also very much smaller than the betatron tunes). To state this criterion another way, the number of "differential" time steps needed for the particle to perform one synchrotron oscillation must be much larger than one for the difference equations to be well modeled by differential equations.

In the above discussion we have assumed an energy below transition, $\eta_\tau < 0$, which is the more "natural" dynamical state—more energetic particles move through the ring faster than less energetic ones. On the other hand, for all high-energy electron synchrotrons and most heavy particle synchrotrons, the machine operates at least part of the time above transition energy (in which case $\eta_\tau > 0$). In this situation, Eq. (4.50) needs to be expanded near $\varphi_0 = 0$, not $\varphi_0 = \pi$, to give stable oscillations. In either case the synchrotron frequency is given by Eq. (4.51). The phase jump in the stable fixed point between π and 0 is one of the major problems encountered in acceleration through **transition energy** (transition crossing).

In order to derive other relevant characteristics of the motion in the circular accelerator, we rewrite Eqs (4.47) and (4.48) in terms of our canonical variables,

$$\frac{d(\delta p)}{dt} = \frac{q\omega_c V_0}{2\pi v_0} \sin(k_z \zeta),$$ (4.53)

and

$$\frac{d\zeta}{dt} = -\frac{v_0 \eta_\tau}{p_0} \delta p,$$ (4.54)

where we have used $k_z \zeta = -\omega\tau + \varphi_0$. These equations can be derived from a Hamiltonian of the same form as Eq. (4.27),

$$H(\zeta, \delta p) = -\frac{\eta_\tau}{2\gamma_0 m_0}(\delta p)^2 + \frac{qV_0}{2\pi h}[\cos(k_z \zeta) - \text{sgn}(\eta_\tau) \cdot 1],$$ (4.55)

where the constant $-\text{sgn}(\eta_\tau) \cdot 1$ is conveniently chosen to be ± 1, above and below transition, respectively. The first term in Eq. (4.55) has the usual positive sign if $\eta_\tau < 0$, and the stable fixed point is at $\varphi = k_z \zeta = \pi$. Above transition, this term is negative (the so-called **negative mass** effect) and the stable fixed point, as mentioned above, moves to $\varphi = 0$.

One can easily derive the properties of the stable region (bucket), which results from this Hamiltonian by looking at the value of the Hamiltonian on the separatrix ($H = qV/2\pi h$). From this information, we notably obtain the bucket height,

$$\frac{\delta p_{\max}}{p_0} = \sqrt{\frac{qV_0}{h\pi|\eta_\tau|p_0 v_0}},$$ (4.56)

and the bucket area in the phase plane,

$$A_b = \frac{8}{k_z}\sqrt{\frac{qV_0\gamma_0 m_0}{h\pi|\eta_\tau|}}.$$ (4.57)

The dynamics of the moving, or accelerating bucket in a circular accelerator are formally identical to those found in the linear accelerator, with the introduction of design phase $\pi/2 < \varphi_0 < \pi$ to accelerate particles below transition, and a design phase $0 < \varphi_0 < \pi/2$ above transition. These changes can be incorporated into the Hamiltonian by including the term $k_z \zeta \sin(\varphi_0)$ inside of the square brackets in Eq. (4.55), in analogy to Eq. (4.40) so that

$$H(\zeta, \delta p) = -\frac{\eta_\tau}{2\gamma_0 m_0}(\delta p)^2 + \frac{qV_0}{2\pi h}[\cos(k_z \zeta) + k_z \zeta \sin(\varphi_0) - \text{sgn}(\eta_\tau) \cdot 1].$$ (4.58)

It is interesting to examine the moving bucket above and below transition. For the case of $\eta_\tau < 0$, the picture looks the same as in Fig. 4.7. Above transition, however, the stable fixed point shifts, as illustrated in Fig. 4.12 (the same as Fig. 4.7 with $\eta_\tau \to -\eta_\tau$ and $\phi_0 \to \pi - \phi_0$) and the fish-shape of the separatrix reverses orientation. In this case the phase flows around the stable fixed point also change direction, from clockwise, to counter-clockwise.

Transition crossing is a difficult phenomenon to quantitatively analyze, so we content ourselves to finish this section with a short conceptual discussion of the major methods of dealing with it. First, we point out that as transition is

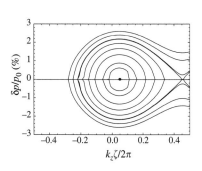

Fig. 4.12 Longitudinal phase plane orbits for the same case of a moving bucket as in Fig. 4.7, but with the analogous above transition case, where $\eta_\tau \to -\eta_\tau$, and $\phi_0 \to \pi - \phi_0$.

approached, $\eta_\tau \to 0$, and the bucket height and area grow large quickly, but not adiabatically. The synchrotron frequency also becomes small at this point. If one tries to lower the rf voltage V_0 to deal with this fact, the acceleration stops, and one cannot negotiate transition at all. Therefore, one typically makes a transition "jump" in which two system parameters simultaneously change state very quickly (a few turns): the phase of the rf wave is changed by π, and the tune of the transverse focusing is raised (with fast-changing quadrupoles) to lower α_c, quickly lowering the transition energy so that it lies below the energy of the beam.

4.8   First-order transverse effects of accelerating fields*

The fields that accelerate the charged particles in an rf linac cavity also give rise to transverse components of the Lorentz force that deflect the particles. The net effects of these transverse components are related to the acceleration by the Panofsky–Wenzel, or deflection, theorem. We will now discuss a general form of the Panofsky–Wenzel theorem relevant to the topics discussed in this text.

This theorem was originally used to give a relationship between the integrated longitudinal and transverse momentum kicks that a particle receives as it traverses an isolated device with an electromagnetic wave excitation in the constant-velocity, paraxial limit. Here we generalize this theorem by including fields that may arise from free charges, assuming that the fields and potentials vanish outside a region of interest. The explicit assumptions leading this generalized form of the Panofsky–Wenzel theorem, state that the particle receiving the kick travels parallel to the z-axis, with a constant velocity.

The net Lorentz force on a charged particle consists of the sum of electric and magnetic components. In general, the electric field $\vec{E}$ may be derived from a scalar potential (ϕ_e) and vector potential ($\vec{A}$), as noted in Eq. (1.6). We can write the Lorentz force equation, with the particle velocity $\vec{v}_0 \equiv \vec{\beta}_0 c = \beta_0 c \hat{z}$, and charge q, to define the net effective force field $\vec{W}$

$$\vec{F} = q(\vec{E} + \vec{v}_0 \times \vec{B}) \equiv q\vec{W}. \tag{4.59}$$

For a particle traveling parallel to the z-axis, we may write

$$\vec{v}_0 \times \vec{B} = v_0 \hat{z} \times (\vec{\nabla} \times \vec{A}) = v_0 \left(\vec{\nabla}(A_z) - \frac{\partial \vec{A}}{\partial z} \right). \tag{4.60}$$

We can obtain an expression for $\vec{W}$ using Eqs (1.6), (4.59), and (4.60), to find

$$\vec{W} = -\frac{\partial \vec{A}}{\partial t} = \vec{\nabla}\phi_e + v_0 \left[\vec{\nabla}(A_z) - \frac{\partial \vec{A}}{\partial z} \right]. \tag{4.61}$$

Using the assumed condition that v_0 is constant (an excellent approximation for relativistic beams, $v_0 \cong c$) we can express the partial derivatives in Eq. (4.61) in terms of the (convective) full derivative in z,

$$\frac{\partial \vec{A}}{\partial t} + v_0 \frac{\partial \vec{A}}{\partial z} = \frac{d\vec{A}}{dt} = v_0 \frac{d\vec{A}}{dz}. \tag{4.62}$$

The full momentum transfer, through a region R, due to the net Lorentz force $q\vec{W}$ can now be rewritten as

$$\Delta \vec{p} = q \int_R \left[-\frac{d\vec{A}}{dz} + \vec{\nabla} \left(A_z - \frac{\phi}{v_0} \right) \right] dz. \tag{4.63}$$

When evaluating the endpoints of integration we find that $\vec{A} = 0$ for an isolated system, where $\vec{E}$ and $\vec{B}$ vanish by the boundary of the region R. This condition makes the first term inside the integrand in Eq. (4.63) vanish, and we obtain

$$\Delta \vec{p} = \vec{\nabla} \left[q \int_R \left(A_z - \frac{\phi}{v_0} \right) dz \right] \equiv -\vec{\nabla} \Omega. \tag{4.64}$$

Since the field of integrated momentum kick, $\Delta \vec{p}$, can be derived from a potential Ω, its curl must vanish,

$$\vec{\nabla} \times (\Delta \vec{p}) = 0. \tag{4.65}$$

This expression is more usefully written to display the relationship between the longitudinal and transverse kicks

$$\vec{\nabla}_\perp (\Delta p_z) = \frac{\partial (\Delta \vec{p}_\perp)}{\partial z}. \tag{4.66}$$

It should be noted that this form of the deflection theorem is much more general than the original Panofsky–Wenzel theorem, where deflections due to the isolated cavities, like those found in synchrotrons, were examined. Equation (4.66) can also be applied to interactions in which free-charges and currents come in contact with the beam particles. This is the case in the beam–plasma and the beam–beam interactions, topics of concern in second volume of this series. Our form of the deflection theorem states that, for a deflection kick to occur, there must be transverse variation of the longitudinal kick imparted by the fields in the isolated system under consideration.

Next, let us examine the possible forms of the longitudinal momentum kicks in a periodic electromagnetic structure. To do so, we must write the off-axis dependence of a cylindrically symmetric solution to Eq. (4.3) in Floquet form,

$$E_z(\rho, z, t) = E_0 \, \text{Im} \sum_{n=-\infty}^{\infty} a_n \exp[i(k_{n,z}z - \omega t)] I_0 [k_{\rho,n}\rho], \tag{4.67}$$

where $k_{z,n} = (2\pi n + \psi)/d$, $k_{\rho,n} = \sqrt{k_{n,z}^2 - (\omega/c)^2}$, and I_0 is the modified Bessel function of order zero.

In the approximation that the Panofsky–Wenzel theorem is valid, we again take v_0 to be constant and approximately equal to $v_{\varphi,0} = \omega/k_{z,0}$ which yields $k_{\rho,0} = \omega/\gamma_0 c$. We then integrate over an integer number of periods M of the structure with the particle at phase $\varphi = k_{z,0}\zeta$ with respect to maximum acceleration, to obtain

$$\Delta p_z(\rho, \varphi) = \begin{cases} qE_0a_0Md\sin(\varphi)I_0(k_{r,0}\rho)/v_0, & n = 0 \\ 0, & n \neq 0. \end{cases} \tag{4.68}$$

The Panofsky–Wenzel theorem then gives

$$\Delta p_\rho(\rho, \varphi) = \int dz \frac{\partial \Delta p_z}{\partial \rho} = -\frac{qa_0E_0Mdk_{\rho,0}}{k_{z,0}v_0}\cos(\varphi)I_1(k_{\rho,0}\rho). \tag{4.69}$$

Equations (4.68) and (4.69) taken together give two results in the ultra-relativistic limit ($v_0 \to c$, and $\gamma_0 \to \infty$), where $k_{z,0} \cong \beta\omega/c \to \omega/c$, and $k_{\rho,0} = \omega/\gamma_0 c \to 0$. First, the acceleration kick is independent of ρ (since $I_0(k_{\rho,0}\rho) \approx 1 + (1/4)(k_{\rho,0}\rho)^2 \to 1$), and dependent only on the phase φ of the particle with respect to the synchronous wave. The second point is that the radial kick disappears in this limit, as $I_1(k_{\rho,0}r) \cong k_{\rho,0}\rho/2 = \omega\rho/2\gamma_0 c \to 0$. These results imply that, for ultra-relativistic particles accelerating in a cylindrically symmetric electromagnetic wave (described by Eq. 4.64), the radial and longitudinal motion is decoupled—there is no focal effect of acceleration. As we shall see in the next section, this is true only to lowest order in the accelerating field strength, and that the alternating gradient, or ponderomotive, focusing appears when we consider the motion to second order in the field strength.

The Panofsky–Wenzel theorem is a powerful tool, but it is also a bit abstract. It is instructive to examine the actual form of the transverse electromagnetic fields in an accelerating structure. To do this, we expand Eq. (1.2) near the axis of assumed cylindrical symmetry, to yield

$$E_\rho \cong -\frac{1}{\rho}\int_0^\rho \left.\frac{\partial E_z}{\partial z}\right|_{\rho=0} \tilde{\rho}\,d\tilde{\rho} = -\frac{\rho}{2}\left.\frac{\partial E_z}{\partial z}\right|_{\rho=0}. \tag{4.70}$$

To evaluate the magnetic component of the force, we also expand Eq. (1.3) about the axis to obtain

$$B_\varphi \cong \frac{1}{\rho c^2}\int_0^\rho \left.\frac{\partial E_z}{\partial t}\right|_{\rho=0} \tilde{\rho}\,d\tilde{\rho} = \frac{\rho}{2c^2}\left.\frac{\partial E_z}{\partial t}\right|_{\rho=0}. \tag{4.71}$$

Equations (4.70) and (4.71) give the lowest order (linear in ρ) contribution of the transverse fields. The further off-axis functional form of the transverse field is in general not linear—for periodic systems it has the form of a modified Bessel function (I_1), as can be deduced from Eqs (1.2), (1.3), and (4.67). Nevertheless, Eqs (4.70) and (4.71) are quite useful, and accurate for paraxial force calculations.

As an example of such a calculation, we can evaluate the radial component of the Lorentz force directly by use of Eqs (4.70) and (4.71),

$$F_\rho = q(E_\rho - v_0B_\varphi) = -\frac{q\rho}{2}\left[\left.\frac{\partial E_z}{\partial z}\right|_{\rho=0} + \frac{\beta_0}{c}\left.\frac{\partial E_z}{\partial t}\right|_{\rho=0}\right]$$

$$\cong -\frac{q\rho}{2}\left[\left.\frac{\partial E_z}{\partial z}\right|_{\rho=0} + \frac{1}{c}\left.\frac{\partial E_z}{\partial t}\right|_{\rho=0}\right] = -\frac{q\rho}{2}\left.\frac{dE_z}{dz}\right|_{\rho=0} \quad (v_0 \to c). \tag{4.72}$$

Thus, in the ultra-relativistic limit ($\beta_0 \to 1$), the net radial Lorentz force is simply proportional to the total z-derivative of the accelerating field. If we integrate the radial forces through an isolated electromagnetic structure with constant velocity and radial offset we obtain

$$\Delta p_\rho \cong \frac{1}{c} \int_{z_1}^{z_2} F_\rho \, \mathrm{d}z = -\frac{q\rho}{2c} \int_{z_1}^{z_2} \frac{\mathrm{d}E_z}{\mathrm{d}z} \bigg|_{\rho=0} \mathrm{d}z = -\frac{q\rho}{2c}[E_z(z_2) - E_z(z_1)] = 0,$$

(4.73)

where we have assumed that z_1 and z_2 are in field-free regions external to the structure. We note the similarity between Eqs (4.73) and (2.37), which applies to the special case of electrostatic fields. Equation (4.73) explicitly shows that the sum of all inward and outward impulses applied within the structure cancel for an ultra-relativistic particle traversing a cylindrically symmetric structure.

On the other hand, an electromagnetic cavity can be used to manipulate relativistic particles if we abandon cylindrical symmetry in the field. Writing only the components of the wave that are synchronous with ultra-relativistic particles, Eq. (4.3) has the same form as Eq. (2.38), with general solution

$$E_z(\rho, \phi, z, t) = E_0 \operatorname{Im} \sum_{m=0}^{\infty} \exp[ik(z - ct)]\rho^m (a_m \sin(m\phi) + b_m \cos(m\phi)).$$

(4.74)

These are **rf multipole** solutions for the wave field. For $m = 0$, we recover the monopole, cylindrically symmetric acceleration of the synchronous wave. In the $m = 1$ case, we have an **rf deflector**, which (assuming $a_1 = 0, b_1 \neq 0$) provides acceleration linear in x. According to Eq. (4.66), this accelerating wave, which imparts a longitudinal kick $\Delta p_z \propto b_1 x \sin(\varphi)$, must deflect the beam as

$$\Delta p_x \propto b_1 \cos(\varphi).$$

(4.75)

This is a transverse dipole kick, which is independent of transverse position (x) but sinusoidal in longitudinal phase φ. Specially designed electromagnetic cavities can allow this type of time-dependent deflection to be used in time-selecting beams to be bent into different trajectories. This type of scheme can be extended to $m = 2$-supporting cavities, which provide **rf quadrupole** focusing.

4.9 Second-order transverse focusing in electromagnetic accelerating fields*

The transverse focal effects experienced by an ultra-relativistic charged particle when it passes through an accelerating cavity have been shown to be negligible to lowest order in the variation of the transverse offset, because they are proportional to γ^{-1}. To clarify what we mean by "lowest order in the variation of the transverse offset," we restate that the charged particle is assumed to travel with constant velocity and transverse position through a period of the harmonic system of electromagnetic fields. Under these conditions, the total integrated transverse forces, which are alternatively strongly focusing and strongly defocusing, cancel nearly perfectly. However, if one relaxes the assumptions of constant velocity and offset from the design orbit and examines the effects of these alternating gradient forces, one finds that they give rise to a second-order secular focusing force of the type investigated in detail in Section 3.4.

For non-relativistic or moderately relativistic particles in rf linear accelerators (in practice, this means all ion beam accelerators), the first order, velocity-dependent effects discussed in the previous section complicate the analysis of second-order effects to some extent. For ultra-relativistic particles accelerating in rf linacs (electrons or positrons), on the other hand, the physical situation is simplified and the analysis is quite straightforward.

Near the axis of a periodic linac structure, the longitudinal electric field is given by Eq. (4.67) and the transverse forces are derived in first order approximation from this expression through use of Eq. (4.72),

$$F_\rho \cong -\frac{q\rho}{2}\frac{dE_z}{dz}\bigg|_{\rho=0} = -\frac{q\rho}{2}E_0 \, \text{Im} \sum_{n=-\infty}^{\infty} ik_{n,z}a_n \exp[i(k_{n,z}z - \omega t)]. \quad (4.76)$$

Here a_n and $k_{z,n} = (2\pi n + \psi)/d$ are the amplitude and the wave number of the nth spatial harmonic, respectively, and the phase shift per period can be written as $\psi = a\pi/b$ (where a and b are integers), as is always so for nearest-neighbor coupled oscillators (see Chapter 7 for more discussion of this point). In this formalism, we assume the $n = 0$ spatial harmonic is synchronous with the relativistic particle motion, or $\omega/k_{z,0} = c$. For a particle located at a phase φ with respect to maximum acceleration, we have

$$E_z = E_0 \left[\text{Im} \sum_{n=-\infty}^{\infty} a_n \, e^{i(2k_0 nz + \varphi_n)} \right], \quad (4.77)$$

and

$$F_\rho \cong -\frac{q\rho}{2}\frac{dE_z}{dz}\bigg|_{\rho=0} = -qE_0\rho \left[\text{Im} \sum_{n=-\infty}^{\infty} ik_0 na_n \, e^{i(2k_0 nz + \varphi_n)} \right]. \quad (4.78)$$

Averaging this expression over a period of the motion must be done with some care, because the particle's effective mass γm_0 is assumed to change due to acceleration. If this effect is ignorably small over a period (requiring $\gamma \gg qE_0 \sin(\varphi)d/m_0c^2$), then one may proceed to find the secular radial equation of motion (as in Section 3.4) by averaging the lowest order oscillatory motion driven by the force in Eq. (4.78) over a structure period:

$$\rho''_{\text{osc}} = \frac{F_\rho|_{\rho=\rho_{\text{sec}}}}{\beta^2 \gamma m_0 c^2} \cong -\frac{qE_0\rho_{\text{sec}}}{\gamma m_0 c^2} \left[\text{Im} \sum_{n=-\infty}^{\infty} ik_0 na_n \, e^{i(2k_0 nz + \varphi_n)} \right], \quad (4.79)$$

with steady-state solution

$$\rho_{\text{osc}} \cong \rho_{\text{sec}} \left[1 - \frac{qE_0}{4\gamma m_0 c^2} \left[\text{Im} \sum_{n=-\infty}^{\infty} \frac{a_n}{ik_0 n} e^{i(2k_0 nz + \varphi_n)} \right] \right]. \quad (4.80)$$

Upon substitution of the value of $\rho_{\rm osc}$ from Eq. (4.80) into Eq. (4.79) and averaging, we have

$$
\rho''_{\rm sec} = -\frac{\bar{F}_\rho}{\beta^2 \gamma m_0 c^2}
$$

$$
\cong -\frac{1}{8}\left[\frac{qE_0}{\gamma m_0 c^2}\right]^2 \rho_{\rm sec} \sum_{n=1}^{\infty}\left(a_n^2 + a_{-n}^2 + 2a_n a_{-n}\sin(2\varphi)\right). \tag{4.81}
$$

This radial focusing force, like that derived from the solenoid (see Section 2.3) provides equal focusing in both x and y, which is second order in applied field strength E_0. In the case of the solenoid, the net radial force is second order due to the accompanying rotation, while in the present case it is of second order because of the fast radial oscillatory motion due to alternating gradient focusing. The factor

$$
\eta(\varphi) \equiv \sum_{n=1}^{\infty}\left(a_n^2 + a_{-n}^2 + 2a_n a_{-n}\sin(2\varphi)\right)
$$

represents a sum over all spatial harmonics that contribute to the alternating gradient force, and is equal to 1 for a pure harmonic standing wave. Note that the synchronous harmonic ($n = 0$) does not contribute to the first or second order force in the ultra-relativistic limit, so that $\eta(\varphi) = 0$ for a pure forward traveling wave ($a_0 = 1$, all other a_n vanishing). It can be seen that the alternating gradient focusing effect arises from the existence of non-synchronous spatial harmonics, and that the (Doppler-shifted) frequencies of the motion due to the nth harmonic are the same as for the $(-n)$th harmonic, and they thus can interfere in the summation given by Eq. (4.79).

Because we have kept the energy constant in the discussion leading to Eq. (4.81), we have not kept the effects of adiabatic damping in the equation of motion. Acceleration can be taken into account by use of the damping term introduced in Eq. (2.57). Because the focusing is symmetric, the resulting equations of motion in x, y and ρ are all equivalent, and are of the form

$$
x'' + \left(\frac{\gamma'}{\gamma}\right)x' + \frac{\eta(\varphi)}{8\sin^2(\varphi)}\left(\frac{\gamma'}{\gamma}\right)^2 x = 0, \tag{4.82}
$$

where $\gamma' = qE_0\sin(\varphi)/m_0 c^2$. The solutions of Eq. (4.82) are obtained in direct analogy to the solutions of Eq. (2.57),

$$
x(z) = x_0 \cos[\alpha(z)] + x'_0\sqrt{\frac{8}{\eta(\varphi)}\frac{\gamma_0}{\gamma'}}\sin(\varphi)\sin[\alpha(z)],
$$

$$
\text{with } \alpha(z) = \left(\frac{\sqrt{\eta(\varphi)/8}}{\sin(\varphi)}\right)\ln\left[\frac{\gamma(z)}{\gamma_0}\right]\gamma_0 = \gamma(z = z_0). \tag{4.83}
$$

The matrix of the transformation corresponding to the solution of the motion during acceleration in an rf linac, given in Eq. (4.81), is thus

$$
\mathbf{M}_{\rm acc} =
\begin{bmatrix}
\cos[\alpha(z)] & \sqrt{\dfrac{8}{\eta(\varphi)}\dfrac{\gamma_0}{\gamma'}}\sin(\varphi)\sin[\alpha(z)] \\[4mm]
\sqrt{\dfrac{\eta(\varphi)}{8}}\dfrac{\gamma'}{\gamma(z)}\dfrac{\sin[\alpha(z)]}{\sin(\varphi)} & \dfrac{\gamma_0}{\gamma(z)}\cos[\alpha(z)]
\end{bmatrix}. \tag{4.84}
$$

Rigorously, one must consider this transformation to be valid over an integer number of rf structure, as it is a period-averaged, or secular, transformation.

In order to give a full treatment of the relativistic charged particle motion through an accelerating linac, one must also take into account the effects of the transient kicks experienced by the particle as it enters and exits the accelerating region. We have already analyzed a similar situation in the context of static electric fields (Section 2.4), but now we must generalize that treatment to include time-dependent field effects. The determination of the entrance (or exit) kick begins by integrating Eq. (4.78) through the transient field region, from zero field up to its initial spatial maximum (or vice versa),

$$\Delta x' = \mp \frac{qE_{\mathrm{m}} \sin(\varphi)}{2\gamma_{0(\mathrm{f})} mc^2} x = \mp \frac{q\gamma'}{2\gamma_{0(\mathrm{f})}} \left[\sum_{n=-\infty}^{\infty} a_n \right] x \equiv \mp \frac{q\gamma'}{2\gamma_{0(\mathrm{f})}} gx. \qquad (4.85)$$

In Eq. (4.85) we have defined g to be the ratio of the maximum accelerating field in the end cell to the average accelerating field experienced by a synchronous particle in the structure. In writing the series form for the longitudinal field in the end cell, we assume that the outlying half of the end cell is identical in field profile to the periodic interior cells. Let us note that Eq. (4.85) could have also been derived by simple application of the Panofsky–Wenzel theorem. It is also important to see that, since these kicks describe the passage of particles from a field-free region to an accelerating region, they are **first order** in the accelerating field amplitude.

In order to construct a transport matrix for these impulsive kicks in the secular particle motion, one must be careful to subtract the angle of the oscillating orbit corresponding to the homogeneous periodic solution to the zeroth order equation of motion given by Eq. (4.80). This angle is, at the maximum of the accelerating field in the entrance (exit) cell,

$$\theta_{\mathrm{osc}} = \mp \frac{x}{2} \frac{\gamma'}{\gamma} \sum_{\substack{n=-\infty \\ n\neq 0}}^{\infty} a_n \equiv \mp \frac{x}{2} \frac{\gamma'}{\gamma} (g - 1). \qquad (4.86)$$

After subtracting the **oscillatory angle** given in Eq. (4.86) from the total angle given in Eq. (4.85), we obtain the **secular** transverse angular kick and the entrance and exit of the structure. The matrix of this impulsive, thin-lens transformation is therefore simply

$$\mathbf{M}_{\mathrm{ent/exit}} = \begin{bmatrix} 1 & 0 \\ \mp \frac{\gamma'}{2\gamma_{0(\mathrm{f})}} & 1 \end{bmatrix} \qquad (4.87)$$

where γ_{f} is the Lorentz factor of the particle at the exit of the structure. Note that since we have assumed that the particle travels with $v \cong c$, that we have ignored possible longitudinal phase dynamics during acceleration.

The total matrix transformation, including interior second-order focusing and adiabatic damping and the first-order entrance and exit kicks can be written in matrix form as

$$\mathbf{M} = \mathbf{M}_{\mathrm{ex}} \mathbf{M}_{\mathrm{acc}} \mathbf{M}_{\mathrm{ent}}$$

$$= \begin{bmatrix} \cos(\alpha) - \sqrt{\frac{2}{\eta(\varphi)}} \sin(\varphi)\sin(\alpha) & \sqrt{\frac{8}{\eta(\varphi)}} \frac{\gamma_0}{\gamma'} \sin(\varphi)\sin(\alpha) \\ -\frac{\gamma'}{\gamma_{\mathrm{f}}} \left[\sqrt{\frac{2}{\eta(\varphi)}} \sin(\varphi) + \sqrt{\frac{\eta(\varphi)}{8}} \frac{1}{\sin(\varphi)} \right] \sin(\alpha) & \frac{\gamma_0}{\gamma_{\mathrm{f}}} \left[\cos(\alpha) + \sqrt{\frac{2}{\eta(\varphi)}} \sin(\varphi)\sin(\alpha) \right] \end{bmatrix}.$$
$$(4.88)$$

4.10   Summary and suggested reading

The acceleration of charged particles based on their interaction with a single traveling electromagnetic wave serves as a powerful model problem for developing an understanding of rf linacs. After introducing general Hamiltonian methods for analyzing this model problem, we divide rf linacs into two categories, gentle and violent acceleration.

The case of violent acceleration, which is typical of modern electron linacs, is in some ways conceptually simpler. We have used various forms of phase space plots, derived from an exact Hamiltonian analysis, to allow visualization of the capture and acceleration process in violent acceleration system. Particles which begin with nonrelativistic velocities may eventually attain an ultra-relativistic final state in such systems.

The case of gentle acceleration, which is applicable to heavy particle (ion) acceleration, is subtler, and has been approached using an approximate Hamiltonian, assuming small relative momentum offsets. This Hamiltonian is formally identical to that of the pendulum, and displays the same characteristics: small amplitude harmonic (synchrotron) oscillations, stable and unstable fixed points of the motion, stable and unstable trajectories in phase space, and a separatrix dividing the stable ("bucket") region from the unstable region. The adiabatic capture of particles into this bucket by slowly increasing the wave amplitude has been explored.

As the maximum momentum deviation imparted to a particle in the bucket is small, acceleration as a whole is accomplished by changing the wave–particle resonance condition slowly, to create a "moving bucket" in which the design particle accelerates. In the moving bucket, it has been noted that the rate of acceleration is limited, and at this limit the stable bucket area vanishes.

The case of gentle acceleration can be naturally generalized to include the longitudinal dynamics of charged particles in the synchrotron, a circular accelerator with a localized rf cavity. This analysis has been formulated as both a difference equation mapping and as an approximate differential equation (derivable from a Hamiltonian). Dynamics both above and below the ring's transition energy have been analyzed.

This chapter is completed by a discussion of both first and second order focusing effects in rf linacs. Analysis of these phenomena leads to a scenario that illustrates acceleration, focusing, and so-called adiabatic damping of transverse motion.

The material in this chapter is presented as a unified treatment of longitudinal motion in electron linear accelerators, ion linear accelerators, and synchrotrons. The reader may also desire a more detailed treatment of one of these subjects, as well as acceleration in devices such as the cyclotron, and induction accelerators. Such treatments are available in the following recommended references to this chapter:

1. T. P. Wangler, *Principles of RF Linear Accelerators* (Wiley, 1993). A complete survey of beam physics in linear accelerators; can act as a reference book. Written at the graduate level.
2. H. Wiedemann, *Particle Accelerator Physics I: Basic Principles and Linear Beam Dynamics* (Springer-Verlag, 1993). A good introduction to rf linacs is contained in this reference.

3. S. Humphries, Jr., *Principles of Charged Particle Acceleration* (Wiley, 1986). A good survey of acceleration by all mechanisms.

4. R.B. Neal, editor, The Stanford Two-Mile Accelerator (W.A. Benjamin, 1968). An extremely complete physics and technology description of the world's highest energy linac, dating from its initial commissioning.

5. W.K.H. Panofsky and W.A. Wenzel, *Rev. Sci. Instrum.* **27** (1956) 967. This article is the original reference for the Panofsky–Wenzel theorem.

6. D. Edwards and M. Syphers, *An Introduction to the Physics of High Energy Accelerators* (Wiley, 1993). A basic introduction to acceleration, especially good for synchrotrons.

7. S.Y. Lee, *Accelerator Physics* (World Scientific, 1996). Very clear treatment of longitudinal dynamics in circular accelerators.

8. P.J. Bryant and K. Johnsen, *The Principles of Circular Accelerators and Storage Rings* (Cambridge University Press, 1993).

9. M. Sands, *The Physics of Electron Storage Rings: An Introduction* (Stanford, 1970). Storage ring longitudinal dynamics, and much more.

Exercises

(4.1) One of the first rf linacs was a device termed the drift tube linac, which is sketched above in Fig. 4.13.

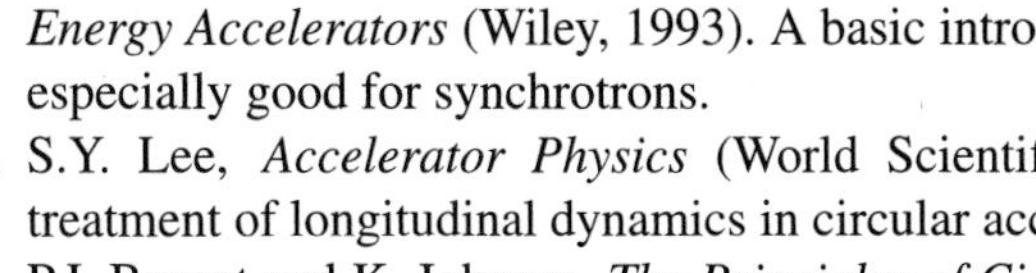

Fig. 4.13 Schematic, cross-sectional view a cylindrically symmetric drift-tube (Alvarez) linac section. Cross-hatched regions are conducting, so the region inside of the drift-tubes is shielded from the applied electric fields (arrowed lines).

(a) A typical field profile in this case can be approximated as step function, with uniform field in between the tubes and no field inside of the tubes, for example, for one period,

$$E_z(z) = \begin{cases} E_0, & 0 \le z < d/2 \\ 0, & d/2 \le z < d. \end{cases}$$

Write this field in terms of a Floquet expansion, noting that this is a **zero-mode** structure, $\psi = 0$, and thus the fundamental spatial harmonic is **not** the synchronous wave.

(b) Assuming the particle is synchronous with the first forward spatial harmonic ($n = 1$), $v_z \cong v_{\phi,1}$, what is the relationship between d and ω?

(c) In ion linacs (where the beam is only partially relativistic), the velocity changes significantly from the beginning to the end of the accelerator, and the structure must change its periodicity d to keep the particles in resonance. For an rf frequency of $f = \omega/2\pi = 500\,\text{MHz}$, find the value of d at the beginning of a proton linac (kinetic energy $T = 5\,\text{MeV}$) and the end ($T = 250\,\text{MeV}$).

(4.2) Evaluate the quantity α_{rf} for the following typical system parameters (assume the phase velocity of the wave is equal to the particle velocity):

(a) A high gradient electron linac, with $f = \omega/2\pi = 2856\,\text{MHz}$, total electron energy of 1 GeV, and accelerating field amplitude of $E_0 = 50\,\text{MV/m}$.

(b) A moderate gradient proton linac, with $f = \omega/2\pi = 805\,\text{MHz}$, proton kinetic energy of 200 MeV (be careful with the velocity!), and accelerating field amplitude of $E_0 = 8\,\text{MV/m}$.

(4.3) In a photocathode electron source, shown schematically in Fig. 4.14, a very short pulse laser (picoseconds, or below) strikes a photocathode, which is embedded in an electromagnetic accelerating structure. The emitted pulse of electrons is trapped, and then **optimally accelerated** in the device, so that its asymptotic phase has the highest acceleration associated with the wave, $\varphi_f = \pi/2$.

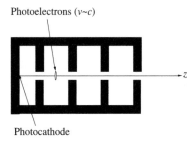

Photoelectrons ($v \sim c$)

Photocathode

Fig. 4.14 Photoinjector schematic.

(a) Assuming (from the picture as drawn, it is obvious that this assumption is not quite correct) that the accelerating structure is a traveling wave device, what is the minimum value of α_{rf} which allows this optimum capture, $\varphi_f = \pi/2$?

(b) What is the initial phase as a function of α_{rf} for any α_{rf} greater than the minimum?

(c) For a photocathode source with the electron linac rf parameters of Problem 4.2(a), what is the initial launch phase which gives $\varphi_f = \pi/2$?

(d) If you launch an electron 1° too early compared to this optimum, what is its final phase? You can do this analysis by direct substitution into the Hamiltonian, or through a more revealing expression derived by evaluating the quantity

$$\partial\varphi_f/\partial\varphi_i = (\partial H/\partial\varphi)_{\varphi_i}/(\partial H/\partial\varphi)_{\varphi_f}.$$

(4.4) For the proton linac parameters given in Exercise 4.2(b), evaluate

(a) The normalized bucket height $\delta p_{max}/p_0$.
(b) The bucket area A_b.
(c) The normalized synchrotron frequency ω_s/ω.

(4.5) Solve for the vibrational motion of a charge particle inside of the bucket in the general case, not just in the small amplitude approximation. Find the frequency of the oscillations, defined as $2\pi/T$, where T is the period of the motion, as a function of amplitude (e.g. maximum normalized momentum offset $\delta p_{max}/p_0$). Hint: The first integral of the motion is obtained by simple use of Eq. (4.27).

(4.6) The final state of an adiabatic capture process leaves the particles in a bucket described by a final maximum momentum excursion $\delta p_{max}/p_0$ and synchrotron frequency ω_s. For an exponential turn-on of the capture potential over five e-foldings, with final ratio of $v_a/\omega_s = 0.02$ and initial phase $\varphi = \pi$, solve the equations of motion numerically. Find the maximum initial momentum offset normalized to $\delta p_{max}/p_0$ which becomes trapped in the final bucket.

(4.7) As a function of synchronous phase φ_0, derive

(a) the bucket height, and
(b) the bucket area,

for moving buckets.

(4.8) As in Exercise 4.5, solve for the vibrational motion of a charged particle inside of a moving bucket (e.g. maximum normalized momentum offset $\delta p_{max}/p_0$). Find the frequency of the oscillations, defined as $2\pi/T$ (where T is the period of the motion), as a function of amplitude (maximum normalized momentum offset $\delta p_{max}/p_0$) and φ_0.

(4.9) For moving buckets, calculate the ratio of fractional momentum change to the design particle during a (small amplitude) synchrotron period $2\pi/\omega_s$ to the bucket height $\delta p_{max}/p_0$, as a function of φ_0. Is it always small?

(4.10) From the smooth approximation arguments explored in Exercise 3.15, we have an estimate of the average horizontal dispersion for the protons and antiprotons circulating in the Fermilab Tevatron.

(a) Using this value of the dispersion, estimate α_c and thus $\gamma_t = \alpha_c^{-1/2}$ for the Tevatron. Injection energy is $U = 150$ GeV in the Tevatron—does it go through transition?

(b) Assume that the applied rf voltage at the Tevatron is $V_0 = 2$ MV, and that the harmonic number $h = 1113$. Find the bucket height and synchrotron frequency for a non-accelerating bucket at 150 GeV and at the final energy of 900 GeV.

(c) Assume that the synchronous phase during acceleration is $\varphi_0 = \pi/6$. How long does it take for the beam to accelerate from 150 to 900 GeV?

(d) How many synchrotron oscillations occur during this acceleration?

(4.11) The differential equation approximation to the motion breaks down for synchrotron tunes that are not small compared to unity. To see this, one needs to plot the results of the mapping given by Eqs (4.47) and (4.48).

(a) For the Tevatron beam (see the previous exercise), plot the motion for ten different stable amplitudes from stable the fixed point to the separatrix non-accelerating bucket at 150 GeV.

(b) Now raise the synchrotron tune to 0.05, and repeat the same exercise. Compare the results.

(4.12) Derive, by using separation of variables, Eq. (4.65) from Eqs (4.3) and (4.4). What happens to the form of the solution when the radial wave-number $k_{\rho,n}$ becomes imaginary (which is guaranteed to happen for large enough n)?

(4.13) Derive Eq. (4.69) from Eq. (4.3) using separation of variables. Can you explain physically why are these fields of the same form as Eq. (2.36)?

(4.14) For the traveling wave fields described by Eq. (4.64),

(a) Write the form of the vector potential component A_z associated with this electric field. In the Lorentz gauge, where the condition

$$\vec{\nabla} \cdot \vec{A} + \frac{1}{c^2} \frac{\partial \phi}{\partial t} = 0$$

is assumed, what must be to the form of A_ρ be to self-consistent with this A_z?

(b) Derive the radial electric and azimuthal magnetic fields associated with the total vector potential.

(c) For the photocathode electron source discussed in Exercise 4.3, the particle is born at the cathode with some canonical radial momentum due to the value of A_ρ at emission. Assume that the "full" cells are pure traveling wave π-mode with speed-of-light phase velocity, the cathode cell is one-half this length, and that the electrons are injected with ultra-relativistic velocity (not a very good assumption!). Re-derive the deflection theorem from Eq. (4.61) forward. What is the final radial kick on the electrons as a function of phase and radial offset?

(d) Derive this radial kick by use of Eq. (4.73), but substituting the correct non-zero value of $E_z(z_1)$.

(4.15) In a long structure, one may be concerned with the long-time response of a particle to a synchronous wave. In this case, we ignore all transients and all fields are assumed to have a longitudinal and temporal dependence that can be expressed solely in terms of $\zeta = z - v_\phi t$, where $v_\varphi \cong v_0$.

(a) Examine the curl of the net field on a particle traveling parallel to the z-axis with speed v_b, $\vec{\nabla} \times \vec{W} = \vec{\nabla} \times (\vec{E} + v_0 \hat{z} \times \vec{B})$, and show that this net field is conservative, so that one may write

a version of the Panofsky–Wenzel theorem,

$$\vec{\nabla}_\perp W_z = \frac{\partial (\vec{W}_\perp)}{\partial \zeta}$$

for synchronous fields.

(b) Explain this result in terms of the picture in the frame of the synchronous wave, using Lorentz transformation of the fields.

(4.16) To further illustrate the dependence of the secular transverse electromagnetic focusing force on variations in the longitudinal acceleration experienced by the particle, derive the relation,

$$\rho_{sec}'' \cong -\frac{\rho_{sec}}{4} \frac{\langle E_z^2 \rangle - \langle E_z \rangle^2}{(\gamma m_0 c^2)^2}.$$

The indicated averages are over the spatial dependence of the field encountered by an ultra-relativistic particle. The expression obtained states that the focusing is proportional to the *variance* of the acceleration. Show that it is equivalent to Eq. (4.81).

(4.17) Assume maximal acceleration ($\varphi = \pi/2$) for a pure standing wave. The focusing in Eq. (4.78) can be thought of as arising from an equivalent solenoidal magnetic field (cf. Eq. (2.57)). For an average acceleration $qE_0 = 50$ MeV/m, what is this field?

(4.18) Evaluate the determinant of the matrix transformation given in Eq. (4.84). Does it give the correct form, which indicates adiabatic damping of phase space (cf. Section 2.6)?

(4.19) Show that the matrix given in Eq. (4.88) can be decomposed into two matrices, $\mathbf{M} = \mathbf{M}_{damp} \mathbf{M}_{foc}$, a focusing matrix, and a matrix describing the adiabatic damping of the angular motion due to acceleration,

$$\mathbf{M}_{damp} = \begin{bmatrix} 1 & 0 \\ 0 & \gamma_0/\gamma_f \end{bmatrix}.$$

5 Collective descriptions of beam distributions

The previous chapters have been concerned with single charged particle motion in both the transverse and longitudinal directions, under the action of types of forces commonly encountered in accelerators. While the methods developed for describing single particle motion form the basis of beam physics, beams are composed not of one, but very many particles, and one must go further. The next set of tools we discuss are concerned, therefore, with multi-particle distributions and description of their dynamics. We begin our treatment by re-introducing the distribution function and the Vlasov equation governing its evolution, which were first mentioned in Chapter 1. By first examining the equilibrium solutions to the Vlasov equation, we then arrive at a common and pedagogically useful form of this distribution function, the bi-Gaussian phase space distribution.

While the Vlasov equation provides a detailed description of the evolution of the distribution function of a non-interacting ensemble of particles acted upon by macroscopic forces, this description is by nature not very compact. We shall see that we can always systematically predict the evolution of the distribution function under linear external forces. However, we discuss these linear transformations mainly to provide the context for introduction of methods that extract less information about the system than the Vlasov equation analysis obtains (i.e. the entire distribution function). These methods aim to provide the lowest-order information of interest, which is the evolution of the beam's spatial and momentum boundaries, or **envelopes**. The boundary of a continuous distribution is not an obviously unambiguous concept, though, and so we proceed by introducing methods for evaluating the motion of the **root-mean-square** (rms, based on second moments of the distribution function) envelope of a beam distribution. The rms formalism is connected to the single particle dynamics analysis by a use of the transport matrices introduced in Chapter 3. The rms envelope description has the additional benefit of allowing a concise definition of the **emittance**, or effective area occupied by the beam particles in a phase plane (or trace space). We shall see that the rms envelope motion is most generally and powerfully evaluated by a nonlinear differential equation. This **rms envelope equation** is most useful when analyzing the effects of beam self-forces (space-charge effects). These self-forces are beyond the discussion of the present chapter, but we prepare for their analysis by introducing the envelope equation in the appropriate context.

The emittance turns out to be very useful concept that, not too surprisingly, allows parameterization of the beam envelope motion in the same way the action (another, related, area in the phase plane) parameterizes the amplitude of a harmonic oscillation. This scaling of the beam envelope with the square root of the beam emittance leads to the introduction of the **Twiss parameter** formalism, which gives insight not only into beam envelope, but single particle dynamics as well. All of the above-mentioned methods, which are initially introduced in the context of transverse dynamics, are also extended to analysis of the longitudinal phase space distribution and longitudinal envelope evolution later in this chapter.

5.1 Equilibrium distributions

The distribution of N particles in phase space can be formally written as

$$f(\vec{x}, \vec{p}) = \sum_{i=1}^{N} \delta^3(\vec{x} - \vec{x}_i)\delta^3(\vec{p} - \vec{p}_i), \tag{5.1}$$

where $\delta^3(\vec{x} - \vec{x}_i)\delta^3(\vec{p} - \vec{p}_i)$ indicates the product of two three-dimensional Dirac delta-functions that locate each particle at a position in phase space $(\vec{x}_i, \vec{p}_i)$. It is more useful at this point to view this distribution in the coarse-grained approximation, in which the minimum relevant volume in phase space $d^3x\,d^3p$ contains many particles. In this case, the distribution can be approximated as a smooth, continuous function $f(\vec{x}, \vec{p})$, so that the number of particles inside a small but finite phase space volume is $f(\vec{x}, \vec{p})\,d^3x\,d^3p$. If one views the distribution purely as a continuous function, there is no way of asking questions about binary or higher-order **microscopic** interactions, and, therefore, the distribution function evolution is described by the Vlasov equation introduced in Chapter 1,

$$\frac{\partial f}{\partial t} + \dot{\vec{x}} \cdot \vec{\nabla}_{\vec{x}} f + \dot{\vec{p}} \cdot \vec{\nabla}_{\vec{p}} f = 0. \tag{5.2}$$

Here the force term $\dot{\vec{p}} = F$ is due only to macroscopic fields, where the source distributions (charge and current densities) are viewed as continuous functions of the coordinates. The effects of other microscopic forces, e.g. binary collisions, can be systematically included on the right-hand side of Eq. (5.2), but we will not address them in this chapter.

The Vlasov equation admits a wealth of solution types, which can represent an infinite variety of physical situations. As such, a detailed discussion of the scope of its solutions is not overly relevant to this book. However, in order to orient ourselves towards further examination of the behavior of particle distributions in accelerators, we first examine the special case of **equilibrium solutions** of the Vlasov equation.

An equilibrium solution of Eq. (5.2) is obviously a function whose partial time derivative vanishes,

$$\dot{\vec{x}} \cdot \vec{\nabla}_{\vec{x}} f + \dot{\vec{p}} \cdot \vec{\nabla}_{\vec{p}} f = 0. \tag{5.3}$$

We next make the assumption that the motion between Cartesian phase planes is uncoupled. In this case, we can write the distribution function in **separable** form as a product of phase plane distributions

$$f(\vec{x}, \vec{p}) = N f_x(x, p_x) f_y(y, p_y) f_z(z, p_z), \tag{5.4}$$

where N is the number of particles, and we choose each phase plane distribution to be normalized to unity, $\int_{-\infty}^{\infty} \int_{-\infty}^{\infty} f_i(x_i, p_i)\,dx_i\,dp_i = 1$. Substitution of

Eq. (5.4) into Eq. (5.3) allows us to write three separate equations of the form

$$\dot{x}\frac{\partial}{\partial x}f_x(x,p_x) + F_x\frac{\partial}{\partial p_x}f_x(x,p_x) = 0. \tag{5.5}$$

From this point forward, through the following six sections, we restrict our discussion to a single transverse phase plane. It will be straightforward to generalize the conclusions we will draw to the other (transverse and longitudinal) phase planes.

If we further make the **ansatz** that the phase plane distribution function is separable, that is, that f_x is separable x and p_x, so that we may write $f_x(x,p_x) = X(x)P(p_x)$, we have

$$\dot{x}\frac{\mathrm{d}X}{\mathrm{d}x}P + F_xX\frac{\mathrm{d}P}{\mathrm{d}p_x} = 0. \tag{5.6}$$

The forces that we have encountered in our discussion to this point have all been dependent on the transverse coordinate in the direction of the force, for example, $F_x(x)$. In these cases, Eq. (5.6) is fully and self-consistently separable, and we have the relations

$$\frac{1}{F_x(x)}\frac{\mathrm{d}X}{\mathrm{d}x} = -\frac{\gamma_0 m_0}{pP}\frac{\mathrm{d}P}{\mathrm{d}p_x} = \lambda_s, \tag{5.7}$$

where λ_s is a separation constant.

We can immediately write the solution of the momentum distribution equation,

$$P(p_x) = C_p \exp\left(-\frac{\lambda_s p_x^2}{2\gamma_0 m_0}\right) \equiv C_p \exp\left(-\frac{p_x^2}{2\sigma_{p_x}^2}\right), \tag{5.8}$$

where C_p is a constant determined by normalization of the distribution. The **Gaussian** momentum spectrum of Eq. (5.8) (which is sometimes referred to as **Maxwellian** in this context) can be put in more familiar terms by noting that the kinetic energy associated with the transverse momentum is, in the paraxial approximation, $p_x^2/2\gamma_0 m_0$. We can then rewrite Eq. (5.8) in terms of the transverse temperature, with $k_B T_\perp = \lambda_s^{-1}$,

$$P(p_x) = C_p \exp\left(-\frac{p_x^2}{2\gamma_0 m_0 k_B T_x}\right) \equiv C_p \exp\left(-\frac{p_x^2}{2\sigma_{p_x}^2}\right). \tag{5.9}$$

This result states that the familiar Maxwellian momentum distribution found in statistical mechanics theory arises as an equilibrium solution to the Vlasov equation, given only the assumption that the forces are independent of momentum.

The macroscopic forces of interest to us can have any spatial (x) dependence allowed by the Maxwell equations,[1] but we begin by examining the mos

[1]The form of the forces becomes quite interesting and nonlinear when the forces arise from the charge and current associated with the distribution itself. When self-electromagnetic forces are important in determining the distribution, the Vlasov equation is often referred to as the **Maxwell–Vlasov equation**.

relevant case, that of a linear (in x) restoring force,

$$F_x(x) = -\gamma_0 m_0 v_0^2 \kappa_0^2 x. \tag{5.10}$$

With this force, we may explicitly solve for the spatial component of the distribution function,

$$X(x) = C_x \exp\left(-\frac{\gamma_0 m_0 v_0^2 \kappa_0^2 x^2}{2k_B T_x}\right) \equiv C_x \exp\left(-\frac{x^2}{2\sigma_x^2}\right), \tag{5.11}$$

where C_x is a normalization constant.

Through Eq. (5.11), we have found that the equilibrium spatial distribution is also of Gaussian form for the case of a linear focusing force, and the total phase plane distribution function, including normalization factors, is

$$f_x(x, p_x) = X(x)P(p_x) = \frac{\kappa_0 v_0}{2\pi k_B T_x} \exp\left[-\frac{1}{2k_B T_x}\left(\gamma_0 m_0 v_0^2 \kappa_0^2 x^2 + \frac{p_x^2}{\gamma_0 m_0}\right)\right]. \tag{5.12}$$

This distribution is termed bi-Gaussian and, by construction, has no coordinate–momentum correlation. Because the bi-Gaussian phase plane distribution function arises naturally from solution of the Vlasov equation, we will use it in subsequent sections to illustrate aspects of phase plane distribution evolution.

Separable distribution functions such as that given in Eq. (5.12) are, of course, a special case, one in which the spatial and momentum projections of the distribution (spatial and momentum densities) are trivially proportional to X and P, as

$$n(x) \equiv \int_{-\infty}^{\infty} f_x(x, p_x) \, dp_x \propto X(x) \quad \text{and} \quad \Psi(p_x) \equiv \int_{-\infty}^{\infty} f_x(x, p_x) \, dx \propto P(p_x). \tag{5.13}$$

These distributions are in thermal equilibrium, as is emphasized by Eq. (5.9). Thermalization in charged particle beam transport is often achieved very slowly, over many revolutions of a circular accelerator, by a combination of damping and heating effects. In fast, transient systems, such as linear accelerators, equilibrating mechanisms are too slow to be relevant, and if equilibria are found, they must be a property of the particle source used.

It is left to the reader to show in Exercises 5.2 and 5.3 that the phase plane distribution function in equilibrium can always be written in the form

$$f_x(x, p_x) = C \exp\left[-\frac{H(x, p_x)}{k_B T}\right], \tag{5.14}$$

where C is a normalization constant and H is the time-independent Hamiltonian. For paraxial motion, where the Hamiltonian governing the transverse motion is approximately

$$H(x, p_x) = \frac{p_x^2}{2\gamma_0 m_0} + V(x), \tag{5.15}$$

the distribution function is also separable in x and p_x, and the momentum distribution is again found to be Maxwellian, in agreement with Eq. (5.9).

An even more general statement about equilibrium solutions to the Vlasov equation can be made, which is that $f_x(x, p_x) = G[H(x, p_x)]$, where G is any differentiable function, is an equilibrium solution to the Vlasov equation. As the value of H is a constant of the motion for a time-independent Hamiltonian, an equilibrium solution labeled only by this value has the property that the contours of constant phase plane probability density, $f_x(x, p_x) = $ constant, are also constant energy curves. Furthermore, these curves follow the trajectory of an individual particle. The reason for this is straightforward: in an equilibrium state, there can be no motion along a gradient in the phase plane (trace space) distribution.

These observations imply that, for a simple harmonic oscillator Hamiltonian of the type governing the paraxial motion, the contours of constant phase plane density are ellipses, as displayed in Fig. 5.1. The use of the ellipse to indicate a relevant area in a phase plane occupied by a beam is often used, even in the general case when the beam distribution function is not in equilibrium. In particular, we recall from Chapter 3 that the motion of a charged particle in a periodic focusing (and therefore manifestly non-equilibrium) system, when plotted in trace space once per period (Poincaré plot), gives an elliptical trajectory. The roles of elliptical regions in trace space will be amplified and clarified considerably in the following sections.

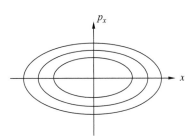

Fig. 5.1 Elliptical contours of constant phase plane distribution density for the simple harmonic oscillator case. The contours also indicate trajectories of particles with a given, constant energy.

5.2   Moments of the distribution function

The full distribution function contains all of the information needed to describe (in a coarse-grained way) the state of a non-interacting ensemble of beam particles. One may not need all of this information, though, and so it is often the case that **moments of the distribution** are used to arrive at a simpler description of the distribution's evolution. These moments are formally written as

$$\int_{-\infty}^{\infty} \int_{-\infty}^{\infty} x^n p_x^m f(x, p_x) \, dx \, dp_x, \tag{5.16}$$

where m and n are equal to zero or a positive integer, and the quantity $m + n$ is referred to as the **order** of the moment.

As we will need to use the apparatus of the matrix description of transverse motion developed in Chapter 3 for the following discussion, we now shift our analysis from the transverse phase plane to the equivalent trace space (x, x'). In trace space, the moments of the distribution are thus

$$\int_{-\infty}^{\infty} \int_{-\infty}^{\infty} x^n (x')^m f_x(x, x') \, dx \, dx'. \tag{5.17}$$

Let us now discuss the lowest-order moments. The zeroth-order moment is simply the normalization condition on the distribution,

$$\int_{-\infty}^{\infty} \int_{-\infty}^{\infty} f_x(x, x') \, dx \, dx' = 1. \tag{5.18}$$

The first-order moments are the **centroids** of the distribution,

$$\langle x \rangle = \int_{-\infty}^{\infty} \int_{-\infty}^{\infty} x f_x(x, x') \, dx \, dx', \tag{5.19}$$

and

$$\langle x' \rangle = \int_{-\infty}^{\infty} \int_{-\infty}^{\infty} x' f_x(x, x') \, \mathrm{d}x \, \mathrm{d}x', \tag{5.20}$$

which vanish when a beam is aligned to its design axis.

The second moments are written in standard notation as the distribution variances,

$$\sigma_{11} \equiv \sigma_x^2 = \langle x^2 \rangle = \int_{-\infty}^{\infty} \int_{-\infty}^{\infty} x^2 f_x(x, x') \, \mathrm{d}x \, \mathrm{d}x', \tag{5.21}$$

$$\sigma_{22} \equiv \sigma_{x'}^2 = \langle x'^2 \rangle = \int_{-\infty}^{\infty} \int_{-\infty}^{\infty} x'^2 f_x(x, x') \, \mathrm{d}x \, \mathrm{d}x', \tag{5.22}$$

as well as the mixed moment,

$$\sigma_{12} = \sigma_{21} \equiv \sigma_{xx'} = \langle xx' \rangle = \int_{-\infty}^{\infty} \int_{-\infty}^{\infty} xx' f(x, x') \, \mathrm{d}x \, \mathrm{d}x', \tag{5.23}$$

which indicates the degree of correlation between x and x'. The two types of notation introduced here belong to two different formalisms for evaluating the evolution of the distribution moments: the notation σ_{ij} for the second moments is used in the **matrix method** of propagating the moments discussed below in Sections 5.3 and 5.4, while the quantities $\sigma_x, \sigma_{x'}$, and $\sigma_{xx'}$ are used in development of the differential equation (rms envelope equation) approach to the moment evolution developed in Section 5.5.

Note also that in discussing the bi-Gaussian equilibrium distribution function in Section 5.1, that the quantities σ_x and $\sigma_{p_x} = p_0 \sigma_{x'}$ were used in the argument of the Gaussian distributions (Eqs (5.11) and (5.8)) to parameterize its spatial and momentum widths, respectively. This defining convention is in agreement with the rms measures of the distribution widths (Eqs (5.21) and (5.22)). There is no confusion concerning the present and previous uses of the designations σ_x and σ_{p_x}, as long as one recognizes that the definitions given in Eqs (5.21) and (5.22) are more general than, and thus supersede, those of Eqs (5.11) and (5.8).

5.3   Linear transformations of the distribution function

To develop the tools needed to follow the evolution of the beam's second moments, we must first formally understand the evolution of the distribution function. The trace space distribution function can be mapped from a known initial condition at one position in the beam transport to another by use of the matrix formalism introduced in Chapter 3. This mapping is accomplished by simply advancing the trace space variables according to the linear transformation **M** that describes the difference between the initial beamline position and the position of interest. This method is straightforward, and its validity is a consequence of the Liouville theorem, which states that the value of the distribution function (i.e. the density in trace space), when followed under the equations of motion generated by a Hamiltonian, is invariant. Therefore, if one knows the initial trace space distribution $f_{x,0}(\vec{x}_i)$, where $\vec{x}_i = (x_i, x_i')$, then to find the value of the distribution function at any point in the trace space plane at a later time, one only needs to evaluate the distribution function at the corresponding initial trace space point, $\vec{x}_i = \mathbf{M}^{-1} \cdot \vec{x}$. Thus, we may write the distribution at a given point in the transport in terms of the initial distribution function $f_{x,0}$, using only the current trace space position $\vec{x}$ and the initial distribution's functional form,

$$f_x(\vec{x}) = f_{x,0}(\mathbf{M}^{-1} \cdot \vec{x}). \tag{5.24}$$

Before we describe the implementation of this general method and its use for propagation of distribution moments, however, it is best to begin with a few examples. These concrete illustrations will demonstrate the use of Eq. (5.24) in simple terms.

Let us take as our first example an initially separable, uncorrelated trace space distribution,

$$f_{x,0}(\vec{x}) = g(x)h(x'). \tag{5.25}$$

If we let the beam drift without applying forces for a length z (x' is unchanged in the transformation, see Eq. (3.18)), we have the final distribution,

$$f_x(\vec{x}) = g(x - x'z)h(x'). \tag{5.26}$$

Likewise, for a thin lens (of focal length f) transformation on the initial distribution, as given by Eq. (3.19), we would have a final distribution in which only the beam angles change,

$$f_x(\vec{x}) = g(x)h\left(x' - \frac{x}{f}\right). \tag{5.27}$$

As a final example, consider the case of a focusing channel with strength κ^2, having the length needed to impart 90° phase advance. In this case, the entire trace space picture rotates by this angle, and we have

$$f_x(\vec{x}) = g\left(-\frac{x'}{\kappa}\right)h(\kappa x). \tag{5.28}$$

These examples are more compelling if accompanied by pictures. In order to create them let us further assume that the initial distribution is, as in the separable Vlasov equilibrium, a bi-Gaussian in trace space,

$$f_x(x, x') = \frac{1}{2\pi\sigma_x\sigma_{x'}} \exp\left[-\frac{x^2}{2\sigma_x^2}\right] \exp\left[-\frac{x'^2}{2\sigma_{x'}^2}\right]. \tag{5.29}$$

In this standard Gaussian notation, we emphasize that the parameters σ_x and $\sigma_{x'}$ in Eq. (5.29) are in fact also the rms beam size $\sqrt{\langle x^2 \rangle}$ and angular spread $\sqrt{\langle x'^2 \rangle}$, respectively.

To visualize the area occupied by the distribution in trace space, we use an ellipse corresponding to a contour of constant probability density. One can pick many useful measures of the area depending on choice of ellipse—it may be interesting to choose an ellipse that describes, say, the half-intensity contour (where f_x is one-half of maximum), or perhaps an ellipse based on enclosed fraction (e.g. 90 per cent), of the beam. The most useful choice from the standpoint of our present discussion, however, is the rms trace space ellipse. For the distribution of Eq. (5.28), the rms trace space ellipse corresponds to the $\exp[-1/2]$ intensity contour, and in this case is described by the expression

$$\frac{x^2}{\sigma_x^2} + \frac{x'^2}{\sigma_{x'}^2} = 1. \tag{5.30}$$

For correlated trace space distributions, the method of determining the rms trace space ellipse is slightly more involved, as discussed further below. While we use the rms trace ellipse to illustrate specific transformations, all of the arguments employed are independent of the actual choice of trace ellipse. The results will, therefore, prove to be useful and valid even in cases when the distribution function is not bi-Gaussian, and/or when the ellipse definition is based on an enclosed fraction of the beam.

Once we have established the ellipse of interest, we can follow it by simply transforming its trace space coordinates using the transport matrix **M**. Most beam transport systems are not operated in equilibrium, which implies that the distribution function will change, and the orientation of a given trace space ellipse will evolve. For example, an uncorrelated distribution may become correlated at another point in the transport, as in Eqs (5.25)–(5.28). It should be emphasized that particles lying inside of the bounding ellipse do not cross it as it deforms under transport (see Figs 5.2–5.4). In fact, no trace space trajectories may cross in this nominally one-dimensional Hamiltonian system. This attribute, combined with the conservation of trace space area derived from Liouville's theorem, gives us two results—the area of the ellipse must be conserved during transport, **and** that the fraction of the beam lying inside of this ellipse is constant. The trace space area inside of the ellipse is written as $A \equiv \pi \varepsilon_x$, where we have introduced the **emittance** ε_x of the beam. For the trace space distribution described by Eq. (5.29), the emittance corresponding to the rms trace ellipse is $\varepsilon_x = \sigma_x \sigma_{x'}$. This is a specific example of an **rms emittance**, which is generally defined, not geometrically in the phase plane but analytically, in terms of distribution moments, as will be seen in Section 5.4.

The examples discussed above, where an initially uncorrelated distribution in trace space is transformed by a drift, a thin lens, or a 90° phase advance focusing

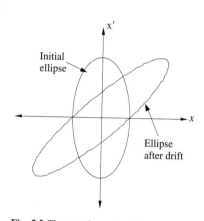

Fig. 5.2 The transformation of an rms trace space ellipse from an initial, uncorrelated state, to a correlated state by a simple drift.

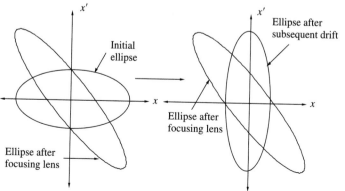

Fig. 5.3 The transformation of an rms trace space ellipse from an initial, uncorrelated state, to a correlated state by application of a thin, focusing lens, and the transformation back to an uncorrelated state by action of a drift.

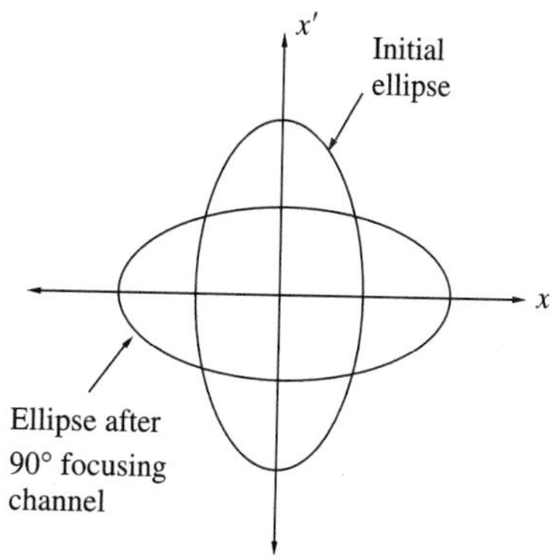

Fig. 5.4 The transformation of an rms trace space ellipse from an initial, uncorrelated state, to a new uncorrelated state by traversal of a 90° phase advance focusing channel.

channel, are illustrated by use of the rms phase ellipse in Figs 5.2–5.4. In the case of the drift (Fig. 5.2), particles with positive angle drift towards the right in trace space, while those with negative angle drift towards the left, skewing the ellipse and producing a correlated distribution. Note that the ellipse by construction has the same area, and thus the same emittance, after the transformation.

In Fig. 5.3, the complementary correlation is produced by the action of a thin lens, in which the position of a particle is held constant while its angle is changed in proportion to its position x. A subsequent drift to focus is shown in Fig. 5.3(a), which again gives rise to an uncorrelated distribution at focus. The distribution has a smaller final spatial extent than it began with, as expected. It can be seen that this focusing is accomplished at the price of increasing the transverse momentum (angular) width of the beam distribution.

Figure 5.4 shows the simple case of the initially uncorrelated distribution being transformed to another uncorrelated distribution by a $\mu = \pi/2$ phase advance focusing channel. In this case, the major and minor axes of the phase ellipse are essentially exchanged, if one uses the scaled coordinate system $(\kappa x, x'/\kappa)$. This transformation can be considered as the thick lens analogue of the two transformations shown in Fig. 5.3.

Let us now investigate the transformation of a trace ellipse in more detail. We begin by writing the correlated distribution function produced from an initially uncorrelated bi-Gaussian distribution by the simple drift shown in Fig. 5.2 as

$$f_x(x, x') = \frac{1}{2\pi\sigma_x\sigma_{x'}} \exp\left[-\left(\frac{(x - x'z)^2}{2\sigma_x^2} + \frac{x'^2}{2\sigma_{x'}^2}\right)\right]. \tag{5.31}$$

The invariance of the (rms, or any other constant density contour) trace space ellipse area is not easily apparent in this form of the distribution. To display the property of area invariance more clearly, it is customary and useful to parameterize a given trace space ellipse as

$$\gamma_x(z)x^2 + 2\alpha_x(z)xx' + \beta_x(z)x'^2 = \varepsilon_x, \tag{5.32}$$

where $\varepsilon_x, \beta_x, \alpha_x$, and γ_x are termed the **Twiss parameters**. It can be seen that this parameterization of a trace space ellipse is useful for, but not at all restricted to, description of the rms trace ellipse. This is because the emittance ε_x on the right-hand side of Eq. (5.32) can be rescaled, with a given β_x, α_x, and γ_x, to define any smaller or larger ellipse (e.g. 90 per cent ellipse) by multiplying ε_x by an arbitrary factor. Note that since we are trying to describe an ellipse of certain major and minor axis sizes, as well as orientation in trace space, that we need only three parameters to accomplish this task, not the four given—$\varepsilon_x, \beta_x, \alpha_x$ and γ_x. In fact, one is redundant, because with the invariance of ε_x the other three parameters can be related by

$$\gamma_x = \frac{1 + \alpha_x^2}{\beta_x}. \tag{5.33}$$

The evolution of the Twiss parameters is simply derived from the trace space transformation $\mathbf{M}$, as will be discussed below. Once one obtains these parameters, the beam size in the spatial and angular dimensions, as well as the distribution's $x - x'$ correlation can be found immediately, as can be seen in Fig. 5.5.

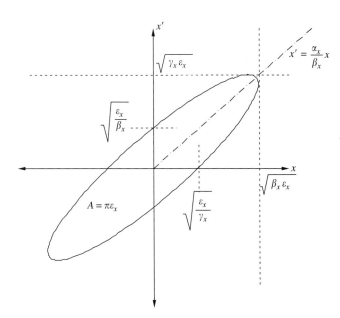

Fig. 5.5 The extrema, intercepts, and correlation of a trace space ellipse, indicated in terms of the Twiss parameters.

Assuming for the sake of discussion that the ellipse pictured in Fig. 5.5 is indeed the rms trace space ellipse, the rms beam envelope boundaries (square roots of the second moments) are seen to occur at

$$\sigma_x = \sqrt{\beta_x \varepsilon_x}, \qquad \sigma_{x'} = \sqrt{\frac{\beta_x}{\varepsilon_x}}, \tag{5.34}$$

while the correlation moment is given by

$$\sigma_{xx'} = -\alpha_x \varepsilon_x. \tag{5.35}$$

In order to recapitulate what we have discussed in this section so far, and clarify the relationship between the distribution, the second moments, and the trace space ellipse, it is instructive to revert to the case of a correlated bi-Gaussian distribution function. With the parameterization of Eq. (5.32), this special case (Eq. (5.31)) can be written in general form using the rms trace space ellipse and associated Twiss parameters as

$$f_x(x, x') = \frac{1}{2\pi\varepsilon_x} \exp\left[-\frac{\gamma_x(z)x^2 + 2\alpha_x(z)xx' + \beta_x(z)x'^2}{2\varepsilon_x}\right]. \tag{5.36}$$

An example of this special distribution function, which is expected in cases where the particle source is in thermal equilibrium, or the distribution evolves towards a bi-Gaussian by thermalizing (e.g. radiation emission, intra-beam binary scattering) effects, is illustrated in Fig. 5.6. The contours of constant trace space density in this case are also shown. In analogy to the case of thermal equilibrium illustrated in Fig. 5.1, they are nested, aligned (directions of the ellipse axes are the same) ellipses, as can be seen by examination of Eq. (5.36). This type of distribution is indeed found in most storage rings, especially strongly synchrotron radiation-damped electron rings, where thermalizing effects have enough time (very many turns) to slowly produce such a distribution.

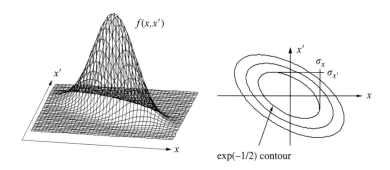

Fig. 5.6 An example of a correlated bi-Gaussian distribution function. The contours of constant trace space density are trace ellipses for this thermalized distribution.

While the correlated bi-Gaussian is, by design, not an equilibrium distribution, if one views only a particular point in a storage ring, a thermalized beam will display this type of distribution. As noted in Chapter 3, trace space ellipses arise from the plotting of Poincaré maps—for a correlated bi-Gaussian distribution these ellipses coincide with the contours of constant density. The trace space ellipses found in Poincaré maps are, in general, not aligned to the trace space axes, and thus the observed distribution must be, in general, correlated. Note that the rms trace ellipse in this case is again the $\exp[-1/2]$ (relative to the peak) density contour, the maximum extents of this ellipse along the x and x' axes are $\sigma_x = \sqrt{\beta_x \varepsilon_x}$ and $\sigma_{x'} = \sqrt{\gamma_x \varepsilon_x}$, respectively, and $\pi \varepsilon_x$ is the area of the rms trace space ellipse.

5.4 Linear transformations of the second moments

The second moments of the beam's trace space distribution function can be formally calculated by using the results above. To begin the discussion of this analysis, we consider the general form of the moment integrals,

$$\sigma_{ij} \equiv \int_{-\infty}^{\infty} \int_{-\infty}^{\infty} x_i x_j f_x(\vec{x}) \, \mathrm{d}^2 \vec{x}, \tag{5.37}$$

where we adopt the notation $\vec{x} \equiv (x_1, x_2) = (x, x')$ and $\mathrm{d}^2 \vec{x} = \mathrm{d}x_1 \, \mathrm{d}x_2 = \mathrm{d}x \, \mathrm{d}x'$. If the initial trace space vector $\vec{x}_0$ is transformed by application of transport matrix representing the portion the beam line under study, $\vec{x}_f = \mathbf{M} \cdot \vec{x}_0$, then the second moments after the transformation can be written as

$$\sigma_{ij_f} \equiv \int_{-\infty}^{\infty} \int_{-\infty}^{\infty} (\vec{x}_f)_i (\vec{x}_f)_j f_x(\mathbf{M}^{-1} \cdot \vec{x}_f) \, \mathrm{d}^2 \vec{x}_f. \tag{5.38}$$

The integrals in Eq. (5.38) can be simplified by replacing the integration variables $\vec{x}_f$ with their initial values $\vec{x}_0$, to obtain

$$\sigma_{ij_f} \equiv \int_{-\infty}^{\infty} \int_{-\infty}^{\infty} (\mathbf{M} \cdot \vec{x}_0)_i (\mathbf{M} \cdot \vec{x}_0)_j f_x(\vec{x}_0) \, \mathrm{d}^2 \vec{x}_0. \tag{5.39}$$

Here we have used the fact that the determinant of the transport matrix is unity, and thus through the Jacobian relation $\mathrm{d}^2 \vec{x}_f = \det(\mathbf{M}) \, \mathrm{d}^2 \vec{x}_0 = \mathrm{d}^2 \vec{x}_0$

Equation (5.39) can be rewritten compactly as

$$\sigma_f = \mathbf{M} \cdot \sigma_0 \cdot \mathbf{M}^T, \tag{5.40}$$

with the **σ-matrix** defined as

$$\sigma \equiv \begin{bmatrix} \sigma_{11} & \sigma_{12} \\ \sigma_{21} & \sigma_{22} \end{bmatrix}, \tag{5.41}$$

We immediately obtain, from Eq. (5.40), that the **σ**-matrix must have an invariant determinant. We can thus explicitly write

$$\det(\sigma) \equiv \sigma_{11}\sigma_{22} - \sigma_{12}^2 = \langle x^2 \rangle \langle x'^2 \rangle - \langle xx' \rangle^2 = \text{constant}, \tag{5.42}$$

where we have used $\sigma_{12} = \sigma_{21}$. It is natural to ask whether this object, which we have seen is a constant of the motion for linear transformations, is related to other known trace space invariants. It is, in fact, related to the **rms emittance** through

$$\varepsilon_{x,\text{rms}}^2 = \det(\sigma) = \langle x^2 \rangle \langle x'^2 \rangle - \langle xx' \rangle^2, \tag{5.43}$$

as the area of the rms trace ellipse is $\pi \varepsilon_{x,\text{rms}}$. This can be verified for the equilibrium case when $\langle xx' \rangle = 0$, and where the area of the rms ellipse can be written $\pi\sqrt{\langle x^2 \rangle \langle x'^2 \rangle} = \pi\sigma_x\sigma_{x'}$.

Since this definition of the rms emittance is consistent with the emittance defined for the uncorrelated rms trace ellipse, it should not be surprising that the Twiss parameters also have analytical definitions, based on the second moments. These definitions are displayed using all of our previously introduced notations,

$$\beta_x \equiv \frac{\langle x^2 \rangle}{\varepsilon_{x,\text{rms}}} = \frac{\sigma_x^2}{\varepsilon_{x,\text{rms}}} = \frac{\sigma_{11}}{\varepsilon_{x,\text{rms}}},$$

$$\alpha_x \equiv -\frac{\langle xx' \rangle}{\varepsilon_{x,\text{rms}}} = -\frac{\sigma_{xx'}}{\varepsilon_{x,\text{rms}}} = -\frac{\sigma_{12}}{\varepsilon_{x,\text{rms}}}, \tag{5.44}$$

$$\gamma_x \equiv \frac{\langle x'^2 \rangle}{\varepsilon_{x,\text{rms}}} = \frac{\sigma_{x'}^2}{\varepsilon_{x,\text{rms}}} = \frac{\sigma_{22}}{\varepsilon_{x,\text{rms}}}.$$

Consistency with the definition of the rms emittance gives

$$\gamma_x \equiv \frac{\langle x'^2 \rangle}{\varepsilon_{x,\text{rms}}} = \frac{\varepsilon_{x,\text{rms}}^2 + \langle xx' \rangle^2}{\langle x^2 \rangle \varepsilon_{x,\text{rms}}} = \frac{1 + \alpha_x^2}{\beta_x}, \tag{5.45}$$

which is also in agreement with the geometric definitions based on the trace space ellipse.

In this section, we have introduced a purely geometric definition of the emittance and the other Twiss parameters, based on ellipses in trace space. In addition, we have given a purely analytical definition of these quantities (Eqs (5.43) and (5.44)) that relies on second moments of the trace space distribution. These two different approaches to description of the Twiss parameters can be connected to each other through the concept of the rms trace space ellipse.

To see this connection, we rewrite the $\boldsymbol{\sigma}$-matrix in terms of the Twiss parameters as

$$\boldsymbol{\sigma} = \varepsilon_{x,\mathrm{rms}} \begin{bmatrix} \beta_x & -\alpha_x \\ -\alpha_x & \gamma_x \end{bmatrix}, \qquad (5.46)$$

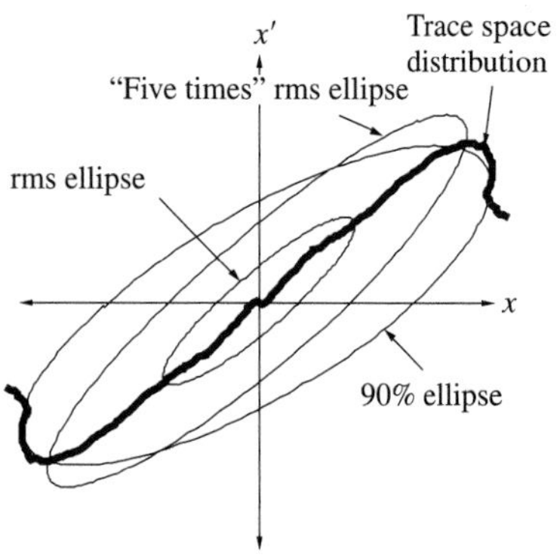

explicity showing the use of the normalized variables α_x, β_x, and γ_x, to describe the aspect ratio and orientation of the rms ellipse, and the scaling factor $\varepsilon_{x,\mathrm{rms}}$ to describe the size of the ellipse. Note that the invariance of both the $\boldsymbol{\sigma}$-matrix determinant and rms the emittance implies that $\beta_x\gamma_x - \alpha_x^2 = 1$, as expected. From the development of the rms formalism, one can see the power of the second moment approach to collective description of the beam. If one desires a more inclusive measure of the beam distribution, it can be obtained by merely scaling the beam emittance upward. For instance, it is common in certain applications to define the beam's distribution "edge" to be at a phase plane ellipse with area equivalent to five times the rms ellipse, $5\varepsilon_{x,\mathrm{rms}}$. The beam "edge" envelope is accordingly found at $x_{\mathrm{edge}} = \sqrt{5}\sigma_x$, as illustrated in Fig. 5.7.

It should also be noted, however, that an emittance alternatively defined through the area of the smallest ellipse containing a given fraction of the beam may **not** yield the same α_x, β_x, and γ_x as the second moment-based (rms) Twiss parameters, as is illustrated in Fig. 5.7. Both the second moment definition and the enclosed fraction definition have some degree of arbitrariness, as can be seen by comparing the ellipses shown in Fig. 5.7 with an example distribution that does not have elliptical symmetry. For a correlated trace space distribution with iso-intensity contours that are nested ellipses, the analytical and geometric emittance definitions give the same α_x, β_x, and γ_x, while with more general, non-thermalized distributions (e.g. the example shown in Fig. 5.7), they will typically not. Thus, some care must be taken in choosing the optimum definition of emittance, in order to describe the distribution characteristics one is interested in. The alert reader will also note that the second-moment definition of Twiss parameters can be anomalously dependent on "tail particles," which are located far from most of the beam population, but still have a strong effect on the calculated rms Twiss parameters. This problem is often encountered in computer simulation of multiple-particle dynamics. In experimental measurements, one must also be careful to handle the tail of the distribution well, as it may be dominated by noise.

We can now begin to discuss the ways in which one implements the Twiss parameter formalisms we have developed. There are a number of simple beam transport examples that one can examine using Eq. (5.40). The most transparent is that of the beam freely expanding from an uncorrelated focus, or **waist** ($\sigma_{12} = 0$ or $\alpha_x = 0$), found at $z = z_0$. In this case one has, from Eq. (5.40),

Fig. 5.7 Distorted trace space distribution along with rms ellipse corresponding to $\varepsilon_{x,\mathrm{rms}}$, ellipse corresponding to $5\varepsilon_{x,\mathrm{rms}}$ and 90% ellipse (minimum area ellipse containing 90% of the particles). Notice that the two different definitions lead to different orientation ellipses, and thus two different Twiss parameter sets, α_x, β_x, and γ_x.

$$\sigma_{11_f} = \sigma_{11_0} + (z - z_0)^2\sigma_{220},$$

$$\sigma_{22_f} = \sigma_{220}, \qquad (5.47)$$

$$\sigma_{12_f} = 2z\sigma_{220},$$

or directly in terms of second moments,

$$\sigma_{xf}^2 = \sigma_{x0}^2 \left[1 + (z - z_0)^2 \frac{\sigma_{x'0}^2}{\sigma_{x0}^2} \right],$$

$$\sigma_{xf}^2 = \sigma_{x0}^2,$$

$$\sigma_{xx'f} = 2(z - z_0)\sigma_{x0}^2. \tag{5.48}$$

In a free drift, the square of the rms beam size, or **envelope**, increases quadratically away from the waist position. The results of Eq. (5.48) can be written in even simpler form by use of the Twiss parameters,

$$\beta_{xf} = \beta_{x0} \left[1 + \left(\frac{z - z_0}{\beta_{x0}} \right)^2 \right] = \beta_x^* \left[1 + \left(\frac{z - z_0}{\beta_x^*} \right)^2 \right],$$

$$\gamma_{xf} = \gamma_{x0},$$

$$\alpha_{xf} = -\gamma_{x0}(z - z_0). \tag{5.49}$$

The first of Eqs (5.49) indicates that the β-function grows quadratically with distance away from the waist, with characteristic length

$$\beta_x^* \equiv \beta_{x0}\big|_{\alpha_{x0}=0} = \frac{\sigma_{x0}^2\big|_{z=z_0}}{\varepsilon_{x,\mathrm{rms}}}, \tag{5.50}$$

the minimum value of the β_x at the waist.

As we have seen in the previous chapters, whenever one introduces bend magnets into the transport system, the horizontal and longitudinal phase planes become coupled. In this and other cases (e.g. coupling of vertical and horizontal motion due to a solenoid), one must use the general six-dimensional phase space vector and associated 6×6 transport matrix (see Eqs (3.83) and (3.84)) to describe the linear transformation of a beam as it traverses a section of beamline. In general, the beam moments can be then mixed from one phase plane to another. For example, if the beam traverses a bend section, and the horizontal momentum dispersion takes on a non-vanishing value, the mixed moments

$$\sigma_{16} \equiv \left\langle x \left(\frac{\delta p_z}{p_0} \right) \right\rangle = \iiint_{\vec{x}} \iint_{\vec{p}} x \frac{(p_z - p_0)}{p_0} f(\vec{x}, \vec{p}) \, d^3\vec{x} \, d^3\vec{p}, \tag{5.51}$$

and

$$\sigma_{26} \equiv \left\langle x' \left(\frac{\delta p_z}{p_0} \right) \right\rangle = \iiint_{\vec{x}} \iiint_{\vec{p}} \frac{p_x(p_z - p_0)}{p_0^2} f(\vec{x}, \vec{p}) \, d^3\vec{x} \, d^3\vec{p} \tag{5.52}$$

are non-zero. The general algorithm for transforming the six-dimensional moments is, in analogy to Eq. (3.83),

$$\mathbf{\Sigma_f} = \mathbf{R} \cdot \mathbf{\Sigma_0} \cdot \mathbf{R}^{\mathrm{T}}, \tag{5.53}$$

where we have introduced the six-dimensional second moment $\mathbf{\Sigma}$-matrix, with $(\mathbf{\Sigma})_{ij} = \sigma_{ij}$.

5.5   The rms envelope equation

In Section 5.4, we constructed a formalism for following the evolution of the second moments of the trace space distribution using the transport matrix **M**. In some cases, however, it is more useful to employ a differential equation approach to deducing the evolution of the second moments, as the physical properties of the system can be more clearly seen by use of this description, termed the **rms envelope equation**. Also, in the case of non-trivial beam self-forces (space charge forces, cf. Ex. 2.14), the rms envelope equation allows the straightforward inclusion of these forces in an envelope analysis.

We begin the analysis by writing the first derivative of the rms beam size as

$$
\frac{d\sigma_x}{dz} = \frac{d}{dz}\sqrt{\langle x^2 \rangle} = \frac{1}{2\sigma_x}\frac{d}{dz}\langle x^2 \rangle
$$

$$
= \frac{1}{2\sigma_x}\frac{d}{dz}\int_{-\infty}^{\infty}\int_{-\infty}^{\infty} x^2 f_x(x,x')\,dx\,dx'
$$

$$
= \frac{1}{\sigma_x}\int_{-\infty}^{\infty}\int_{-\infty}^{\infty} xx' f_x(x,x')\,dx\,dx' = \frac{\sigma_{xx'}}{\sigma_x}. \tag{5.54}
$$

The second derivative of the rms beam size is then given by

$$
\frac{d^2\sigma_x}{dz^2} = \frac{d}{dz}\frac{\sigma_{xx'}}{\sigma_x} = \frac{1}{\sigma_x}\frac{d\sigma_{xx'}}{dz} - \frac{\sigma_{xx'}^2}{\sigma_x^3}
$$

$$
= \frac{1}{\sigma_x}\frac{d}{dz}\int_{-\infty}^{\infty}\int_{-\infty}^{\infty} xx' f_x(x,x')\,dx\,dx' - \frac{\sigma_{xx'}^2}{\sigma_x^3}
$$

$$
= \frac{\sigma_{x'}^2 + \langle xx'' \rangle}{\sigma_x} - \frac{\sigma_{xx'}^2}{\sigma_x^3} \tag{5.55}
$$

or simply

$$
\sigma'' = \frac{\sigma_x^2\sigma_{x'}^2 - \sigma_{xx'}^2}{\sigma_x^3} - \frac{\langle xx'' \rangle}{\sigma_x} = \frac{\varepsilon_{x,\mathrm{rms}}^2}{\sigma_x^3} - \frac{\langle xx'' \rangle}{\sigma_x}. \tag{5.56}
$$

For the linear transport conditions we have been discussing up until this point that all forces give rise to the general relation

$$
x'' + \kappa_x^2 x = 0, \tag{5.57}
$$

where κ_x^2 can be positive, negative, or vanishing, and need not be piece-wise constant. We, thus, can substitute Eq. (5.57) into Eq. (5.58), to arrive at the rms envelope equation,

$$
\sigma_x'' + \kappa_x^2\sigma_x = \frac{\varepsilon_{x,\mathrm{rms}}^2}{\sigma_x^3}. \tag{5.58}
$$

The envelope evolution is controlled by an expression, Eq. (5.58), which has a homogenous portion of the equation identical to that of the single particle motion. The inhomogeneous term on the right-hand side of Eq. (5.58) can be interpreted mathematically as the outward forcing of the beam envelope by the rms spread in trajectory angle, which is parameterized by the non-vanishing rms emittance.

It can also be interpreted physically in terms of the outward pressure in the beam region due to the thermal nature of the collection of particles in the beam. To see this connection clearly, we examine the case of a distribution in thermal equilibrium with the externally applied linear focusing channel, as discussed initially in Section 5.1. In this simple case, the net inward force per unit area towards the symmetry plane on the distribution must be equal and opposite to the outward force per unit area away from the symmetry plane due to the pressure gradient $\vec{\nabla}(nkT)$ within the distribution,

$$\frac{dF_{x,\text{out}}}{dA} = \hat{x} \cdot \vec{\nabla}(nkT) = \Sigma_b kT_x \frac{\partial}{\partial x} f_x(x, p_x) = \Sigma_b kT_x \frac{\partial}{\partial x} f_x(x, p_x) = \Sigma_b kT_x \frac{\partial X}{\partial x},$$
(5.59)

where Σ_b is the number of beam particles per unit area, as projected onto the y–z plane. As this force is nonlinearly dependent on x, it does not locally cancel the applied linear force, but, as stated above, these forces must cancel when integrated from $x = 0$ to a limit of the distribution's range. For the integrated outward directed force per unit area away from the symmetry plane, we have

$$\int_0^\infty \frac{dF_{x,\text{out}}}{dA} dx = \Sigma_b k_B T_x \int_0^\infty \frac{\partial X}{\partial x} dx = \Sigma_b k_B T_x X(0).$$
(5.60)

Since we are assuming thermal equilibrium, we know from solution of the Vlasov equation that the function X is of Gaussian form, $X(x) = [\sqrt{2\pi}\sigma_x]^{-1} \exp(-x^2/2\sigma_x^2)$, and

$$\int_0^\infty \frac{dF_{x,\text{out}}}{dA} dx = \frac{\Sigma_b k_B T_x}{\sqrt{2\pi}\sigma_x}.$$
(5.61)

Also, it is straightforward to evaluate the total inward directed force on one side of the distribution,

$$\int_0^\infty \frac{dF_{x,\text{in}}}{dA} dx = \frac{\Sigma_b \kappa_0^2 \beta_0^2 \gamma_0 m_0 c^2}{\sqrt{2\pi}\sigma_x} \int_0^\infty x X(x)\, dx = \frac{\Sigma_b \kappa_0^2 \beta_0^2 \gamma_0 m_0 c^2 \sigma_x}{\sqrt{2\pi}}.$$
(5.62)

By requiring balance between the integrated forces in Eqs (5.61) and (5.62), we have for the equilibrium value of σ_x

$$\sigma_{\text{eq}}^2 = \frac{k_B T_x}{\kappa_0^2 \beta_0^2 \gamma_0 m_0 c^2}.$$
(5.63)

The temperature in Eq. (5.63) is not a constant for a given system (if one raises the focusing strength κ^2, the temperature increases as well, where $T_x = \sigma_{x,\text{rms}}^2/2\gamma_0 m_0 c^2$, and $p_{x,\text{rms}} = \beta_0 \gamma_0 \sigma_{x'} = \beta_0 \gamma_0 \varepsilon_{x,\text{rms}}/\sigma_{\text{eq}}$), so we rewrite it in terms of the conserved rms emittance,

$$\sigma_{\text{eq}}^2 = \frac{\varepsilon_{x,\text{rms}}}{\kappa_0}.$$
(5.64)

This simple result could have been deduced by cursory examination of the equilibrium solution ($\sigma_x'' = 0$) to the envelope equation, Eq. (5.58), but deriving it in this way illustrates well the physical meaning of the emittance term in this equation.

The effects of acceleration can also be straightforwardly included in the envelope equation by noting that, with an instantaneous spatial rate of change of momentum $p_0' = (\beta_0\gamma_0)' m_0 c^2$,

$$\frac{d^2\sigma_x}{dz^2} = \frac{d}{dz}\frac{\sigma_{xx'}}{\sigma_x} = \frac{1}{\sigma_x}\frac{d\sigma_{xx'}}{dz} - \frac{\sigma_{xx'}^2}{\sigma_x^3}$$

$$= \frac{1}{\sigma_x}\left[\sigma_{x'}^2 - \kappa_x^2\sigma_x^2 - \frac{(\beta_0\gamma_0)'}{\beta_0\gamma_0}\sigma_{xx'}\right] - \frac{\sigma_{xx'}^2}{\sigma_x^3}. \tag{5.65}$$

In more standard form, this equation is written as

$$\frac{d^2\sigma_x}{dz^2} + \frac{(\beta_0\gamma_0)'}{\beta_0\gamma_0}\frac{d\sigma_x}{dz} + \kappa_x^2\sigma_x = \frac{\varepsilon_{n,x}^2}{(\beta_0\gamma_0)^2\sigma_x^3}, \tag{5.66}$$

where we have introduced the **normalized emittance**, $\varepsilon_{n,x} \equiv \beta_0\gamma_0\varepsilon_{x,\text{rms}}$, which is explicitly

$$\varepsilon_{n,x}^2 \equiv (\beta\gamma)^2[\langle x^2\rangle\langle x'^2\rangle - \langle xx'\rangle^2] = (m_0c)^{-2}[\langle x^2\rangle\langle p_x^2\rangle - \langle xp_x\rangle^2]. \tag{5.67}$$

The normalized emittance, and not the rms emittance defined by Eq. (5.43), is, in fact, invariant under the combined effects of linear transverse forces, where $F_x \propto x$, and longitudinal acceleration (see Ex. 5.14). This result is a direct consequence of the adiabatic damping of the beam particle angles under acceleration, which causes the emittance defined in trace space to be diminished. The invariant normalized emittance, on the other hand, is an effective area occupied by the beam in the phase plane, not the trace plane. It is analogous, in the uniform focusing and acceleration case discussed in Section 2.6, to the single particle action of a particle with maximum amplitude equal to the rms beam size σ_x.

5.6 Differential equation description of Twiss parameter evolution

The rms envelope equation is but one way to write a differential equation governing the rms beam size in a linear transport system, especially when one wishes to include the effects of self-forces. At other times, it is often more useful, as will be seen below, to write an equivalent system of equations based on the Twiss parameters. If we begin by differentiation of the rms expressions given in Eq. (5.44), we have

$$\beta_x' = 2\frac{\langle xx'\rangle}{\varepsilon_{x,\text{rms}}} = -2\alpha_x,$$

$$\alpha_x' = -\frac{\langle x'^2\rangle + \langle xx''\rangle}{\varepsilon_{x,\text{rms}}} = -\frac{\langle x'^2\rangle - \kappa_x^2\langle x^2\rangle}{\varepsilon_{x,\text{rms}}} = -\gamma_x + \beta_x\kappa_x^2, \tag{5.68}$$

$$\gamma_x' = \frac{2\langle x'x''\rangle}{\varepsilon_{x,\text{rms}}} = -\frac{2\kappa_x^2\langle xx'\rangle}{\varepsilon_{x,\text{rms}}} = 2\kappa_x^2\alpha_x.$$

These equations are particularly straightforward when one examines the behavior of a beam in a force-free drift. In this case $\kappa_x^2 = 0$, and we immediately have

$$\gamma_x = \text{constant} = \gamma_{x0} \quad \text{and} \quad \alpha_x = \alpha_{x0} - \gamma_{x0}(z - z_0), \qquad (5.69)$$

where $\gamma_{x0} = \gamma_x(z_0)$ and $\alpha_{x0} = \alpha_x(z_0)$. The behavior of the β-function is then obtained by a final integration,

$$\beta_x = \beta_{x0} - 2\alpha_{x0}(z - z_0) + \gamma_{x0}(z - z_0)^2, \qquad (5.70)$$

with $\beta_{x0} = \beta_x(z_0)$. If we define z_0 to be at a waist ($\alpha_{x0} = 0, \beta_x \equiv \beta_x^*$), then we have simply, using $\gamma_{x0} = \beta^{*-1} = \beta_{x0}^{-1}$,

$$\beta_x = \beta_{x0} + \gamma_{x0}(z - z_0)^2 = \beta_{x0} + \frac{(z - z_0)^2}{\beta_{x0}} = \beta_{x0}\left[1 + \left(\frac{z - z_0}{\beta_{x0}}\right)^2\right], \quad (5.71)$$

in agreement with Eqs (5.48) and (5.49).

Equation (5.68) also allow a simple analysis of thin-lens focusing, in which the parameters of importance in the problem become immediately apparent. If one places a thin focusing lens of focal length f at a longitudinal position z_0, Eqs (5.68) become, after integrating through the lens,

$$\beta_{xf} = \beta_{x0}, \alpha_{xf} = \alpha_{x0} + \frac{\beta_{x0}}{f} \quad \text{and} \quad \gamma_{xf} = \frac{1 + \left(\alpha_{x0} + \frac{\beta_{x0}}{f}\right)^2}{\beta_{x0}}. \qquad (5.72)$$

We can deduce directly from Eqs (5.69) both the size of the beam at its subsequent waist (if the waist exists) and the distance to this waist. Since in a drift $\gamma_x = \text{constant} = \beta^*$, the ratio of the beam size at the focus to its initial size at z_0 is

$$\frac{\sigma_x^*}{\sigma_{x0}} = \sqrt{\frac{\beta^*}{\beta_{x0}}} = \sqrt{\frac{1}{\gamma_{xf}\beta_{x0}}} = \sqrt{\frac{1}{1 + \left(\alpha_{x0} + \frac{\beta_{x0}}{f}\right)^2}}. \qquad (5.73)$$

The distance to this focus can be found by noting from Eqs (5.68) that the α_x changes linearly in a drift, so the distance L^* to the waist, where α_x vanishes, is simply

$$L^* = -\frac{\alpha_{xf}}{\gamma_{xf}} = \frac{\beta_{x0}\left(\frac{\beta_{x0}}{f} + \alpha_{x0}\right)}{1 + \left(\alpha_{x0} + \frac{\beta_{x0}}{f}\right)^2}. \qquad (5.74)$$

Examination of relevant quantities in Eqs (5.73) and (5.74) illustrates several aspects of the physics involved. First, we note that if $-\alpha_{x0} > \beta_{x0}/f$, then the beam is too divergent to be focused by the lens of the chosen strength, and the distance to the next waist is negative in Eq. (5.74) (there is no subsequent waist). In the case where $-\alpha_{x0} \ll \beta_{x0}/f$, however, the beam is strongly focused, and the ratio of the beam size at the next waist to its initial size is

$$\frac{\sigma^*}{\sigma_{x0}} \underset{-\alpha_{x0} \ll \beta_{x0}/f}{\cong} \frac{f}{\beta_{x0}}. \qquad (5.75)$$

Thus, we see that the ratio β_{x0}/f governs the impact of the lens on the beam size. When this ratio greatly exceeds unity, the position of the next waist with respect to the lens becomes

$$L^* \underset{-\alpha_{x0} \ll \beta_{x0}/f}{\cong} \frac{\frac{\beta_{x0}^2}{f}}{1 + \left(\frac{\beta_{x0}}{f}\right)^2} \underset{\beta_{x0} \gg f}{\cong} f, \qquad (5.76)$$

as would be naively expected from single particle ray optics. This limit is obtained because the β-function near the focus is a measure of how quickly the beam diverges. If the focal length is much shorter than the β-function, then the tendency of the beam to diverge can be neglected, and the focal position is essentially independent of beam spreading.

The first-order equations given in (5.68), in addition to being useful for direct integration, form the building blocks of further analysis. Differentiating the first of these equations, and substituting in the second two, we have

$$\beta_x'' = -2\alpha_x' = 2\gamma_x - 2\kappa_x^2\beta_x, \quad \text{or} \quad \beta_x'' + 2\kappa_x^2\beta_x = \frac{2}{\beta_x} + \frac{(\beta_x')^2}{2\beta_x}. \quad (5.77)$$

This second-order differential equation is obviously nonlinear. This is as expected, because Eq. (5.77) is equivalent to as the rms envelope equation, which is also nonlinear. The equilibrium value of the β-function ($\beta_x'' = \beta_x' = 0$) in the case of a uniform focusing channel $\kappa_x^2 = \kappa_0^2$ can be found by inspection of Eq. (5.77) to be

$$\beta_{\text{eq}} = \frac{1}{\kappa_x}. \quad (5.78)$$

This result which states that the "matched" β-function is simply the inverse of the focusing wave number, could have also been noted from Eq. (5.64).

In a drift ($\kappa_x^2 = 0$), the second-order differential equation is, in fact, linear, however, as γ_x is a constant

$$\beta_x'' = 2\gamma_x. \quad (5.79)$$

Other linear forms of the second-order differential equation describing the evolution of the β-function are examined in Section 5.7.

5.7   Oscillations about equilibrium

The equilibrium value of the rms beam size in a uniform focusing channel is, from inspection of Eq. (5.58), or the physical derivation of Eq. (5.64),

$$\sigma_{\text{eq}} = \sqrt{\frac{\varepsilon}{\kappa_0}}. \quad (5.80)$$

Deviations from this equilibrium, $\delta\sigma \equiv \sigma_x - \sigma_{\text{eq}}$, can be examined by linearizing the envelope equation in the neighborhood of the equilibrium. This process consists of expanding the terms in the equation to first order in a Taylor series about the equilibrium point as follows:

$$\delta\sigma_x'' + \kappa_0^2\delta\sigma_x = -\frac{3\varepsilon_{x,\text{rms}}^2}{\sigma_x^4}\delta\sigma_x, \quad (5.81)$$

which becomes, upon substitution of Eq. (5.80),

$$\delta\sigma_x'' + \kappa_0^2\delta\sigma_x = -3\kappa_0^2\delta\sigma_x,$$

or simply

$$\delta\sigma_x'' + 4\kappa_0^2\delta\sigma_x = 0. \quad (5.82)$$

The term of the right-hand side of Eq. (5.81) adds an additional "restoring" force to the envelope dynamics near equilibrium, since the pressure forcing on

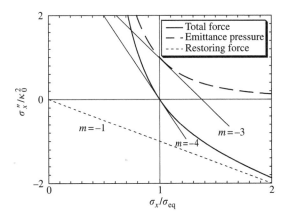

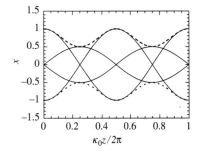

Fig. 5.8 The contributions of each term in the envelope equation to the total forcing of the envelope. In this plot the linear focusing appears as a line of slope $m = -1$, the forcing due to emittance or pressure effects at the equilibrium beam size as a line of slope $m = -3$, and the total forcing as line of slope $m = -4$.

the envelope falls rapidly with increasing beam size, as seen in Fig. 5.8. The slope of the emittance force term near equilibrium is seen to be three times that of the linear restoring force due to the external focusing, as also explicitly seen in the first equation of (5.82). The total effective linear restoring force near the equilibrium beam size is thus four times that of the external focusing alone.

Because of this combination of effects, the frequency of oscillations about the equilibrium is twice that of the single particle betatron frequency κ_0, with general solution of the form,

$$\delta\sigma_x'' = \delta\sigma_m \cos(2\kappa_0 z + \theta_0). \tag{5.83}$$

Here, $\delta\sigma_m$ is the amplitude of the oscillation (which in order for the analysis to be valid is assumed to be small, $\delta\sigma_m \ll \sigma_{eq}$), and θ_0 is an arbitrary phase. This doubling of the betatron frequency is easy to explain with a physical illustration, given in Fig. 5.9. The envelope, shown in bold dashed lines, oscillates downward when the largest amplitude particles move through their zero crossing, and oscillates outward whenever these particles are near their extrema. Both of these conditions happen twice per betatron period, thus driving the frequency of envelope oscillations to twice the betatron frequency.

It can be deduced from Fig. 5.9 that the oscillations of the beam envelope about equilibrium always have periodicity one-half that of the betatron oscillations of the particles in the channel. It is also clear from the form of Eq. (5.58) that they are not simply sinusoidal for all amplitudes $\delta\sigma_m$, as the linear analysis leading to Eq. (5.80) is no longer valid for large amplitude oscillations. For such large amplitude envelope motion, therefore, one expects that the oscillations must have a different form, but the same periodicity as predicted by linear arguments. These large amplitude oscillations are best explored by use of the Twiss parameter formalism.

We begin this analysis by noting that a linear differential equation may be obtained by differentiating Eq. (5.77) once again, to yield

Fig. 5.9 Four representative betatron trajectories in a beam traversing a uniformly focusing channel (in solid line), two cosines of opposing sign, and two sines of opposing sign, with half the amplitude of the cosines. A covering, sinusoidally varying envelope of these trajectories is shown by the bold dashed line, which is periodic with twice the betatron frequency κ_0.

$$\beta_x''' = -2\alpha_x'' = 2\gamma_x' - 2\beta_x'\kappa_x^2 - 2\beta_x(\kappa_x^2)' \quad \text{or} \quad \beta_x''' + 4\kappa_x^2\beta_x' + 2\beta_x(\kappa_x^2)' = 0. \tag{5.84}$$

In order to examine to the behavior of the β-function in a uniform focusing channel, we must proceed carefully. We can connect Eq. (5.84) to the equivalent second-order differential equation within the focusing channel by integrating over the singularity in the function $(\kappa_x^2)' = \kappa_0^2 \delta(z)$ at the start of the channel, assumed to be located at $z = 0$. This analysis gives a discontinuity in the second derivative of the β-function, $\Delta \beta_x'' = -2\kappa_0^2 \beta_x$. We can, thus, construct a second-order biased oscillator equation with the correct integration constant on its right-hand side,

$$\beta_x'' + 4\kappa_0^2 \beta_x = 2\kappa_0^2 \beta_{x0} + 2\gamma_{x0}, \tag{5.85}$$

where $\gamma_{x0} = \gamma_x(z_0)$ and $\beta_{x0} = \beta_x(z_0)$. This expression gives oscillations in the β-function at twice the betatron frequency, $2\kappa_0$, about a particular solution,

$$\beta_p = \frac{\beta_{x0}}{2} + \frac{\gamma_{x0}}{2\kappa_0^2}. \tag{5.86}$$

This is a curious result (arising from the fact that Eq. (5.84) is of third order), as it indicates that the particular solution to the differential equation is not a constant equilibrium value, but one that is dependent on initial conditions. There is still a well-defined equilibrium (matched solution) associated with this particular solution, however, as setting $\beta_p = \beta_{x0} = \beta_{eq}$, $\alpha_{x0} = \alpha_x(0) = 0$, we have, in agreement with previous results,

$$\beta_{eq} = \frac{\gamma_{x0}}{\kappa_0^2} = \frac{1}{\beta_{eq}\kappa_0^2} \quad \text{or} \quad \beta_{eq} = \frac{1}{\kappa_0}. \tag{5.87}$$

5.8   Collective description of longitudinal beam distributions

The methods developed in this chapter for describing transverse beam distributions and their second moment evolution can be adapted well to longitudinal distributions. We begin a survey of these adapted methods by examining the equilibrium solutions to the Vlasov equation in the longitudinal phase plane. Following the methods of Section 5.1, we examine the form of the longitudinal distribution function, $f_z(\zeta, \delta p)$, with $\delta p = p_z - p_0$ and $\zeta = z - v_\varphi t$, as before. For this case, since we have invested considerable effort in development of the Hamiltonian governing longitudinal motion, we use the results of Exercise 5.2 to immediately write the solution to the Vlasov equation as

$$f_z(\zeta, \delta p) = C \exp \left[-\frac{H(\zeta, \delta p)}{k_B T_z} \right]. \tag{5.88}$$

When the longitudinal Hamiltonian in an rf linear accelerator or synchrotron is written in the case of gentle accelerating forces (which is the relevant case, as violent acceleration cases do not easily produce equilibria), it is, in fact, additively separable. In this case, the exponential form given in Eq. (5.88) guarantees that the distribution function is multiplicatively separable, $f_z(\zeta, \delta p) = Z(\zeta)P(\delta p)$. In this small amplitude Hamiltonian (e.g. Eq. (4.37)), the momentum enters in as a quadratic term, and the momentum distribution, as before, is simply a Gaussian,

$$P(\delta p_z) = \frac{1}{\sqrt{2\pi}\sigma_{\delta p}} \exp\left(-\frac{\delta p^2}{2\sigma_{\delta p}^2}\right). \tag{5.89}$$

On the other hand, the confining potential in the longitudinal dimension is not linear, but sinusoidal (or a linearly biased sinusoid in the case of a moving bucket), and the distribution is not, in contrast to the transverse case, generally Gaussian in ζ. We begin our discussion of this situation with the stationary bucket in an rf linear accelerator, in which case we can write

$$f_z(\zeta, \delta p) = C \exp\left[-\frac{m_0 c^2}{k_B T_z}\left(\frac{\beta_0^2}{2\gamma_0 p_0^2}(\delta p^2) + \alpha_{RF}[\cos(k_z\zeta) + 1]\right)\right]. \tag{5.90}$$

There is a single, unitless parameter that controls the longitudinal (or **beam current**) distribution in this expression,

$$\Gamma_1 \equiv \frac{H_{sep}}{k_B T_z} = \frac{2\alpha_{RF} m_0 c^2}{k_B T_z}. \tag{5.91}$$

When Γ_1 is much less than one, most of the particles in the distribution have momentum deviations from the design larger than the maximum found in the bucket, and the distribution is almost entirely outside of the separatrix. In this limit, the distribution is approximately

$$f_z(\zeta, \delta p) = C \exp\left[-\frac{\Gamma_1 \beta_0^2}{4\alpha_{RF}\gamma_0 p_0^2}(\delta p^2)\right] \exp\left[-\frac{\Gamma_1}{2}([\cos(k_z\zeta) + 1])\right]$$

$$\cong \frac{N'}{\sqrt{2\pi}\sigma_{\delta p}} \exp\left[-\frac{\delta p^2}{2\sigma_{\delta p}^2}\right]\left[1 - \frac{\Gamma_1}{2}\cos(k_z\zeta)\right], \quad \Gamma_1 \ll 1, \tag{5.92}$$

where $\sigma_{\delta p}^2/p_0^2 = 2\alpha_{RF}\gamma_0/\Gamma_1\beta_0^2$. For this scenario, which is typical of the beginning of the adiabatic capture process discussed in Section 4.5, the beam's longitudinal spatial profile is simply a constant $N' = I/qv_0$ (I is average value of the beam current over the rf period) with a small amplitude ($\Gamma_1/2$) density modulation arising from the weak longitudinal forces.

On the other hand, when the beam has been captured tightly inside the bucket, so that $\Gamma_1 \gg 1$, the beam distribution is quite localized in the region of the stable fixed point at $k_z\zeta = \pi$. In this case, expanding the argument of the exponential in Eq. (5.90) about $\varphi_0 = \pi$ in the distance $\delta\zeta = \zeta - \varphi_0/k_z$, we have

$$f_z(\zeta, \delta p) = C \exp\left[-\frac{\Gamma_1 \beta_0^2}{4\alpha_{RF}\gamma_0 p_0^2}(\delta p^2)\right] \exp\left[-\frac{\Gamma_1}{2}([\cos(k_z\zeta) + 1])\right]$$

$$\cong \frac{N'}{\sqrt{2\pi}\sigma_{\delta p}} \exp\left[-\frac{\delta p^2}{2\sigma_{\delta p}^2}\right] \exp\left[-\frac{\delta\zeta^2}{2\sigma_{\delta\zeta}^2}\right], \quad \Gamma_1 \gg 1. \tag{5.93}$$

Here again $\sigma_{\delta p}^2/p_0^2 = \alpha_{RF}\gamma_0/\Gamma_1\beta_0^2$, and the longitudinal beam distribution is approximately Gaussian, with rms phase extent

$$\sigma_\varphi \equiv k_z\sigma_{\delta\zeta} = \sqrt{\frac{2}{\Gamma_1}}. \tag{5.94}$$

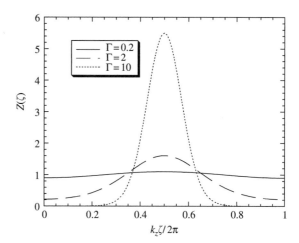

Fig. 5.10 Equilibrium longitudinal distributions for cases with $\Gamma_1 = 0.2$, 2, and 10.

From Eq. (5.94), we can see that the beam's rms length is self-consistently predicted to be well within a period of the wave. The density distributions corresponding to several different values of cases with Γ_1 are illustrated in Fig. 5.10, including an intermediate state, $\Gamma_1 = 2$, in which the density modulation is deep, but the beam is not quite completely located inside of the separatrix.

The equilibrium cases we have examined allow us to introduce the notion of the rms longitudinal emittance, which is defined using second moments, just as in the case of the rms transverse emittance first specified by Eq. (5.43),

$$\varepsilon_{\zeta,\mathrm{rms}} = \sqrt{\langle \delta\zeta^2 \rangle \left\langle \left(\frac{\delta p}{p_0}\right)^2 \right\rangle - \left\langle \delta\zeta \frac{\delta p}{p_0} \right\rangle^2}. \tag{5.95}$$

This definition of the longitudinal emittance is appropriate, as we shall see, for the case of violent acceleration, as it will tend to be conserved in such a system. In the case of gentle acceleration, however, this definition of the longitudinal emittance is "damped" by the acceleration process, in which p_0 grows. Thus, a more useful measure of the occupied longitudinal phase plane (as opposed to a trace space) area in a gentle acceleration system is the **normalized longitudinal emittance**,

$$\varepsilon_{\zeta,\mathrm{n}} \equiv \beta_0 \gamma_0 \varepsilon_{\zeta,\mathrm{rms}} = (m_0 c)^{-1} \sqrt{\langle \delta\zeta^2 \rangle \langle \delta p^2 \rangle - \langle \delta\zeta \delta p \rangle^2}. \tag{5.96}$$

The roles of these two different emittances in describing envelope dynamics in linear and circular accelerators will become more apparent when we discuss specific examples.

One of the most useful aspects of the rms transverse emittance is that it remains invariant under linear transformations, those induced by forces that are linear in distance away from the design orbit. It is clear that this is never the case for longitudinal forces, in which the applied force is in the form of a harmonic wave. Nevertheless, once particles are trapped inside a bucket into a region near the stable fixed point, the forces are approximately linear, and the longitudinal emittance plays the role of an approximate invariant just as in the transverse case. The "trick" of adiabatic capture is to take an initially unbunched particle distribution and bunch it without dilution of the phase plane density, or alternatively, growth of the longitudinal emittance. It is instructive to look at this problem from these alternative points of view.

The distribution function of an initially unbunched thermal beam of N_λ particles located within an rf wavelength is $f_z(\zeta, \delta p) = (k_z N_\lambda / (2\pi)^{3/2}\sigma_{\delta p,0}) \exp[-\delta p^2/2\sigma_{\delta p,0}^2]$, where $\sigma_{\delta p,0} = \sqrt{\langle \delta p^2 \rangle}$. After bunching, assuming that the adiabatic capture process produces a thermal momentum distribution in its final state, the phase plane distribution function has the form $f_z(\zeta, \delta p) = (N_\lambda/2\pi\sigma_{\delta p,b}\sigma_{\zeta,b}) \exp[-\delta\zeta^2/2\sigma_{\zeta,b}^2] \exp[-\delta p^2/2\sigma_{\delta p,b}^2]$, where the subscript b indicates a bunched distribution. Further, it can be seen from the Vlasov equation that in the vicinity of the fixed point at $(\delta\zeta, \delta p) = (0,0)$, the distribution function (peak) must be constant, and therefore we have that the final momentum spread is

$$\sigma_{\delta p,b} = \frac{\sqrt{2\pi}\sigma_{\delta p,0}}{k_z\sigma_{\zeta,b}}. \tag{5.97}$$

As a perfectly thermal state is not guaranteed (this obviously depends on the degree of adiabaticiy of the bunching), it is helpful to also look at this process using the normalized longitudinal emittances. In the initial unbunched state, the normalized rms emittance is

$$\varepsilon_{z,n} = \frac{\pi\sigma_{\delta p,0}}{\sqrt{3}k_z}. \tag{5.98}$$

Here $\sigma_{\delta p,0} = \sqrt{\langle \delta p^2 \rangle}$ is the initial thermal rms momentum spread. After bunching, the longitudinal emittance is simply

$$\varepsilon_{z,n} = \sigma_\zeta\sigma_{\delta p,b}, \tag{5.99}$$

and, assuming no rms emittance growth in the bunching process, the rms momentum spread after bunching is

$$\sigma_{\delta p,b} = \frac{\pi\sigma_{\delta p,0}}{\sqrt{3}k_z\sigma_\zeta}. \tag{5.100}$$

The estimate one obtains from Eq. (5.100) is quite similar to that given by Eq. (5.97), differing by a constant factor of 1.38. This difference can be attributed to the false assumption that there is no rms emittance growth in the bunching process—as it employs nonlinear forces, the rms emittance cannot be conserved.

The notion of longitudinal emittance can also be used to develop an rms longitudinal envelope equation, which allows us to move beyond equilibrium situations, and analyze the evolution of a longitudinal distribution. This derivation of this equation does not differ conceptually from that leading to Eq. (5.58), but it is complicated slightly in the case of bending systems by the non-trivial relationship between the momentum deviation and the apparent longitudinal velocity, relative to the design particle, including path length effects. This relationship is given formally in Eq. (3.77), with the explicit calculation of the path length parameter α_c detailed in Eq. (3.79). We first examine the case of rectilinear motion in a linac, where a longitudinally focusing force is constantly applied to the bunched distribution, and we can write the following second moment relationships:

$$\frac{\mathrm{d}\sigma_\zeta}{\mathrm{d}z} \equiv \frac{\mathrm{d}}{\mathrm{d}z}\sqrt{\langle\delta\zeta^2\rangle} = \frac{1}{\sigma_\zeta}\frac{\mathrm{d}}{\mathrm{d}z}\langle\delta\zeta^2\rangle = \frac{1}{\gamma_0^2}\frac{\sigma_{\zeta\delta p}}{\sigma_\zeta} \tag{5.101}$$

where $\sigma_\zeta \equiv \sqrt{\langle\delta\zeta^2\rangle}$ and $\sigma_{\zeta\delta p} \equiv \sqrt{\langle\delta\zeta\cdot\delta p\rangle}/p_0$, The second derivative of the rms bunch length is then derived to be

$$\frac{\mathrm{d}^2\sigma_\zeta}{\mathrm{d}z^2} = \frac{1}{\gamma_0^2}\frac{\mathrm{d}}{\mathrm{d}z}\frac{\sigma_{\zeta\delta p}}{\sigma_\zeta} = \frac{1}{\gamma_0^2}\left[\frac{1}{\sigma_\zeta}\frac{\mathrm{d}\sigma_{\zeta\delta p}}{\mathrm{d}z} - \frac{1}{\gamma_0^2}\frac{\sigma_{\zeta\delta p}^2}{\sigma_\zeta^3}\right],$$

$$= \frac{1}{\gamma_0^2}\left[\frac{1}{\gamma_0^2}\frac{\sigma_{\delta p}^2}{\sigma_\zeta} + \frac{\langle\delta\zeta\cdot\delta p'\rangle}{p_0\sigma_\zeta} - \frac{1}{\gamma_0^2}\frac{\sigma_{z\delta p}^2}{\sigma_\zeta^3}\right], \tag{5.102}$$

or

$$\sigma_\zeta'' - \frac{1}{\gamma_0^2}\frac{\langle\delta\zeta\cdot\delta p'\rangle}{p_0\sigma_\zeta} = \frac{1}{\gamma_0^4}\frac{\varepsilon_{z,\mathrm{rms}}^2}{\sigma_\zeta^3}. \tag{5.103}$$

The second term on the left-hand side of Eq. (5.102) can be approximated as having a linear ζ dependence for a tightly bunched beam ($k_z\sigma_\zeta \ll 1$), as

$$\langle\delta\zeta\cdot\delta p'\rangle = \frac{\langle\delta\zeta\cdot\dot p\rangle}{p_0 v_0} \simeq \frac{\langle\delta\zeta^2\rangle}{v_0}\alpha_{\mathrm{RF}}k_z^2\cos(\varphi_0), \tag{5.104}$$

where for a stable bucket, $\cos(\varphi_0) < 0$. With Eqs (5.102)–(5.104), we may rewrite the longitudinal envelope equation succinctly as

$$\sigma_\zeta'' + k_\mathrm{s}^2\sigma_\zeta = \frac{\varepsilon_{z,\mathrm{rms}}^2}{\gamma_0^4\sigma_\zeta^3}. \tag{5.105}$$

Here we have reintroduced the (spatial) synchrotron frequency, $k_\mathrm{s} = \omega_\mathrm{s}/v_0$, the frequency of small amplitude oscillations about the stable fixed point as given by Eq. (4.40).

For circular accelerators, there are several changes needed to make the envelope analysis more appropriate. The first is that the path length term in the time dispersion is no longer ignorable, and one must use $-\eta_\tau$ instead of simply γ_0^{-2} in the relationship given by Eq. (5.102). Another is that the longitudinal forces are not applied continuously, but in a local region around the ring. Because of this difference, it makes sense to write the envelope equation using time (only approximately a continuous variable, as it advances in quanta of the revolution period) as the independent variable. The envelope equation then becomes

$$\ddot\sigma_\zeta + \omega_\mathrm{s}^2\sigma_\zeta = \eta_\tau^2 v_0^2\frac{\varepsilon_{z,\mathrm{rms}}^2}{\sigma_\zeta^3}, \tag{5.106}$$

where v_0 is the synchronous velocity, and we now use the definition of the synchrotron frequency, ω_s, for circular machines given in Eq. (4.49).

Up to this point in the present section, we have been discussing the approximately thermal distributions produced in gently accelerating systems such as ion linacs or circular accelerators. In the linear motion near a stable fixed point in such a system there is little dilution of the coarse-grained phase plane density, and thus the normalized longitudinal emittance (phase plane area) is nearly constant—especially after the capture process has been completed. In a violently accelerating system, however, the particles do not rotate about the fixed point, being essentially frozen in longitudinal position after they become ultra-relativistic. In this scenario, the emittance is not thermal in nature, and is due almost entirely to the nonlinearity of the applied rf wave. As a consequence of this, the phase plane area in the rms sense is not conserved, but the longitudinal trace space area is roughly constant. These attributes of the distribution under violent acceleration are illustrated in Fig. 5.11, in which the synchronous (speed-of-light) rf wave is plotted, along with the distribution in the longitudinal phase plane, with momentum normalized to the maximum achievable by running at the maximum acceleration phase in the wave $\varphi_0 = \pi/2$. The distribution is localized in longitudinal coordinate, and because the phase slippage is ignorable in the system, the momentum distribution is given to good approximation by

$$p_z(\zeta) \cong p_{\max} \sin(k_z \zeta) \cong p_0 \left[1 - \cot(\phi_0) k_z \zeta - \tfrac{1}{2}(k_z \zeta)^2 \right], \qquad (5.107)$$

where, for a radio-frequency linac of length L_{acc}, the maximum achievable momentum is approximately $p_{\max} c \cong q E_0 L_{\mathrm{acc}}$. If we assume further that the longitudinal (current) profile is Gaussian, $n_\zeta(\zeta) = \int f_\zeta(\zeta, \delta p)\, \mathrm{d}p = \exp(-\delta\zeta^2/2\sigma_\zeta^2)/\sqrt{2\pi}\sigma_\zeta$, then the following second moments can be calculated, expanding the assumed momentum distribution to second order in ζ,

$$\langle \delta\zeta^2 \rangle = \sigma_\zeta^2, \qquad (5.108)$$

$$\frac{\langle \delta p^2 \rangle}{p_0^2} = \frac{(k_z \sigma_\zeta)^4}{\sqrt{2}} + \cot^2(\phi_0)(k_z \sigma_\zeta)^2, \qquad (5.109)$$

and

$$\frac{\langle \delta\zeta \cdot \delta p \rangle}{p_0} = -\sigma_\zeta \cot(\phi_0)(k_z \sigma_\zeta). \qquad (5.110)$$

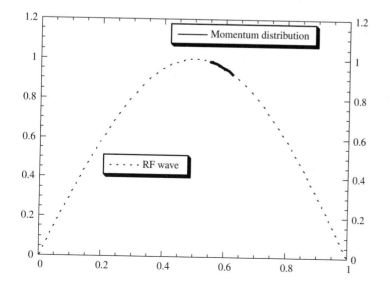

Fig. 5.11 Plot of the synchronous (violently accelerating) wave in a linac, with the accompanying distribution in the longitudinal phase plane, with momentum normalized to the maximum achievable.

From Eqs (5.108)–(5.110), the normalized rms longitudinal emittance for this system can be calculated approximately as

$$\varepsilon_{\zeta,n} \equiv \beta_0\gamma_0 \cdot \sqrt{\langle\delta\zeta^2\rangle\frac{\langle\delta p^2\rangle}{p_0^2} - \frac{\langle\delta\zeta\cdot\delta p\rangle^2}{p_0}} = \beta_0\gamma_0\sigma_\zeta\frac{(k_z\sigma_\zeta)^2}{\sqrt{2}}. \tag{5.111}$$

This emittance clearly grows linearly with mean momentum, $p_0 = \beta_0\gamma_0 m_0 c$, and is thus not a useful invariant. On the other hand, the trace space emittance (Eq. (5.95)) is, because the longitudinal momentum distribution relative to the mean momentum remains invariant. Thus, we have

$$\varepsilon_{\zeta,\mathrm{rms}} = \frac{\sigma_\zeta}{\sqrt{2}}(k_z\sigma_\zeta)^2 \tag{5.112}$$

as a useful measure of the longitudinal **trace space** area in violently accelerating systems.

This emittance can be used to write down Twiss parameters for the second moments of the longitudinal distribution as

$$\beta_\zeta = \frac{\langle\delta\zeta^2\rangle}{\varepsilon_{\zeta,\mathrm{rms}}} = \frac{\sqrt{2}\sigma_\zeta}{(k_z\sigma_\zeta)^2}, \tag{5.113}$$

$$\alpha_\zeta = \frac{\langle\delta\zeta\cdot\delta p\rangle}{p_0\varepsilon_{\zeta,\mathrm{rms}}} = -\sqrt{2}\frac{\cot(\phi_0)}{k_z\sigma_\zeta} \tag{5.114}$$

and

$$\gamma_\zeta = \frac{\langle\delta p^2\rangle}{p_0^2\varepsilon_{\zeta,\mathrm{rms}}} = \frac{1+\alpha_\zeta^2}{\beta_\zeta} = \frac{(k_z\sigma_\zeta)^2 + 2\cot^2(\phi_0)}{\sqrt{2}\sigma_\zeta}. \tag{5.115}$$

In this type of system (e.g. electron linac), one can correlate the distribution by running at an off-crest ($\cot\varphi \neq 0$) acceleration phase, and remove this correlation by use of a magnetic chicane, to **compress** the bunch. The Twiss parameter treatment of the distribution is particularly fitting for analyzing this process, as it has analogies to the thin-lens transverse focusing discussion introduced in Section 5.6. The analogy is as follows: the trace-space correlation is introduced by the longitudinal "lens" of the linac run off-crest, with correlation parameter α_ζ as given in Eq. (5.114). The longitudinal "drift" needed to bring the distribution to an uncorrelated "focus" is introduced by the magnetic chicane, which has a large momentum compaction, as examined in Exercise 3.19. The idea behind the chicane is relatively simple—larger-momentum particles pass through the system with a shorter path length than smaller momentum particles, and thus a beam with a negative α_ζ can be compressed, by bringing all particles toward the same longitudinal position, as shown in Fig. 5.12. The changes made to the longitudinal trace space distribution by the traversal of the bunch through the chicane are displayed in Fig. 5.13, with the final compressed beam having a "comma" shape in trace space.

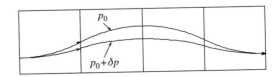

Fig. 5.12 Schematic of the design trajectory and a higher momentum trajectory in a magnetic chicane (cf. Fig. 3.14).

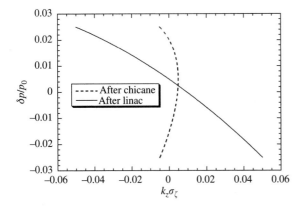

Fig. 5.13 Typical longitudinal trace space distribution after violent acceleration in the linac (before chicane) and after traversal of the chicane.

The matrix element $R_{56} = -\eta_\tau \cdot \Delta s$ for the magnetic system displayed in Fig. (5.12) is the parameter analogous to the drift length in the transverse thin lens. This quantity (the calculation of which is left for Ex. 5.18) is a function of the bend angle, the radius of curvature, and, weakly in the ultra-relativistic limit, on the design momentum. With longitudinal dynamics we have, by analogy to Eq. (5.68), $\Delta\alpha_\zeta = -\gamma_\zeta R_{56} = \gamma_\zeta \cdot \eta_\tau \cdot \Delta s$ where γ_ζ is constant in the acceleration-free region. Thus, the required R_{56} in the chicane to remove the applied correlation ($\alpha_\zeta = 0$) is

$$R_{56} = \frac{\alpha_\zeta}{\gamma_\zeta} = \frac{2\sigma_\zeta \cot(\phi_0)}{(k_z\sigma_\zeta)^3 + (k_z\sigma_\zeta)2\cot^2(\phi_0)}. \tag{5.116}$$

Once the correlation has been removed by traversal of the chicane, the compressed, uncorrelated rms bunch length, can be deduced from the invariance of γ_ζ, in terms of the initial bunch length σ_ζ,

$$\frac{\sigma_\zeta^*}{\sigma_\zeta} = \sqrt{\frac{\beta^*}{\beta_\zeta}} = \sqrt{\frac{1}{\gamma_\zeta\beta_\zeta}} = \sqrt{\frac{1}{1+\alpha_\zeta^2}} = \sqrt{\frac{(k_z\sigma_\zeta)^2}{(k_z\sigma_\zeta)^2 + 2\cot^2(\phi_0)}}. \tag{5.117}$$

In the limit of large correlation, this compression ratio becomes

$$\frac{\sigma_\zeta^*}{\sigma_\zeta} \underset{\alpha_\zeta \gg 1}{\Rightarrow} \frac{1}{\alpha_\zeta} = \frac{k_z\sigma_\zeta}{\sqrt{2}\cot(\phi_0)}. \tag{5.118}$$

Equation (5.118) illustrates two points well: first, that a beam that is already short compared to the radio-frequency wavelength ($k_z\sigma_\zeta \ll 1$) is easy to compress further, due to the strong dependence of the longitudinal emittance on the bunch length (the energy spread is quadratic in $k_z\sigma_\zeta$). The second point is the somewhat obvious comment that in order to compress the beam, it is necessary to run off-crest, $\cot(\varphi_0) > 0$.

5.9 The Twiss parameters in periodic transverse focusing systems

Periodic transverse focusing systems are of primary importance in beam physics, due to their application in circular accelerators. The analysis of single particle motion in this type of system was discussed extensively in Chapter 3, so we naturally take up the discussion of the collective description of particle beams here. We will begin by examining periodic solutions to the transverse beam envelope as described by the Twiss parameters. In the context of a periodic focusing system, this type of solution is termed matched, in the sense that the beam envelope has the same periodicity and symmetry as the focusing system. This is but one solution corresponding to a specific boundary condition on the differential equation governing the system, that is, Eq. (5.77). For this solution to be physically realized, the beam distribution must be either injected with the correct initial conditions upon entry into the ring, or must relax towards this distribution through dissipative or nonlinear processes. These considerations will be elucidated further at the end of this section.

We start the analysis with a simple examination of a beam in a focusing array which consists of drifts of length L_p separating thin lenses of focal length f. In the drift sections, the solution must have the form of Eq. (5.71). Further, a periodic solution will be required to have $\alpha_x = 0$ at both the lens position, where the beam size is at a maximum, and at the mid-point between the two lenses, where we define $\beta_x = \beta_{min}$. We thus have, at the lens positions,

$$\beta_{x,lens} = \beta_{max} = \beta_{min}\left[1 + \left(\frac{L_p}{2\beta_{min}}\right)^2\right] \quad \text{and} \quad \alpha_{x,lens} = -\frac{\beta'_{x,lens}}{2} = -\frac{L_p}{2\beta_{min}} \tag{5.119}$$

After application of the lens, $\beta'_{x,lens}$, and thus $\alpha_{x,lens}$, must reverse sign to allow symmetry of the β-function about the lens position, as illustrated in Fig. 5.14. This reversal gives a relation for the minimum β-function between the lenses,

$$\Delta\alpha_x = -2\alpha_{x,lens} = \frac{L_p}{\beta_{min}} = \frac{\beta_{max}}{f} = \frac{\beta_{min}}{f}\left[1 + \left(\frac{L_p}{2\beta_{min}}\right)^2\right], \tag{5.120}$$

which can be solved to yield both extrema,

$$\beta_{min} = \sqrt{L_p\left(f - \frac{L_p}{4}\right)} \quad \text{and} \quad \beta_{max} = \frac{fL_p}{\beta_{min}} = f\sqrt{\frac{L_p}{f - (L_p/4)}}. \tag{5.121}$$

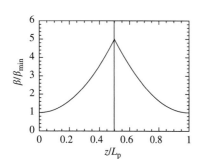

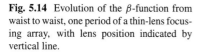

Fig. 5.14 Evolution of the β-function from waist to waist, one period of a thin-lens focusing array, with lens position indicated by vertical line.

Note that the minimum β-function (at the waist) approaches zero, and the maximum β-function becomes unbounded as $f \Rightarrow L_p/4$ from above. This is entirely expected from the matrix analysis introduced in Chapter 3, because $f = L_p/4$ is the limit of single particle stability, where the eigenvector rays cross the axis at the mid-point between the lenses.

With less ideal, more complicated systems than the simple focusing lens array, the direct construction of symmetric, periodic differential equation solutions for the β-function evolution is not practical. In this situation, as with single particle optics, we turn to matrix methods to solve for the Twiss parameters in a given periodic focusing array.

In order to proceed with this analysis, we need to go back and re-examine the single particle motion in a periodic system, with methods that include Twiss parameters in a systematic way. This is done by noting that the second-order differential equation with periodic focusing, Eq. (3.10), is of a general form

known as Hill's equation. It has been shown by Floquet that the following
substitution allows a useful (and, by now, familiar) recasting of the differential
equation:

$$x(z) = \sqrt{\varepsilon} \cdot \sqrt{\beta(z)} \cos(\psi(z)). \qquad (5.122)$$

We shall see below that the parameter suggestively termed ε (the subscript x
has been left out here to avoid identifying this constant with the rms emittance) corresponds to a single particle invariant—the action, or **single particle
emittance**—and the amplitude function $\beta(z)$ corresponds under **matched conditions** to the familiar Twiss parameter β-function. Before undertaking this
discussion, we first examine the results of differentiating Eq. (5.122), which
gives

$$x'(z) = \sqrt{\varepsilon} \cdot \left[\frac{\beta'}{2\beta^{1/2}} \cos(\psi) - \beta^{1/2}\psi' \sin(\psi) \right] \qquad (5.123)$$

and

$$x''(z) = \sqrt{\varepsilon} \cdot \left[\left(\frac{\beta''}{2\beta^{1/2}} - \frac{(\beta')^2}{4\beta^{3/2}} - \beta^{1/2}(\psi')^2 \right) \cos(\psi) \right.$$
$$\left. - \left(\beta^{1/2}\psi'' + \frac{\beta'\psi'}{\beta^{1/2}} \right) \sin(\psi) \right]. \qquad (5.124)$$

Further, upon substitution of Eq. (5.124) into the single particle equation of
motion, Eq. (3.10), we have

$$\left(\frac{\beta''}{2\beta^{1/2}} - \frac{(\beta')^2}{4\beta^{3/2}} - \beta^{1/2}(\psi')^2 + \kappa_x^2 \beta^{1/2} \right) \cos(\psi)$$
$$- \left(\beta^{1/2}\psi'' + \frac{\beta'\psi'}{\beta^{1/2}} \right) \sin(\psi) = 0. \qquad (5.125)$$

It is clear that the coefficients of the sine and cosine terms in Eq. (5.125) must
both vanish, and thus two separate equations emerge. We first examine the sine
coefficient, rewriting it as

$$\frac{\psi''}{\psi'} = -\frac{\beta'}{\beta}. \qquad (5.126)$$

The solution to Eq. (5.126), which has logarithmic derivatives on either side of
the expression, is

$$\ln(\psi') = -\ln(\beta) + \text{const.} \qquad (5.127)$$

or

$$\psi' = \frac{C}{\beta}. \qquad (5.128)$$

The integration constant C in Eq. (5.128) can be chosen arbitrarily, as the
consequence of changing it is simply to rescale ε. We chose $C = 1$, to make the
definition of $\beta(z)$ as consistent as possible with the Twiss parameter definition.
We thus have the solution for the **Floquet phase**

$$\psi(z) = \int_{z_0}^{z_f} \frac{dz}{\beta(z)} + \psi_0, \qquad (5.129)$$

which generalizes the notion of phase advance per period μ, first introduced in
Section 3.2, to a continuously increasing quantity that, after one period, must
arrive at $\Delta\psi = \mu$.

With the aid of Eq. (5.128), the second equation arising from this analysis becomes

$$\beta'' + \kappa_x^2 \beta = \frac{(\beta')^2}{2\beta} + \frac{2}{\beta}, \qquad (5.130)$$

which is identical in form to Eq. (5.77). At this point there may be some confusion between the two definitions of β, the one defined by rms quantities and the one defined by the assumed periodic solution to Hill's equation. This confusion is resolved by noting that they are the same function only for **periodic** systems, that is, the rms definition of β is equivalent to the definition given in Eq. (5.122) only for periodic solutions to the rms equation. The definition of β as a periodic envelope function in Eq. (5.122) allows us to assign the β-function to be a property of the periodic focusing system itself, however, free from considerations of the beam distribution.

The trajectory in trace space given formally by Eqs (5.122) and (5.123), which is specified by the correct, periodic β-function belonging to the periodic focusing system, can be plotted once per period, to give a trace space ellipse. Such elliptical trajectories have a given orientation and major-to-minor radius ratio, and an overall size parameterized by $\varepsilon = A/\pi$, where A is the area of a given ellipse, as illustrated in Fig. 5.15. An actual beam distribution with the same orientation and aspect ratio in trace space is termed **matched**, and remains the same after each period. Any other type of distribution is mismatched to the focusing lattice, and after each period, the distribution changes. An example of a **mismatched distribution** is given as well in Fig. 5.15. This distribution does not remain invariant after each period, as each particle rotates along its trace space curve by a phase angle $\Delta \psi = \mu$ each period. An example of this is shown in Fig. 5.16, in which the distribution of Fig. 5.15 is advanced in phase by $\mu = 90°$.

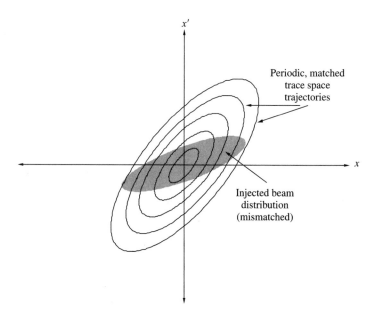

Fig. 5.15 Poincaré plot of typical periodic trace space trajectories of different trace space action ε are given by nested ellipses. Mismatched beam distribution shown by shaded region.

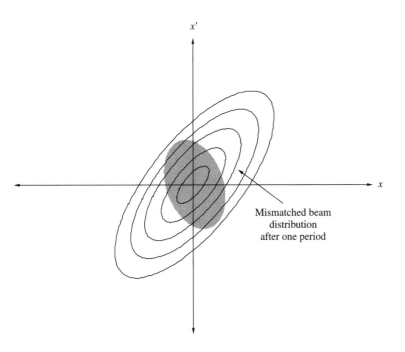

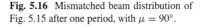

Fig. 5.16 Mismatched beam distribution of Fig. 5.15 after one period, with $\mu = 90°$.

This mismatch may seem harmless at first glance, as one may envision the beam oscillating for all time about the matched β-function. This is not quite the case, however, as in any real circular accelerator, for instance, there are small nonlinearities (e.g. sextupole fields) that produce non-negligible effects on long time scales. The phenomenon produced that is most relevant to the present discussion is **amplitude dependent tune**. As one travels away from the trace space origin, the tune at each ellipse (corresponding to an increasing value of ε) changes slightly, that is, $\nu_x(\varepsilon) \neq$ constant. If particles at different amplitudes rotate about the origin in trace at different rates, eventually the particles **decohere**, and the entire trace space inside of the largest amplitude particle orbit is eventually filled by particles, albeit at a lower trace space density. This phenomenon is known as decoherence, or **phase mixing**, and eventually it results in damping of the mismatch oscillation. Decoherence also, significantly, produces in the final state a beam with larger rms emittance that is matched to the focusing system. This situation is illustrated in Fig. 5.17, which shows the long-term final state of the beam distribution shown in Figs 5.15 and 5.16 after it has decohered.

If, instead of a generating a Poincaré plot from Eq. (5.122), we look at the form of the solutions given by Eq. (5.122) as a continuous function of z, we can identify it as giving an envelope function,

$$x_{\mathrm{env}}(z) = \sqrt{\varepsilon} \cdot \sqrt{\beta(z)}. \qquad (5.131)$$

In Eq. (5.131), the envelope is defined in terms of the scaling parameter $\sqrt{\varepsilon}$, and the amplitude function $\sqrt{\beta(z)}$. A given trace space ellipse, like those shown in Figs 5.15–5.17 is parameterized by a value of the trace space action ε, which is well defined for a periodic system. At a particular z, this ellipse is populated away from the envelope edge by particles of phase $\psi \neq 0, \pi$.

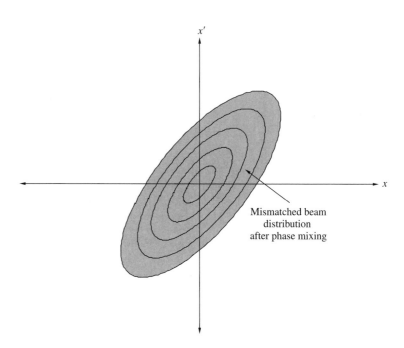

Mismatched beam
distribution
after phase mixing

Fig. 5.17 Mismatched beam distribution of Fig. 5.15 after many periods, with amplitude dependent tune-driven phase mixing causing all regions of trace space inside of largest amplitude particle to be populated.

The Floquet formalism can be used to deduce aspects of the Twiss parameter behavior in periodic systems. To see this, we start by rewriting Eqs (5.122) and (5.123) in slightly different, but equivalent form,

$$x(z) = x_0 \sqrt{\frac{\beta(z)}{\beta_0}} \cos(\psi(z)) + (\beta_0 x_0' + \alpha_0 x_0) \sqrt{\frac{\beta(z)}{\beta_0}} \sin(\psi(z)) \qquad (5.132)$$

and

$$x'(z) = x_0 \frac{-(1 + \alpha_0^2)}{\sqrt{\beta(z) \cdot \beta_0}} \sin(\psi) + x_0' \sqrt{\frac{\beta(z)}{\beta_0}} [\cos(\psi) - \alpha_0 \sin(\psi)], \qquad (5.133)$$

where $\alpha_0 = \alpha(0)$, $\beta_0 = \beta(0)$ and $\psi(0) = 0$. With Eqs (5.132) and (5.133), the matrix for transport of the trace space vector (x, x') through **one period** of the system is,

$$\mathbf{M} = \begin{bmatrix} \cos(\mu) + \alpha \sin(\mu) & \beta \sin(\mu) \\ -\gamma \sin(\mu) & \cos(\mu) - \alpha \sin(\mu) \end{bmatrix}, \qquad (5.134)$$

where we have used the fact that in the periodic solutions (we are examining the system after one period $z = L_p$) $\alpha = \alpha(L_p) = \alpha_0$ and $\beta = \beta(L_p) = \beta_0$.

By simply writing the transport matrix $\mathbf{M}$ as outlined in Chapter 3 and calculating μ, therefore, we can discover the solutions for the Twiss parameters by comparing the matrix elements we obtain to the form given in Eq. (5.134), and solving for the unknowns, α and β. For the cases of envelope maxima or minima, the positions of which can often be deduced from symmetry considerations, hese matrix elements simplify even further,

$$\mathbf{M} = \begin{bmatrix} \cos(\mu) & \beta_{\text{max}/\text{min}} \sin(\mu) \\ -\frac{\sin(\mu)}{\beta_{\text{max}/\text{min}}} & \cos(\mu) \end{bmatrix}, \tag{5.135}$$

and the extrema of the β-function in a periodic system can be trivially found.

The utility of this method is illustrated by the example of the periodic thin-lens focusing array discussed previously in this section. At a distance Δz past the midpoint between lenses, the matrix transformation through one period is

$$\mathbf{M} = \begin{bmatrix} \left(1-\frac{1}{f}\right)\left(\Delta z + \frac{L_{\text{p}}}{2}\right) & \left(L_{\text{p}} - \frac{1}{f}\right)\left(\frac{L_{\text{p}}^2}{4} - \Delta z^2\right) \\ -\frac{1}{f} & \left(1-\frac{1}{f}\right)\left(\frac{L_{\text{p}}}{2} - \Delta z\right) \end{bmatrix}, \tag{5.136}$$

which simplifies for the symmetric point ($z = 0$) to give

$$\mathbf{M} = \begin{bmatrix} 1 - \frac{L_{\text{p}}}{2f} & L_{\text{p}}\left(1 - \frac{L_{\text{p}}}{4f}\right) \\ -\frac{1}{f} & 1 - \frac{L_{\text{p}}}{2f} \end{bmatrix}. \tag{5.137}$$

We thus have

$$\cos(\mu) = 1 - \frac{L_{\text{p}}}{2f} \qquad \sin(\mu) = \sqrt{\frac{L_{\text{p}}}{f}\left(1 - \frac{L_{\text{p}}}{4f}\right)} \tag{5.138}$$

and

$$\beta_{\text{min}} = f\sin(\mu) = \frac{L_{\text{p}}}{\sin(\mu)}\left(f - \frac{L_{\text{p}}}{4}\right) = \sqrt{L_{\text{p}}\left(f - \frac{L_{\text{p}}}{4}\right)}, \tag{5.139}$$

in agreement with Eq. (5.121). The maximum in the β-function is found in a similar way, by choosing $\Delta z = L_{\text{p}}/2$ (which is **not** a point where $\alpha = 0$, as one must traverse half of the thin lens to achieve this condition again) to give

$$\mathbf{M} = \begin{bmatrix} 1 - \frac{L_{\text{p}}}{f} & L_{\text{p}} \\ -\frac{1}{f} & 1 - \frac{L_{\text{p}}}{f} \end{bmatrix}, \tag{5.140}$$

yielding

$$\beta_{\text{max}} = \frac{L_{\text{p}}}{\sin(\mu)} = f\sqrt{\frac{L_{\text{p}}}{f - \frac{L_{\text{p}}}{4}}}, \tag{5.141}$$

as expected.

The parameterization of the transport matrix in terms of phase advance and the Twiss parameters allows insight into the eigenvectors of the transport matrix introduced in Chapter 3. Using Eq. (5.134), we can rewrite the eigenvalue equation as

$$\mathbf{M} - \lambda_{1,2}\mathbf{I} = \sin(\mu) \begin{bmatrix} \alpha \mp i & \beta \\ -\gamma & -\alpha \mp i \end{bmatrix} \cdot \vec{d}_{1,2}. \tag{5.142}$$

The eigenvectors that solve this equation are simply

$$\vec{d}_{1,2} = \begin{pmatrix} \beta \\ -\alpha \pm i \end{pmatrix} = \begin{pmatrix} \beta \\ -\alpha \end{pmatrix} \pm i \begin{pmatrix} 0 \\ 1 \end{pmatrix}. \tag{5.143}$$

The eigenvectors have the same real part, which defines the new coordinate axis for the observed simple harmonic motion,

$$\mathrm{Re}\, \vec{d} = \begin{pmatrix} \beta \\ -\alpha \end{pmatrix}. \tag{5.144}$$

In order to align the trace space to the eigenvector axes, one can, therefore, define a new trace space angle $\tilde{x}' = x' + \alpha x/\beta$. Note that this simple transformation could have been deduced simply from the geometry shown in Fig. 5.5.

This transformation is a related one that rigorously takes an arbitrary ellipse in trace space, and maps it to a circle. The resulting transformation is a generalized form of the action-angle space discussed in Chapter 1. If we take the new coordinate to be the Floquet angle ψ and momentum ε (the action), then we may write $x(z) = \sqrt{\varepsilon\beta(z)}\cos(\psi(z))$, as in Eq. (5.122), and write a transformation of the beam angle $x'(z)$ inspired by Eq. (5.123),

$$\tilde{x}'(z) = x'(z) + x(z)\frac{\alpha(z)}{\beta(z)}. \tag{5.145}$$

The transformation in Eq. (5.146) makes $\tilde{x}'(z)$ (Eq. (5.123)) dependent only on $\sin(\psi(z))$, not on $\cos(\psi(z))$. When constructing a Poincaré plot of trace space trajectories, one may use $(x, \tilde{x}')$ to compactly illustrate the motion. This is especially useful when subtle effects due to nonlinear forces (e.g. magnetic multipoles of higher order than quadrupole) are present.

To obtain an action-angle description of the motion, one make the transformation in Eq. (5.145) first, and proceed in analogy to Eqs (1.26)–(1.28). If we take x' to be the old momentum, a Hamiltonian (with z as the independent variable) can be written that reproduces Eqs (5.122) and (5.123),

$$G = \tfrac{1}{2}(x')^2 + \tfrac{1}{2}\kappa_0^2(z)x^2. \tag{5.146}$$

This Hamiltonian can be written in terms of the new coordinate and momentum by use of the generating function

$$F_1(x, \psi) = \frac{x^2}{\beta}[\tan(\psi(z)) + \alpha], \tag{5.147}$$

to give a new "action-angle" Hamiltonian

$$\tilde{G} = G + \frac{\partial F_1}{\partial z} = \frac{\varepsilon}{\beta(z)}. \tag{5.148}$$

Equation (5.148) indicates that ε is constant (it is the action, or single particle "emittance"), and that $\psi' = 1/\beta$, in agreement with Eq. (5.128). The action-angle formulation of the motion is also very useful in analyzing the effects of nonlinear forces on the motion.

5.10   Summary and suggested reading

In this chapter, we have generalized the description of charged particle motion from a single particle viewpoint to one concerned with distributions in phase space. The first classes of distributions we have examined are simple equilibria. We have found equilibrium distribution functions, by examining time-independent solutions of the Vlasov equation in the transverse phase plane. Such distribution functions are found to be Gaussian in momentum and, when the applied focusing is linear, Gaussian in the coordinate as well. We have investigated statistical mechanics aspects of beam distributions in equilibrium, including temperature and pressure.

Since the distribution function is unwieldy for looking at the general case of non-equilibrium beam transport, we have introduced a way of describing the properties of the beam through moments of the distribution. The moment approach leads, through examination of the second moments of a trace space distribution, to an rms envelope description of the beam. The evolution of a beam's second moments has been shown to be derivable from the matrix formulation of linear transformations. The second moments also form the basis for defining the rms emittance, which is a useful analytical formulation of the effective trace space area of the beam. An alternative way of defining such an area based on ellipses (defined by bi-Gaussian distribution function contours, or by the fraction of the beam contained inside of the ellipse) was also introduced. We have introduced a parameterization of such ellipses through use of the Twiss parameters, which have slightly different meaning depending on whether they are defined by second moments or by geometric considerations.

While the matrix method of transforming the second moments is powerful, other formalisms have their use in beam analyses. Among them we have explored differential equations, such as the rms envelope equation, and various approaches to solving for the evolution of the Twiss parameters.

The oscillations of beam envelopes about their equilibrium value in a constant focusing channel have been examined, and shown by a perturbed envelope equation analysis to proceed at twice the betatron frequency. Such oscillations are found to be exactly described using a β-function analysis.

The approaches to collective descriptions of beam distributions introduced in the context of transverse phase space have been generalized to apply to longitudinal phase space. Longitudinal envelopes, emittances, and other Twiss parameters have been defined in analogy to their transverse counterparts. Equilibrium distributions that arise from application of nonlinear (sinusoidal) rf forces have been introduced. The transformation of such thermal distributions, and their description by envelope equations has been examined. A Twiss parameter analysis was shown to be useful for describing the compression of a beam in a chicane system.

The power and elegance of the Twiss parameter system become apparent when it is applied in the context of periodic systems such as circular accelerators. In periodic systems, the solutions for β-functions are demanded to be periodic, in contrast to the initial value problem we have assumed to this point. For the periodic case, in fact, one can assign the Twiss parameters to a property of the focusing lattice. One must inject the beam into this lattice with the correct, matched Twiss parameters in order to have the desired periodic beam envelopes, and avoid phase space dilution in real systems. In simple cases, one may exploit symmetry to derive salient characteristics of the β-functions in periodic systems. For more complex cases, we have introduced matrix-based tools for deducing relevant aspects of the β-function evolution. We ended the chapter's discussions by connecting the β-function formalism to mathematical transformations that may be used to aid analysis and visualization of particle motion in periodic focusing systems.

The material in this chapter is intended to cover in the broadest possible way most of the commonly encountered descriptions of beam distribution and envelope evolution. In order to do this, we have minimized the discussion of the periodic systems, which would be emphasized in a text concentrating on circular accelerators. More information on these and other subjects relevant to this chapter are contained in the following references:

1. D. Edwards and M. Syphers, *An Introduction to the Physics of High Energy Accelerators* (Wiley, 1993). A very straightforward discussion of the Twiss parameters.
2. M. Reiser, *Theory and Design of Charged Particle Beams* (Wiley, 1994). Good examples of the use of the envelope equation, and an informative investigation of the temperature in relativistic beam systems.
3. H. Wiedemann, *Particle Accelerator Physics I: Basic Principles and Linear Beam Dynamics* (Springer-Verlag, 1993). More mathematically explicit than other treatments of the Twiss parameters.
4. S.Y. Lee, *Accelerator Physics* (World Scientific, 1996).
5. M. Sands, *The Physics of Electron Storage Rings: an Introduction* (Stanford, 1970). This monograph discusses the mechanisms for creation of equilibria in storage rings through synchrotron radiation damping (see Chapter 8 of this text), quantum fluctuations, and intrabeam scattering.
6. J. D. Lawson, *The Physics of Charged Particle Beams* (Clarendon Press, 1977). Physically insightful discussion of envelope equations and Twiss parameters.

Exercises

(5.1) Find the normalization constants in Eqs (5.9) and (5.11) in terms of the transverse temperature T_x and the rms measures of the distribution, σ_x and σ_{px}.

(5.2) The distribution function given by Eq. (5.12) can be written formally as

$$f_x(x, p_x)$$

$$= \frac{\kappa v_0}{2\pi k_B T_x} \exp\left[-\frac{1}{2\gamma_0 m_0 k_B T_x}(p_x^2 + \gamma_0^2 m_0^2 v_0^2 \kappa_0^2 x^2)\right]$$

$$= \frac{\kappa v_0}{2\pi k_B T_x} \exp\left[-\frac{H(x, p_x)}{2k_B T_x}\right]$$

where $H(x, p_x) = (1/2\gamma_0 m_0)(p_x^2 + (\gamma_0 m_0 \kappa_0 v_0)^2 x^2)$ is the time-independent transverse Hamiltonian governing simple harmonic oscillations in system.

(a) Show that for this Hamiltonian, the correct paraxial equations of motion are obtained, that is, one has simple harmonic motion of particles with effective mass $\gamma_0 m_0$ and oscillation frequency $\omega = \kappa_0 v_0$.

(b) Show that for any time-independent Hamiltonian the corresponding distribution function

$$f_x(x, p_x) = C \exp\left[-\frac{H(x, p_x)}{k_B T_x}\right],$$

where C is a normalization constant, is always an equilibrium solution to the Vlasov equation.

(c) For an octupole-like oscillator potential, $V(x) = Ax^4$, $A = $ constant, write and plot (contours of constant f_x, see Fig. 5.1) the correctly normalized distribution function $f_x(x, p_x)$ given by the expression in (b).

(d) The form of the distribution function given in part (b) is not the most general form possible for the solution to the Vlasov equation in equilibrium. Show that

$$f_x(x, p_x) = G[H(x, p_x)]$$

is an equilibrium solution to the Vlasov equation, where G is **any** differentiable function of the Hamiltonian.

(5.3) It can be seen from Exercise 5.2 that the equilibrium momentum component of the distribution function is Maxwellian only because the Hamiltonian is of non-relativistic form, with quadratic momentum dependence, $H \propto p_x^2$.

(a) Show that the Hamiltonian providing the transverse equations of motion has quadratic momentum dependence, if the paraxial condition $p_x \ll p_0$ is obeyed.

(b) Now consider a one-dimensional, relativistic simple harmonic oscillator Hamiltonian

$$H(x, p_x) = \sqrt{(p_x c)^2 + (m_0 c^2)^2} + \tfrac{1}{2} Kx^2.$$

Plot the equilibrium momentum distribution, in the form suggested by Exercise 5.1(b), for the cases $k_B T/m_0 c^2 = 0.1, 1$, and 10, and compare the results.

(c) What happens to the distribution in x for these cases? How do these cases differ from the expectations of a non-relativistic Hamiltonian,

$$H(x, p_x) = \frac{p_x^2}{2m_0} + \frac{1}{2}Kx^2.$$

(5.4) Show that the fraction of particles contained inside of the rms trace space ellipse, for the bi-Gaussian distribution given by Eq. (5.29), is $\exp[-1/2]$.

(5.5) Show an ellipse in trace space always transforms into another ellipse under linear transport, when the final trace space vector is obtained from the initial vector by $\vec{x} = \mathbf{M} \cdot \vec{x}_i$.

(5.6) It is implied by Eq. (5.24) that a bi-Gaussian phase plane distribution always transforms into another (generally correlated) bi-Gaussian. Show that the projections of a correlated bi-Gaussian distribution onto the spatial and angular axes are also Gaussian function.

(5.7) Show, by using Eqs (5.12) and (5.36), that for a transverse beam distribution in equilibrium ($\alpha_x = 0$) in a linear focusing channel with wave number κ:

(a) One obtains the following relationship between the normalized rms emittance and the temperature of the distribution:

$$\varepsilon_n = \frac{k_B T}{m_0 c^2} \left(\frac{1}{\kappa \beta_0}\right).$$

(b) Using this result, show that the equilibrium beam size is given, in terms of the temperature, by

$$\sigma_{eq} = \frac{1}{\kappa}\sqrt{\frac{k_B T}{\beta_0 m_0 c^2}}.$$

(5.8) Consider the case of a beam of emittance $\varepsilon_{x,rms}$, propagating in a uniform focusing channel of strength κ_0^2. Assume the initial conditions $\alpha_x = 0$ and $\beta_x = \beta_x^*$ at $z = z_0$.

(a) Write the transport matrix for the horizontal motion for $z > z_0$.

(b) Find the evolution of the second moment σ_{11} using Eq. (5.40). Use this result to write the evolution of β_x.

(c) Show that $\sigma_{11} = \varepsilon_{x,rms}/\kappa_0$ is an equilibrium condition for the beam size. What is the value of β_x corresponding to this equilibrium condition?

(5.9) Using Eqs (5.40) and (5.46), one obtains an algorithm for propagating the Twiss parameters once the transport matrix $\mathbf{M}$ is known. Another common transformation is written by viewing the Twiss parameters not as matrix, but vector elements $(\beta_x, \alpha_x, \gamma_x)$. Show, by use of the trace space vector transformation $\vec{x}_f = \mathbf{M} \cdot \vec{x}_0$ (with $\vec{x} = (x, x')$) and Eq. (5.32), that the Twiss parameter transformation can be written as

$$\begin{pmatrix} \beta_x \\ \alpha_x \\ \gamma_x \end{pmatrix}_f = \begin{bmatrix} M_{11}^2 & -2M_{11}M_{12} & M_{12}^2 \\ -M_{11}M_{12} & M_{11}M_{22}+M_{12}M_{21} & -M_{12}M_{22} \\ M_{21}^2 & -2M_{12}M_{22} & M_{22}^2 \end{bmatrix} \times \begin{pmatrix} \beta_x \\ \alpha_x \\ \gamma_x \end{pmatrix}_0.$$

(5.10) In Exercise 5.9, one obtains the simple relation between the initial beam state and final beam size,

$$\beta_{x,f} = \beta_{x,0}M_{11}^2 - 2M_{11}M_{12}\alpha_{x,0} + M_{12}^2\gamma_x,$$

or multiplying by $\varepsilon_{x,rms}$,

$$\sigma_{11,f} = \sigma_{11,0}M_{11}^2 + M_{11}M_{12}\sigma_{12,0} + M_{12}^2\sigma_{22,0}.$$

Construct a beam transport matrix $\mathbf{M}$ from a simple thin lens of variable focal length f followed by a drift of length L. Show how to determine, from three separate measurements of $\sigma_{11} = \langle x^2 \rangle$ corresponding to three different values of f, the initial Twiss parameters β_x, α_x (and thus γ_x), and the emittance $\varepsilon_{x,\text{rms}}$.

(5.11) Show, using the methods outlined in this section, that:

    (a) Eq. (5.53) gives the transformation of the general second moment Σ-matrix.

    (b) Explicitly write the generalized rms emittance invariant for the six-dimensional phase space (the determinant of the Σ-matrix).

(5.12) Find, by direct integration of the envelope equation in a force-free drift ($\kappa_x^2 = 0$), the result previously obtained in Eq. (5.48).

(5.13) In differentiating the second moments, for example, Eq. (5.54), the differential operation is performed on the weighting variables x and x', but not on these same symbols as contained in the distribution function $f_x(x, x')$. Why is this? Hint: consider the more direct representation of the beam distribution given in Eq. (5.1).

(5.14) Prove, by direct differentiation of Eq. (5.67), that the normalized emittance $\varepsilon_{n,x} \equiv \beta_0\gamma_0\varepsilon_{x,\text{rms}}$ is conserved under acceleration and linear transverse forces.

(5.15) For any second-order, alternating gradient (ponderomotive) force, such as is generally discussed in Section 3.4 (see also the discussion leading to Eq. (4.80)), the applied force is of the form $F_x = -(C/(\beta_0\gamma_0)^2)x$. Show that this form leads to a constant equilibrium beam size during acceleration.

(5.16) Show that the full solution to Eq. (5.85) is

$$\beta_x(z) = \beta_p + \Delta\beta_0 \cos(2\kappa_0 z) - \frac{\Delta\alpha_0}{\kappa_0} \sin(2\kappa_0 z),$$

with $\Delta\beta_0 = \beta_{x0} - \beta_p = \beta_{x0}/2 - \gamma_{x0}/2\kappa_0^2$, and $\Delta\alpha_0 = \alpha_{x0}$.

(5.17) Prove that the β-function found in Exercise 5.16 is always positive. Hint: examine its value at the minimum, where $\alpha_x = 0$.

(5.18) For the chicane shown in Fig. 5.12 (also Fig. 3.14), it is instructive to calculate the value of $R_{56} = -\eta_\tau \Delta s = (\alpha_c - (1/\gamma_0^2))\Delta s$ not through the dispersion integral in Eq. (3.79) (see Ex. 3.19), but directly, using geometric analysis.

    (a) First, write the total length of the path length Δs of the design trajectory through the chicane in terms of the individual magnet lengths L and radius of curvature R, leaving the bend angle θ_b implicit.

    (b) Noting that logarithmic differentiation with respect to R is equivalent to the negative of the same operation on the momentum p, determine the quantity

$$\frac{\partial(\Delta s)}{\partial R}\frac{R}{\Delta s} = -\frac{\partial(\Delta s)}{\partial p}\frac{p}{\Delta s} = -\alpha_c,$$

and from this obtain R_{56}.

(5.19) In the moving bucket case, since acceleration is occurring an equilibrium can only be approximated, much as the existence of the bucket itself is defined approximately in terms of a slowly varying design momentum. In this case of a synchronous phase of $\varphi_0 = 3\pi/4$, plot the equilibrium density $Z(\zeta)$ for the values $\Gamma = 0.2, 2$, and 10, and compare the results to Fig. 5.10.

(5.20) Find the relationship between the longitudinal temperature T_z and the normalized longitudinal emittance (e.g. Eq. (5.99)) for a bunched beam in equilibrium, with $\Gamma \gg 1$.

(5.21)   (a) For a tightly bunched beam that is not accelerating, $\varphi_0 = \pi$, find the equilibrium rms longitudinal beam size in an radio-frequency linac from Eq. (5.105) in terms of the accelerating field, radio-frequency wavelength, and longitudinal emittance.

    (b) For a tightly bunched beam that is accelerating, $\varphi_0 \neq \pi$, show by direct differentiation of Eq. (5.96) that the longitudinal emittance is conserved in the linear field limit where Eq. (5.104) applies.

    (c) Derive the longitudinal rms envelope equation for the case of an accelerating bunch, in analogy to Eq. (5.66).

    (d) Can you find an equilibrium rms longitudinal beam size in an radio-frequency linac for the case of $\varphi_0 \neq \pi$, in terms of the accelerating field, radio-frequency wavelength, and longitudinal emittance and φ_0?

(5.22) In circular accelerators, one often performs "phase space rotations" to manipulate the beam after injection or before extraction. To proceed with an example of this, imagine first that the beam is first matched to a certain voltage V_0.

    (a) What is the equilibrium rms longitudinal beam size deduced from Eq. (5.106) in terms of V_0, harmonic number, rf wavelength, and longitudinal emittance and η_τ.

    (b) If we change the rf voltage instantaneously as $V_0 = 4V_0$, the beam will be mismatched. Follow, by numerical integration, the evolution of the rms beam size for a time $\omega_s \Delta t = \pi/2$, where the beam size should be at a minimum. What is the relationship between minimum and maximum beam sizes in the system?

    (c) The general relationship between the maximum and minimum sizes in this system can be found by analytically integrating Eq. (5.106) once through multiplication of the equation by $\dot\sigma_\zeta$, to obtain a relation between σ_ζ and $\dot\sigma_\zeta$. Setting σ_ζ to be a maximum, find the minimum rms size for arbitrary mismatch parameter V/V_0.

5.23) In a practice design of a chicane compressor, consider the following electron beam and linac parameters: $U_0 = 20\,\text{MeV}$, $f = \omega/2\pi = 2856\,\text{MHz}$, and $\sigma_\zeta = 1\,\text{mm}$. You will set the linac at 25° forward of the wave peak, and the magnets available to you are to be run at $B_0 = 0.3\,\text{T}$.

(a) Calculate the initial longitudinal Twiss parameters for this scenario.

(b) Specify the value of R_{56} that one needs for compression to an uncorrelated minimum beam size. Using the results of Ex. (5.18), find the bend angle in the chicane magnets corresponding to this value. What is the compressed beam size?

5.24) For the previous exercise, the beam as compressed no longer has a Gaussian longitudinal profile after compression. Conservation of longitudinal probability density gives a relation between the initial and final distribution,

$$n_\zeta(\zeta_f)\,d\zeta_f = n_\zeta(\zeta_0)\,d\zeta_0.$$

Thus, if one knows the transformation between initial and final position, $\zeta_f(\zeta_0)$, then the final distribution is given by

$$n_\zeta(\zeta_f) = n_\zeta(\zeta_0)\left(\frac{d\zeta_f}{d\zeta_0}\right)^{-1}.$$

(a) For the case of the Gaussian beam compressed to an rms uncorrelated longitudinal distribution, write the final density function. There will be two problems associated with this calculation. The first is that there are two values of ζ_0 that can give rise to a final value of ζ_f, as shown in Fig. 5.13. Simply add the densities associated with each contributing ζ_0. The second is that the derivative $d\zeta_f/d\zeta_0$ vanishes at a certain point in the distribution, making the density unbounded at that point.

(b) The problem of an unbounded linear density can be avoided if one assumes that there is a finite spread in the initial momentum distribution at a given value of ζ_0. Thus, the phase plane distribution after the linac can be written as

$f_\zeta(\zeta, p_z)$

$$\cong \left[\begin{array}{l}\exp\left(-\left[p - p_0\left(1 - \cot(\varphi_0)\,k_z\delta\zeta\right.\right.\right.\\\left.\left.\left.-\tfrac{1}{2}\left(k_z\delta\zeta\right)^2\right)\right]^2 / 2\sigma_{p,\text{th}}^2\right)\\\times \exp\left(-\delta\zeta^2/2\sigma_\zeta^2\right)\end{array}\right]^{-(2\pi\sigma_{p,\text{th}}\sigma_\zeta)}$$

where $\sigma_{p,\text{th}}$ is a small thermal spread in the momentum that is not correlated with ζ_0. For an

assumed value $\sigma_{p,\text{th}} = 0.1(k_z\sigma_\zeta)^2 p_0$, transform (using the methods of Section 5.3) this distribution through the chicane of Exercise (5.23) and recalculate $n_\zeta(\zeta_f)$.

5.25) Consider the FODO system introduced in Section 3.2. The periodic β-function in this system has a minimum at the defocusing lens position, and a maximum at the focusing lens position.

(a) For equal focal length lenses (f) and separation (L_p) between lenses, find the minimum and maximum values of the β-function found at the lenses. Note that in order to create a symmetry point inside the defocusing (focusing) lens to find the minimum (maximum) β-function, the thin lens must be split into two equal parts.

(b) Write these β-function extrema in terms of the phase advance per FODO period, μ.

5.26) For the alternating focus–defocus (FD) system introduced in Exercise 3.6, find the extrema in the β-function:

(a) for the first stability region;

(b) for the second stability region;

(c) plot these extrema as a function of κ_0^2; and

(d) in the case of $\kappa_0 L_p = 3\pi$, in the second stability region, find the β-function everywhere in the period.

5.27) Find a transformation that takes an arbitrary Poincaré plot, and maps it, by using Eq. (5.146) and appropriate rescaling of axes, to a circle of unit radius. How do the scale factors depend on the single particle "emittance" ε?

5.28) Note that the transformation matrix through a focusing period (Eq. (5.134)) can be written as

$$\mathbf{M} = \mathbf{I}\cos(\mu) + \mathbf{J}\sin(\mu), \quad \text{where } \mathbf{J} = \begin{bmatrix} \alpha & \beta \\ -\gamma & -\alpha \end{bmatrix}.$$

This breakdown of the matrix has similar characteristics to the function $\exp(i\mu)$, because of the unique characteristics of $\mathbf{J}$.

(a) What is $\mathbf{J}^2$?

(b) Show that

$$[\mathbf{I}\cos(\mu_1) + \mathbf{J}\sin(\mu_1)][\mathbf{I}\cos(\mu_2) + \mathbf{J}\sin(\mu_2)]$$
$$= \mathbf{I}\cos(\mu_1 + \mu_2) + \mathbf{J}\sin(\mu_1 + \mu_2).$$

(c) What is the mapping $\mathbf{M}^n$ that represents n applications of the same periodic focusing, in terms of the total phase advance $n\mu$?

Accelerator technology I: magnetostatic devices

[1] This statement is a obviously a bit too sweeping to be strictly correct—particles can be accelerated by magnets (e.g. induction acceleration in a betatron), and particles can be bent and focused by electromagnetic wave fields in non-axisymmetric configurations.

The interplay between technology and charged particle dynamics in the field of particle beam physics is apparent from the previous chapters' discussions. In short, charged particles are focused and bent by use of magnets, and accelerated by use of electromagnetic waves in cavities or guides.[1] In this chapter, we flesh out the discussion of particle manipulation through external forces by first examining the design principles and practical aspects of magnetostatic devices. In Chapter 7, we undertake a discussion of the ideas important in the bounded electromagnetic systems that are used in accelerators.

6.1 Magnets based on current distributions

Conceptually, the simplest magnets used in accelerator laboratories essentially contain only currents, with no ferromagnetic or permanent magnet material. In fact, the strongest magnets in use in today's accelerators are of this type, but because they employ ultra-high current densities to achieve high fields, they must be superconducting. Thus conceptual simplicity gives way to the technological complexity associated with cryogenics and superconducting materials, among other issues. These issues are beyond the scope of this text, and so we will concentrate on the physical principles associated with the proper choice of current distribution in such magnets. Since magnets are classified in terms of their multipole character (e.g. dipole, or quadrupole, or combined function), the choice of current configuration can be examined with a particular pure multipole symmetry in mind.

In Section 2.5, we introduced a magnetic scalar potential written as a sum of multipole components that form a basis set of solutions to the Laplace equation in cylindrical coordinates. While the scalar potential approach to analysis is appropriate for iron-dominated magnets, it should be slightly modified here because the scalar potential is only valid where the current vanishes. For the type of two-dimensional magnet model considered here, all current densities $\vec{J}_z$ are in the z direction (in or out of the plane of a picture like Fig. 2.8). Therefore it suffices to define the transverse magnetic field in terms of the longitudinal component of the **vector potential**, as in the Coulomb gauge we have a single Poisson (scalar) equation which governs the relevant potential,

$$\vec{\nabla}^2 A_z = -\mu_0 J_z. \tag{6.1}$$

This potential, in turn, generates transverse magnetic field through the relation

$$\vec{B}_\perp = \frac{\partial A_z}{\partial y}\hat{x} - \frac{\partial A_z}{\partial x}\hat{y} = \frac{1}{\rho}\frac{\partial A_z}{\partial \phi}\hat{\rho} + \frac{\partial A_z}{\partial \rho}\hat{\phi}. \tag{6.2}$$

The solutions to Eq. (6.1) for regions inside of a cylinder containing no current are identical in form to those of the scalar potential ψ given by Eq. (2.39). A more general form of the solution in arbitrary current-free regions is given by,

$$A_z = \text{Re} \sum_{n=1}^{\infty} [A_n \rho^n + B_n \rho^{-(n+1)}] \exp(in\phi). \tag{6.3}$$

This expression indicates that there are two components to the vector potential: one corresponding to fields that are due to source currents inside of the observation radius ρ, and a second outside of this radius. In general, the most powerful tools to connect these solutions to the source current distributions are: (i) finding and employing a Green function giving the solution to Eq. (6.1) for a δ-function source located at a given (ρ, ϕ); and (ii) directly summing the source distributions with appropriate weight to find multipole moments. The first approach produces results that are not very compact, and the second requires a fairly involved formal derivation. To effectively introduce multipole moments into the discussion here we instead undertake a heuristic approach, that of finding a localized current sheet distribution placed at a certain radius a, which has the correct symmetry to give rise to the fields in question.

As an example, we examine the case of a dipole field with strength $B_0 = -A_1$, oriented along the y direction, so that

$$A_z = \begin{cases} A_1 \rho \cos(\phi), & \rho < a \\ B_1 \rho^{-2} \cos(\phi), & \rho > a. \end{cases} \tag{6.4}$$

The potential is continuous at the radius where the currents are located, $\rho = a$, giving $B_1 = A_1 a^3$. This implies that a dipole magnet of this sort will have a field immediately outside of its current distribution which is proportional to its size a. The field associated with the potential in Eq. (6.4) is

$$\vec{B} = B_0 \begin{cases} \sin(\phi)\hat{\rho} + \cos(\phi)\hat{\phi}, & \rho < a \\ (a/\rho)^3 [\sin(\phi)\hat{\rho} - 2\cos(\phi)\hat{\phi}], & \rho > a. \end{cases} \tag{6.5}$$

Note that the tangential field B_ϕ is discontinuous at $\rho = a$ due to the current sheet located there. Use of the Stokes theorem with Eq. (1.3) specifies the surface current (current sheet) distribution $K_z(\phi)$ needed to give this discontinuity,

$$\Delta B_\phi = 2B_0 \cos(\phi) = \mu_0 K_z(\phi), \tag{6.6}$$

or in terms of a current density,

$$J_z(\rho, \phi) = K_z(\phi)\delta(\rho - a) = \frac{2B_0}{\mu_0} \cos(\phi)\delta(\rho - a). \tag{6.7}$$

This distribution, not too surprisingly, has the same azimuthal dependence as the vector potential component which gives rise to the desired multipole. This condition could have been deduced from Eq. (6.1) for a pure multipole solution.

The current distribution of Eq. (6.7) is the ideal upon which superconducting dipoles, the enabling technology that high-energy hadron circular accelerators, such as the Tevatron (Fermilab), RHIC (Brookhaven National Laboratory near New York), and the Large Hadron Collider (LHC, currently under construction at the CERN laboratory near Geneva, Switzerland), are built upon. This type of magnet design is termed, for obvious reasons, a "$\cos(\phi)$" magnet.

The distribution of Eq. (6.7) is unsatisfactory from the viewpoint of actually designing magnets because of the singular (sheet) current distribution. One can easily imagine a non-singular current density being built up out of a sum of sheet distributions,

$$J_z(\rho, \phi) = j(\rho) \cos(\phi), \tag{6.8}$$

which gives a pure dipole field inside of the radius $\rho_{\min}$ where the currents vanish, $j(\rho < \rho_{\min}) = 0$. In fact, with this more general dipole distribution, we can view Eq. (6.8) as specifying a type of simple Green function, which allows us obtain the value of the field in the current-free, near-axis region,

$$B_0 = \frac{\mu_0}{2} \int_{\rho_{\min}}^{\rho_{\max}} j(\tilde{\rho}) \, d\tilde{\rho}. \tag{6.9}$$

It can be seen from this expression that the current distribution giving a pure near-axis dipole field is not unique, but can have infinite variety.

Equation (6.8), while of conceptual value, is not typically used in practice, because magnets are usually wound from series of tightly packed coils made up of a conductor with constant transverse dimensions and current carried. Another, more practical way of approximating the $\cos(\phi)$-magnet can be constructed by noting that the sum of two overlapping, opposing cylinders of uniform current density also gives a pure dipole field, as shown in Fig. 6.1. This is a case exactly solvable by superposition of sources and their associated fields. Note that superposition of the current density implies that in the region of overlap the currents from each cylinder cancel, leaving a current distribution very much akin to the $\cos(\phi)$ distribution, as illustrated in Fig. 6.1.

The magnetic field arising from a uniform current density, J_z, cylinder centered at a point $\vec{a} = (a_x, a_y)$, evaluated at a point $\vec{\rho} = (x, y)$, is trivially obtained from Ampere's law,

$$\vec{B} = \frac{\mu_0 J_z}{2} (\vec{\rho} - \vec{a}) \times \hat{z}. \tag{6.10}$$

The superposition of the fields from the two opposing current densities $\mp J_z$ centered at $\vec{a} = \pm(d/2)\hat{x}$, respectively, is therefore simply

$$\vec{B} = \frac{\mu_0 J_z}{2} \left(\vec{\rho} + \frac{d}{2}\hat{x} \right) \times \hat{z} - \frac{\mu_0 J_z}{2} \left(\vec{\rho} - \frac{d}{2}\hat{x} \right) \times \hat{z}$$

$$= \frac{\mu_0 d J_z}{2} \hat{x} \times \hat{z} = -\frac{\mu_0 d J_z}{2} \hat{y}, \tag{6.11}$$

which is a pure, vertically oriented dipole field in the "overlap" region. The configuration shown in Fig. 6.1 well approximates many existing superconducting magnets, where the windings are based on cryogenically cooled niobium alloys.

It is interesting to compare the practical geometry of Fig. 6.1 with the ideal dipole-generating current sheet configuration of Eq. (6.7). This can be done by taking the limit that the displacement d of the current cylinders is much smaller than their diameter D. In this case, the surface current density associated with the

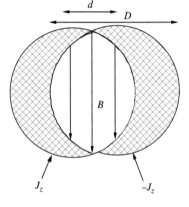

Fig. 6.1 Realistic current density distribution for approximating a pure $\cos(\phi)$-magnet design for dipole field.

current distribution can be approximated as

$$K_z(\phi) \cong J_z \Delta\rho(\phi) \cong J_z \, \mathrm{d} \cos(\phi), \tag{6.12}$$

where $\Delta\rho(\phi)$ is the thickness of the layer of current as a function of ϕ. According to Eq. (6.7), this surface current distribution should give rise to a field

$$B_0 = \frac{\mu_0 K_z(\phi)}{2\cos(\phi)} = \frac{\mu_0 d J_z}{2}, \tag{6.13}$$

in agreement with Eq. (6.11).

For more arbitrary or complicated types of magnets, with no pure multipole characteristic, one may find the magnetic field from the currents by using the Biot–Savart integral form,

$$\vec{B} = \frac{\mu_0}{4\pi} \iiint_V \frac{\vec{J} \times \vec{r}}{r^3} \, \mathrm{d}V. \tag{6.14}$$

One practical limitation on the strength of superconducting magnets arises from the self-forces generated by the current. The force per unit volume on an element of current carrying wire can be written as

$$\frac{\mathrm{d}\vec{F}}{\mathrm{d}V} = \vec{J} \times \vec{B}. \tag{6.15}$$

If we wish to estimate the total pressure on the windings in a magnet at a location along the azimuth, we may view the current distribution as a sheet, as in Eq. (6.12), which allows Eq. (6.15) to be written as a force per unit area, or pressure,

$$\frac{\mathrm{d}\vec{F}}{\mathrm{d}A} \cong \tfrac{1}{2}\vec{K} \times \vec{B} = \frac{B_0^2}{\mu_0} \cos(\phi)\hat{x}. \tag{6.16}$$

For the ultra-high field superconducting dipole magnet designed for the LHC at the European high-energy physics laboratory CERN located in Geneva, Switzerland, the design magnetic field is over 6 T. This indicates that the windings experience a maximum pressure of nearly 300 atms! As one can imagine, dealing with these pressures requires a serious engineering effort in order to have a mechanically stable device.

6.2   Magnets based on ferromagnetic material

The use of ferromagnetic (mainly iron-based) material in electromagnets is by far the most common design tool in magnetostatic charged particle optics devices. Its utility lies in the fact that iron is highly permeable, with low-field values of μ/μ_0 in the neighborhood of 10^4. We know from Eq. (1.3) that the tangential component of $\vec{H} = \vec{B}/\mu$ is continuous at a boundary, while we also know from Eq. (1.1) that the longitudinal component of $\vec{B}$ is continuous at an interface containing iron on one side. Because of these constraints, at an iron/non-permeable region interface, the tangential component of the $\vec{B} = \mu_0\vec{H}$ must nearly vanish on the non-permeable side of such a boundary. The existence of boundary surfaces with the magnetic fields forced to be normal to such surfaces implies that they are **equipotential surfaces**. Here the potential under consideration must be a scalar, so we revert to the approach of Section 2.5, in which we introduced the potential ψ, with $\vec{B} = -\vec{\nabla}\psi$, or for regions with constant permeability, $\vec{H} = -\mu^{-1}\vec{\nabla}\psi$.

We have already seen in Chapter 2 that this scalar magnetic potential has simple forms corresponding to the possible multipoles under consideration. The equipotential surfaces, which we now explicitly see are in fact the surfaces of the magnet iron, are therefore planes in the case of dipoles, and hyperbolae in the case of quadrupoles, as illustrated in Fig. 6.2. Note that the ideal quadrupole surface is very similar to the geometry of the real quadrupole ferromagnet shown in Fig. 2.10.

In designing magnets, one must mathematically connect the desired field in the magnet gap (iron-free region) with the inner dimensions of the magnet pole and the current used to excite the field. The simplest illustration of this process can be understood by analyzing the "C-magnet" geometry displayed in Fig. 6.3. This picture shows the transverse cross-section of a bend magnet, with a dipole field excited in the gap by the coils that carry a certain total current. This current is usually contained in N windings wrapped around the iron, so that one often writes the total current as NI, the number of windings times the current I in each, resulting in a total indicated by the unit **Ampere-turns**. The planar surfaces at the bottom and top of the gap set the dipole field inside of the gap. In addition, the iron also provides a return path for the magnetic flux, which must close on itself, since $\vec{B}$ is a divergence-less field. The iron portion of the magnet that lies away from the poles is termed the **yoke**.

The relationship between the total excitation current and the field in the gap can be simply estimated by use of the integral form of Ampere's law,

$$\oint \vec{H} \cdot d\vec{l} = I_{\text{enc}}. \tag{6.17}$$

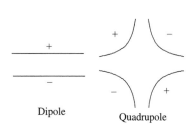

Fig. 6.2 Equipotential surfaces for simple magnetic multipoles.

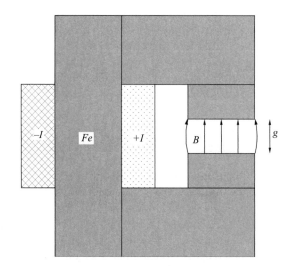

Fig. 6.3 Transverse cross-section of a C-magnet dipole.

For the geometry shown in Fig. 6.4, Eq. (6.17) can be written approximately as the sum of two components,

$$\frac{B_0}{\mu_0}g + \frac{1}{\mu}\int_{\text{Fe}} \vec{B} \cdot d\vec{l} = I_{\text{enc}}, \tag{6.18}$$

where B_0 is the dipole field strength in the gap. Because the magnitude of $\vec{B}$ inside of the iron is not too different from B_0, while the permeability is orders of magnitude larger in the iron, the second term on the left-hand side of Eq. (6.18) is typically ignored, and we have

$$B_0 \cong \mu_0 I / g. \tag{6.19}$$

For other multipoles, we note that, if we ignore the contribution of the integral in Eq. (6.18) in the iron, the relation between field integrals and currents can then be written simply as

$$\frac{1}{\mu_0}[\psi_1 - \psi_2] = I_{\text{enc}}, \tag{6.20}$$

where ψ_1 and ψ_2 are the magnetostatic potentials at adjacent pole tips, and I_{enc} is the current enclosed by a path through the iron and between the pole tips. For a pure quadrupole, referring to Fig. 2.10, this prescription yields

$$\frac{2b_2}{\mu_0}\left[\frac{a^2}{2} - \left(-\frac{a^2}{2}\right)\right] = \frac{B'}{\mu_0}a^2 = 2I_{\text{pole}}, \quad \text{or} \quad B' = \frac{2\mu_0 I_{\text{pole}}}{a^2}, \tag{6.21}$$

where a is the radial distance of the closest point on the pole tip to the origin, and I_{pole} is the current circulating around each pole tip.

The design model we have introduced relies on two idealizations of the iron pole surfaces. The first is that the internal pole dimensions extend to infinity in ρ, so that a pure multipole is obtained from poles of the type shown in Fig. 6.2. This is of course untrue, and the corrections due to finite pole dimension can be accounted for by analytical models, such as **conformal mapping**, which is based on a complex representation of the potential and its equipotential boundaries, or, much more commonly, through the use of computer codes (see Fig. 6.6). The second idealization is that of effective infinite permeability of the iron, which allows the pole surfaces to be viewed as equipotentials, and also gives

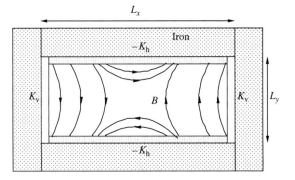

Fig. 6.4 Iron, current, and field configurations for a Panofsky quadrupole.

rise to the approximation that the entire magnetic potential drop $\mu_0 I_{enc}$ occurs in the gap between poles. Once again this problem is best dealt with through computational modeling.[2]

[2] Before the easy availability of computers, it was common to make a two-dimensional "analog" computer model of magnet geometries, on circuit board material, where the relative resistivity of a surface could be made to mimic the relative permeability of the iron. Equivalent magnetic potentials were then determined by a voltage map. This method addressed only the effects of constant permeability, not those of saturation, which does not have a simple resistive analogy.

One of the more novel configurations of magnet that uses iron is the Panofsky quadrupole, which is shown in Fig. 6.4. The iron in this case, has dipole-like symmetry, and is obviously not used to set the form of the potential inside of the magnet directly. Instead, the current distributions (which are approximated as current sheets in the z direction) located inside of the iron, interface directly with the quadrupole field distribution, and cancel all components tangential to their surface before the $\vec{B}$-field enters the iron. Thus the field is purely normal entering the iron, and the iron surfaces are again magnetic equipotentials.

The role of the current sheets in the Panofsky quadrupole can be quantified as follows. The magnetic field, written as

$$\vec{B} = B'(y\hat{x} + x\hat{y}), \qquad (6.22)$$

clearly has components tangential to the "box" surfaces of the quadrupole interior which are independent of the coordinate along the surface and proportional to the distance of the surface from the origin. The tangential field at the vertical surfaces in Fig. 6.4 is simply $B_\| = \pm B'L_x/2$, which must be cancelled by a current sheet with constant surface current density $K_v = \mp B_\|/\mu_0 = B'L_x/2\mu_0$.

Likewise, the current sheets on the horizontal surfaces must cancel fields of strength $B_\| = \mp B'L_y/2$, and therefore have a constant surface current density equal to $K_h = B'L_y/2\mu_0$. Thus, the simple rule for constructing a Panofsky quadrupole: the ratio of surface current density on the sides of the magnet to that found on the top and the bottom is $K_v/K_h = L_x/L_y$.

Panofsky quadrupoles were invented by W. Panofsky in order to fabricate a quadrupole with large apertures that can be chosen to be very different in the horizontal and vertical dimensions, to manipulate charged particles emanating from fixed-target high-energy physics experiments. This type of device is especially useful in momentum spectrometer lines, in which the beam may be much larger horizontally than vertically.

6.3   Saturation and power loss in ferromagnets

The performance of a ferromagnet is strongly influenced, and ultimately limited by, the saturation of the iron. This effect can be understood qualitatively as follows: ferromagnetic material aids in the flow of the B-field by aligning its microscopic dipole moments to the field, and "relaying" the field without causing loss of the integrated H-field. Once all of the dipole moments are optimally aligned, the B-field in the iron does not increase with additional H in the same proportion as for low values. This means that the effective incremental permeability,

$$\mu_\delta \equiv \frac{dB}{dH}, \qquad (6.23)$$

falls from values as high as $10^4 \mu_0$ when $H \ll H_{sat}$, to a value that descends ultimately to near μ_0 for $H \gg H_{sat}$. This implies that a magnet employing iron to shape the magnetic field and guide the magnetic flux will not behave

he same at large excitations. In particular, there is some value of the magnetic
ield, as illustrated by a typical $B(H)$ curve shown in Fig. 6.5, which is not easily
exceeded in a ferromagnetic material, that is known as the **saturation** field. For
common low carbon steels, this saturation field is in the region of 1.5–2.0 T.

Avoidance of saturation places a straightforward constraint on the design of
electromagnets. To ensure that the field in the iron does not approach saturation,
the iron in the yoke must not have too small of a cross-sectional area. This can
be seen by noting that the total flux throughout a pole tip

$$\Phi = \iint_{\text{pole}} \vec{B} \cdot d\vec{A} \tag{6.24}$$

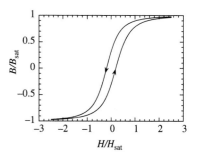

Fig. 6.5 $B(H)$ curve for a iron, with hysteresis
effect illustrated (direction of H-excitation
indicated by arrows).

s a conserved quantity (see Eq. (1.16)) throughout the iron, assuming that no
ield lines leave the iron. We thus have

$$B_{\text{Fe}} \cong B_{\text{pole}} \frac{A_{\text{pole}}}{A_{\text{Fe}}}, \tag{6.25}$$

where A_{Fe} is the total cross-sectional area of the iron which the return flux from
the pole is allowed to pass through. Note that in many situations the flux splits
into two legs of the yoke, as illustrated by the example in Exercise 6.6. If we
restrict the field in the iron to be much smaller than saturation, we have the
design constraint

$$A_{\text{Fe}} \gg A_{\text{pole}} \frac{B_{\text{pole}}}{B_{\text{sat}}}. \tag{6.26}$$

In practice, one uses Eqs (6.20), (6.24) and (6.26) as guidelines to begin a
design, and then employ computer programs to numerically solve the field
or potential equations on a computational grid. An example of this type of
program's output is shown in Fig. 6.6, which illustrates an H-magnet geometry
and associated field lines, calculated by the two-dimensional magnetostatic
simulation code POISSON. These types of codes, which also permit three-
dimensional geometries to be simulated, allow optimization of pole tip shapes,
minimization of flux leakage in the yokes, and can predict the portions of the
iron that may saturate.

The $B(H)$ curve shown in Fig. 6.5 also displays a phenomenon known as
hysteresis, in which the magnetization of iron does not disappear as one moves

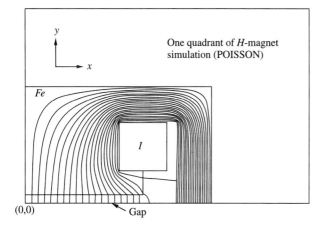

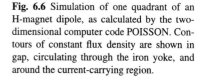

Fig. 6.6 Simulation of one quadrant of an
H-magnet dipole, as calculated by the two-
dimensional computer code POISSON. Contours of constant flux density are shown in
gap, circulating through the iron yoke, and
around the current-carrying region.

from an extremum in the applied H to zero excitation. The value of B which is left over in the absence of any applied H-field is dependent on the direction and magnitude of the recent history of H-excitation, and it is known as a **remanent field**. It appears because the detailed state of the microscopic dipole moment alignment is dependent on the history of nearby re-orientations of these magnetic moments. In order to remove these remanent fields, one must slowly cycle the magnet through ever decreasing amplitude excitations. If one excites the windings with a current having a damped sine-wave time dependence, then such a **degaussing** of the magnet can be accomplished.

Hysteresis effects not only cause a remanent field problem, which affects low-excitation performance of electromagnets, but they also are a source of energy deposition in magnets that must be cycled, such as betatron or synchrotron magnets. This can be seen by noting that the magnetic energy density in a permeable medium is

$$u_{\mathrm{M}} \cong \tfrac{1}{2} \vec{B} \cdot \vec{H}. \tag{6.27}$$

The existence of a non-vanishing area enclosed by the $B(H)$ curve indicates that work must be done to execute a traversal of such a loop in the $B(H)$ diagram. If we indicate this area (which represents an energy density) by measuring it as a fraction of the maximum, $A_{BH} = \alpha_{\mathrm{m}} B_{\max}^2/2\mu$, then the power density deposited in the magnet as it cycles with repetition rate f is

$$\frac{\mathrm{d}P}{\mathrm{d}V}\bigg|_{\mathrm{hysteresis}} \cong f \alpha_{\mathrm{m}} \frac{B_{\max}^2}{2\mu}. \tag{6.28}$$

The fractional loss per cycle of the stored magnetic energy in the iron is α_{m}, and this energy ultimately is transferred to heat. This power loss played a major role in the betatron's decline in favor as a device for creating large fluxes of MeV electrons needed by applications such as medical X-ray production. The total particle flux in such a device is clearly proportional to the repetition rate, which is strictly limited by the heat power handling capabilities of the iron, a material with poor heat conductivity.

Hysteresis is not the only potential source of power dissipation in a cycling magnet containing iron. As the magnetic field rises, an induction electric field arises which is proportional to the temporal rate of rise in B. The iron is typically sliced into narrow cross-sections termed **laminations**, which are electrically insulated from each other. This is done in order to prevent the induced electric field from driving excessive **eddy currents** in the iron, which produce an ohmic power loss

$$\frac{\mathrm{d}P}{\mathrm{d}V}\bigg|_{\mathrm{ohmic}} = \vec{J}_{\mathrm{eddy}} \cdot \vec{E}_{\mathrm{induction}} = \sigma_{\mathrm{c}} \vec{E}_{\mathrm{induction}}^2. \tag{6.29}$$

These slices are long in the direction in which little of the circulating induced electric field projects. This procedure therefore mitigates the ohmic power losses due to eddy currents. It should also be noted that these currents are **diamagnetic**, in the sense that they will configure themselves to oppose the change in the magnetic field. Such diamagnetic fields are as undesirable as the power losses themselves, because they may degrade the quality of the field inside of the magnet gap.

The other significant power loss associated with electromagnet operation is found in the conducting windings, which are typically made up of high conductivity ($\sigma_{\mathrm{c}} = 5.8 \times 10^5 \ \Omega^{-1}\mathrm{cm}^{-1}$) copper that are insulated on the exterior. The windings, in the case of large ohmic losses, may be cooled using pressurized water flow through an internal cooling channel. This allows current densities in excess of 750 A cm^{-2} (ohmic power density of approximately 1 Wcm^{-2}) to be used. In high field magnets, where ohmic power losses can be excessive, one may be forced to abandon electromagnets based on iron in favor of superconducting designs. As noted in Section 6.1, superconducting magnets often use no iron, which we now note is due to a desire to achieve fields well in excess of saturation, for example, the 6 T LHC dipole.

6.4 Devices using permanent magnet material

In some experimental scenarios, it is not practical to rely on electromagnets or superconducting currents. This is particularly true when the magnetic device must be compact. In such cases, the magnet designer may employ permanent magnet materials to achieve high fields. The **magnetization** $\vec{M}$ of a material is related to the $\vec{H}$ and $\vec{B}$ fields by the expression

$$\vec{M} = \frac{\vec{B}}{\mu_0} - \vec{H}. \qquad (6.30)$$

While in a ferromagnet the magnetization is considered to be a function of applied $\vec{H}$ field, $\vec{M} = \vec{H}[(\mu/\mu_0) - 1]$, in a permanent magnet, $\vec{M}$ is idealized as a constant characteristic of the material. For the most common configurations, $\vec{M}$ is thought of as constant in amplitude and direction within a macroscopic region. This is in reality only a good approximation; a permanent magnet does not have constant magnetization (independent of $\vec{H}$), but is actually a material with a very large residual magnetization, and thus remanent field. Stated mathematically, a permanent magnet has both a large $\vec{M}$ at $\vec{H} = 0$, and a weak dependence of $\vec{M}$ on $\vec{H}$ near $\vec{H} = 0$. The nearly constant magnetization obtained in neodymium-based compounds can be in range of 1.2 T, while more common (and inexpensive) ferrites may have smaller magnetizations. It should also be noted that magnetization is not completely "permanent" due to other practical effects; demagnetization may occur due to excessive heating or high radiation levels.

The analysis of permanent magnet arrays can be simplified by introducing fictitious "**magnetic charges**" or electric currents. To employ these useful concepts, their roles must first be clarified. A fictional distribution of magnetic charges is applied if one wishes to calculate the $\vec{H}$-field. This distribution is derived from Eq. (1.1) through

$$\vec{\nabla} \cdot \vec{B} = \mu_0(\vec{\nabla} \cdot \vec{H} + \vec{\nabla} \cdot \vec{M}) = 0, \qquad (6.31)$$

or

$$\vec{\nabla} \cdot \vec{H} = \rho_m, \quad \text{with } \rho_m = -\vec{\nabla} \cdot \vec{M}. \qquad (6.32)$$

For uniformly magnetized pieces, the divergence of $\vec{M}$, and thus the magnetic charge density, vanishes everywhere except the boundary surfaces that may be normal to the magnetization. At these boundaries, we can derive from Gauss' law an effective magnetic surface from the discontinuity in $\vec{M}$,

$$\sigma_{\mathrm{m}} = \vec{M} \cdot \hat{n}, \tag{6.33}$$

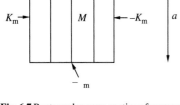

Fig. 6.7 Rectangular cross-section of permanent magnet piece, showing the magnetization
vector $\vec{M}$, the magnetic surface charge σ_{m},
and the Amperian current sheets $\vec{K}_{\mathrm{m}}$.

where $\hat{n}$ is the unit normal to a surface. The existence of effective magnetic
surface charges implies (through Eq. (6.32)) that the normal component of $\vec{H}$
is discontinuous at such a boundary, which is in fact implied by the continuity
of $\vec{B}$ and the discontinuity of $\vec{M}$ (cf. Eq. (6.30)) These fictitious surface charges
are illustrated in Fig. 6.7, which shows a rectangular cross-section of a long
permanent magnet piece. This piece can be thought of as one of the components
of an undulator magnet of the type discussed in Section 2.8 (see Fig. 2.12). Once
the effective magnetic charge distribution is found, Eqs (6.32) and (6.33) can be
solved to give the $\vec{H}$-field directly, or indirectly through a magnetic potential.

In other situations, it may be more useful to represent the permanent magnet
in terms of effective **Amperian currents**. These arise from consideration of
Eq. (1.3) in the static, current-free case ($\vec{J}_{\mathrm{e}} = 0$), so that we write

$$\vec{\nabla} \times \vec{B} = \mu_0 \vec{\nabla} \times (\vec{H} + \vec{M})$$

$$= \mu_0 \vec{\nabla} \times \vec{M} \equiv \mu_0 \vec{J}_{\mathrm{m}}. \tag{6.34}$$

The Amperian current density $\vec{J}_{\mathrm{m}}$ also vanishes in a region with uniform magnetization, and one is left only with a contribution due to the boundaries that
are parallel to $\vec{M}$. These boundaries have an Amperian current sheet of surface
current density

$$\vec{K}_{\mathrm{m}} = \vec{M} \times \hat{n}. \tag{6.35}$$

Amperian currents can be used to find the magnetic field directly by standard
methods from magnetostatics (e.g. use of the Biot–Savart law, or in cases of
exploitable symmetry, use of Stokes theorem and Eq. (6.35)). Alternatively,
these currents may be used to find the magnetic vector potential $\vec{A}$.

As an example of these methods, let us examine the permanent magnet-based
undulator shown in Figs 2.10 and 6.8, which is known as a **Halbach undulator**, in honor of its inventor, Klaus Halbach (who initially also developed
the POISSON code used to create Fig. 6.6). The permanent magnet pieces are
square with side dimension a in (y–z) cross-section and, for ease of analysis,
infinitely long in the x-dimension. In this case we begin by defining the magnetic charge density which, as seen in Fig. 6.7, consists of merely two sheets,
infinitely long in x, of opposite sign surface charge density at the poles of the

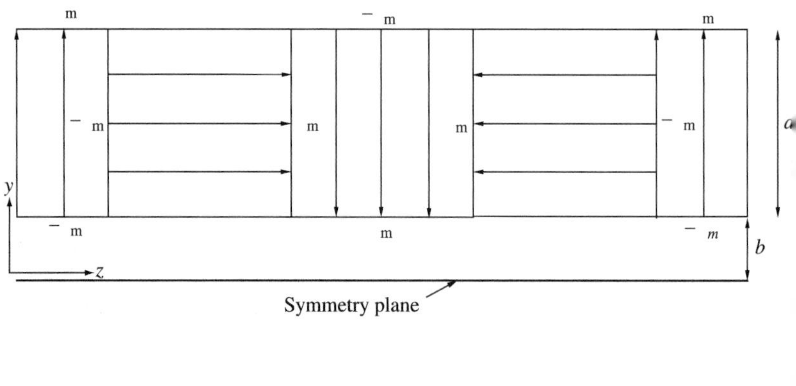

Fig. 6.8 Cross-section of one period (upper
half only) of a permanent magnet-based undulator, with magnetization direction indicated
by arrowhead lines, and magnetic surface
charges specified at appropriate boundaries.

Symmetry plane

piece. Because it is often simpler to use potentials, we examine a magnetic potential ψ_H, where $\vec{H} = -\vec{\nabla}\psi_H$. This scalar potential, which is related to the source charge distribution through a Poisson equation,

$$\vec{\nabla}^2\psi_H = -\rho_m. \tag{6.36}$$

can be found by many methods. The most obvious is based upon integration over the bulk and sheet magnetic charge density distributions using, in analogy to electrostatics, Green function solutions to Eq. (6.36),

$$\psi_H(\vec{r}) = \iiint_V \frac{\rho_m(\vec{r}')}{4\pi|\vec{r} - \vec{r}'|}\,dV' + \sum_j \iint_{S_j} \frac{\sigma_m(\vec{r}')}{4\pi|\vec{r} - \vec{r}'|}\,dA'. \tag{6.37}$$

For the case of present interest, the volume integral in Eq. (6.37) vanishes, and we must only consider the surface integrals over the separate surfaces[3] enumerated by the index j (an infinite number for a purely periodic system, as shown in Fig. 6.8). Because this procedure involves summation over the sheets, we do not work out the detailed solution formally given by Eq. (6.37), but instead examine an alternative formulation where we explicitly take advantage of the periodicity of the magnetic charge distribution. This formulation arises naturally from Eq. (6.36) in the case of periodic solutions, because it is solved straightforwardly using harmonic functions.

We begin our periodic analysis by noting that the periodic magnetic surface charge density shown in Fig. 6.8 can be described by a Fourier series in z. The Fourier representation of this surface charge density distribution breaks down naturally into two separate components—the magnetic charge on vertical ($z = $ constant) surfaces, and the magnetic charge on horizontal ($y = $ constant) surfaces. The potential corresponding to each component can be found separately, and the full potential obtained by linear superposition of the component potentials. We first examine the vertical surfaces, in which case the Fourier series describing the magnetic charge distribution is given by

$$\rho_m = \sum_{n=1,\text{odd}}^{\infty} c_n \cos(k_n z), \, b < |y| < b+a, \quad \text{with } k_n = nk_u = \frac{n\pi}{2a} \tag{6.38}$$

and

$$c_n = -\frac{2M}{a}\cos\left(\frac{n\pi}{4}\right), \quad n \text{ odd.} \tag{6.39}$$

The distribution described by Eqs (6.38) and (6.39) is then substituted into Eq. (6.35), to give

$$\vec{\nabla}^2\psi_H = -\rho_m = -\sum_{n=1,\text{odd}}^{\infty} c_n \cos(k_n z). \tag{6.40}$$

The solution to Eq. (6.40) has a particular part,

$$\psi_{H,p} = \sum_{n=1}^{\infty} c_n \left(\frac{2a}{n\pi}\right)^2 \cos(k_n z) = -\frac{8aM}{\sqrt{2}\pi^2}\sum_{n=1}^{\infty}\left(\frac{1}{n}\right)^2 \cos\left(\frac{n\pi}{4}\right)\cos(k_n z). \tag{6.41}$$

[3] The second integral on the right hand side of Eq. (6.37) is of course formally the same as the first integral, if we view the surface charge density distributions as singular bulk charge distributions. The second integral then arises from integrating over the Dirac δ-functions, which describe the surfaces the singularities reside upon.

There is also a homogeneous part, which can be written in terms of the periodic functions,

$$
\psi_{H,h} =
\begin{cases}
\sum_{n=1,\text{odd}}^{\infty} d_n \cosh\left(k_n\left(y-\left(b+\frac{a}{2}\right)\right)\right)\cos(k_n z), \\
\quad b < y < b+a, \\
\sum_{n=1,\text{odd}}^{\infty} e_n \exp\left[-k_n\left|y-\left(b+\frac{a}{2}\right)\right|\right]\cos(k_n z), \\
\quad y < b \text{ and } y > b+a,
\end{cases}
\tag{6.42}
$$

if we concentrate for the moment on only the upper side, $y \geq 0$, of the symmetry plane of the of the magnet array. The solutions given in Eq. (6.41) are by construction symmetric about the middle of the magnet array, $y = b + a/2$. The coefficients d_n and e_n are obtained by requiring continuity of the potential and its y derivative at the boundaries, $y = b$, $y = b + a$,

$$
\psi_H = -\frac{8aM}{\sqrt{2}\pi^2}
\begin{cases}
\sum_{n=1,\text{odd}}^{\infty} \frac{\cos(n\pi/4)}{n^2}\left[1-\exp\left(-\frac{n\pi}{4}\right)\right] \\
\quad \times \cosh\left(k_n\left|y-\left(b+\frac{a}{2}\right)\right|\right)\cos(k_n z), \quad b < y < b+a, \\
\sum_{n=1,\text{odd}}^{\infty} \frac{\cos(n\pi/4)}{n^2}\sinh\left(\frac{n\pi}{4}\right) \\
\quad \times \exp\left[-k_n\left|y-\left(b+\frac{a}{2}\right)\right|\right]\cos(k_n z), \quad y < b, \ y > b+a.
\end{cases}
\tag{6.43}
$$

Now that we have obtained the magnetic potential due to the magnetic surface charges on vertical surfaces in the region $b < y < b+a$, we next examine the potential due to horizontal surface distributions at $y = b$ and $y = b + a$. These surface distributions (in the region $y > 0$) have the longitudinal Fourier representation

$$
\rho_m = \sum_{n=1,\text{odd}}^{\infty} C_n[\delta(y-(b+a))-\delta(y-b)]\cos(k_n z),
\tag{6.44}
$$

with

$$
C_n = \frac{4M}{n\pi}\sin\left(\frac{n\pi}{4}\right), \quad n \text{ odd.}
\tag{6.45}
$$

Only the homogeneous solutions to Eq. (6.36) (of the same form as Eq. (6.42)) need be examined for this distribution, and they are connected by integration through the magnetic charge sheets at $y = b$ and $y = b+a$ using the prescription

$$
\Delta\psi_H' = -\sigma_m,
\tag{6.46}
$$

where the derivative in Eq. (6.46) is normal to the boundary containing the charge sheet. Thus, we obtain the potential due to the horizontal magnetic surface charge distributions,

$$
\psi_H = \frac{4Ma}{\sqrt{2}\pi^2}\sum_{n=1,\text{odd}}^{\infty}\frac{\sin(n\pi/4)}{n^2}\cos(k_n z)
$$

$$
\times
\begin{cases}
\exp(k_n(y-(b+a)))-\exp(k_n(y-b)), & y < b, \\
\sinh\left(k_n\left[y-\left(b+\frac{a}{2}\right)\right]\right), & b < y < b+a, \\
\exp(-k_n(y-(b+a)))-\exp(-k_n(y-b)), & y > b+a.
\end{cases}
\tag{6.47}
$$

Note that only odd values of n contribute to the series found in Eqs (6.43) and (6.47).

The final step in the analysis of the Halbach undulator magnet requires that we superimpose the potential from a magnet array like that shown in Fig. 6.8 with its associated pair located in the region $-b > y > -(b + a)$. The magnet array in this region has the identical magnetization configuration and potential, with the entire picture translated down in y by a distance $2b + a$. Thus we can find the potential in the gap (the region of interest for beam dynamics analyses),

$$\psi_{\mathrm{H}} = \frac{4Ma}{\sqrt{2}\pi^2} \sum_{n=1,5,9,\ldots}^{\infty} \frac{1}{n^2} \left(\exp\left(-\frac{n\pi}{2} \right) - 1 \right) \exp(-k_n b) \sinh(k_n y) \cos(k_n z),$$

(6.48)

or, in terms of the undulator wave number, k_{u},

$$\psi_{\mathrm{H}} = \frac{8M}{\sqrt{2}\pi k_{\mathrm{u}}} \sum_{n=1,5,9,\ldots}^{\infty} \frac{1}{n^2} \left(\exp\left(-\frac{n\pi}{2} \right) - 1 \right)$$

$$\times \exp(-nk_{\mathrm{u}}b) \sinh(nk_{\mathrm{u}}y) \cos(nk_{\mathrm{u}}z).$$

(6.49)

From Eq. (6.49), we obtain the magnetic field inside of the gap,

$$\vec{B} = -\mu_0 \vec{\nabla} \psi_{\mathrm{H}} = \mu_0 M \frac{8}{\sqrt{2}\pi} \sum_{n=1,5,9,\ldots}^{\infty} \frac{1}{n} \left(1 - \exp\left(-\frac{n\pi}{2} \right) \right)$$

$$\times \exp(-nk_{\mathrm{u}}b) \cosh(nk_{\mathrm{u}}y) \cos(nk_{\mathrm{u}}z)\hat{y}$$

$$- \mu_0 M \frac{8}{\sqrt{2}\pi} \sum_{n=1,5,9,\ldots}^{\infty} \frac{1}{n} \left(1 - \exp\left(-\frac{n\pi}{2} \right) \right)$$

$$\times \exp(-nk_{\mathrm{u}}b) \sinh(nk_{\mathrm{u}}y) \sin(nk_{\mathrm{u}}z)\hat{z}.$$

(6.50)

Equation (6.50) indicates that the undulator field has many longitudinal harmonics, but that these higher harmonics, which only begin at $n = 5$, diminish rapidly in amplitude as n becomes large. Note also the exponentially decaying dependence of the field amplitude in the gap as a function of b. This is due to the tendency of the magnetic flux to circulate without crossing the $y = 0$ plane when the gap height is not ignorably small compared to the period of the undulator. The application of the Halbach undulator in free-electron lasers may dictate that the gap be of non-negligible height. Since in this case, the higher harmonics in the field produce undesirable harmonics in the induced electron wiggle motion (see Sections 2.8 and 8.9), one must be careful not to choose too small of a gap.

In most cases, analytical investigation of a given permanent magnet geometry using magnetic charge densities yields the magnetic field more straightforwardly than the use of Amperian surface currents to represent permanent magnet pieces. On the other hand, Amperian currents can be extremely useful in computational modeling, as they provide a simple way of extending the power of a simulation code such as POISSON, which provides for magnetic sources based only on free currents. As an example of this, we show a POISSON simulation of the Halbach undulator geometry in Fig. 6.9, which uses Amperian current sheets as prescribed by Eq. (6.35). This simulation shows a final interesting aspect of the Halbach undulator—the magnetic flux is returned almost perfectly through the permanent magnet material, with almost no leakage of the field into regions

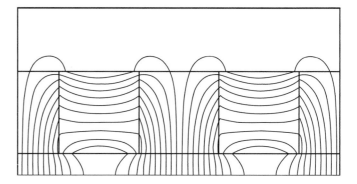

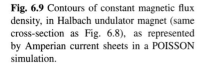

Fig. 6.9 Contours of constant magnetic flux density, in Halbach undulator magnet (same cross-section as Fig. 6.8), as represented by Amperian current sheets in a POISSON simulation.

outside of the magnet ($|y| > b + a$). Remarkably, this is accomplished without the use of any permeable material.

In practice, three-dimensional effects in undulator magnets give rise to the need for three-dimensional tools. As analytical approaches, even perturbative schemes, are much more difficult in three dimensions than in two, one now typically relies on three-dimensional **magnet modeling codes**, which are modern, practical extensions of two-dimensional codes such as POISSON. These codes can model the effects of three-dimensional geometry, saturation, power dissipation, permanent magnet material, electromagnetic stress, and other relevant physical phenomena.

Magnets, like other components in laboratories, are subject to much testing during the fabrication process. Magnetic field maps can be made locally using **Hall probes**, which utilize the Hall effect to give a voltage proportional to the magnetic field. For systems like quadrupoles, or undulators where the field changes very rapidly in space, this approach may not give the most accurate result. For quadrupoles and other nominally pure multipole devices, a **rotating coil** measurement can yield the multipole content of a magnetic field (cf. Exercise 6.4). In the case of an undulator magnet, the field itself is perhaps not as important as the trajectory information (first and second integrals of the field along the axis). By pulsing current through a taut wire that is stretched through the undulator and measuring the mechanical waves induced in the wire by the resulting magnetic force impulse, one can deduce these integrals.

6.5   Summary and suggested reading

In this chapter, we have given an overview of the physics and engineering principles behind the types of magnets employed in accelerators and beam transport lines. These types have been broken down as follows:

1. Magnets that are based on currents only, such as superconducting dipoles (e.g. $\cos(\phi)$ magnet). Methods of analysis based on multipoles, the Biot–Savart law, and Ampere's law (exploiting symmetry) have been introduced. While ignoring technical aspects of superconducting materials and cryogenics, we have noted the significant self-forces such magnets may experience at high field.

2. Electromagnets based on current excitation applied to set field strength and ferromagnetic materials employed to give the desired field shape.

The choice of ferromagnetic material geometry, currents, and desired multipole strength has been explored using a magnetic "circuit" model based on a scalar magnetic potential. Relevant aspects of such magnets such as saturation, eddy currents and power loss have been introduced.

3. Magnets that employ permanent magnet material. We have discussed how this type of magnet can analyzed using two equivalent source models—magnetic "charges" for calculating the H-field, and Amperian currents for deriving the B-field. The magnetic charge method has been illustrated in the detailed example of the Halbach undulator magnet.

The design of magnets is a sophisticated field, and all three types of magnets we have discussed possess many more physics and engineering details than are included in this chapter. For more information on magnet design, see the following references:

1. R. Wangsness, *Electromagnetic Fields* (Wiley, 1986). This text includes introductory discussions of electromagnets, including magnetic circuit analysis, and saturation effects.
2. A. Chao and M. Tigner, editors, *Handbook of Accelerator Physics and Engineering* (World Scientific, 1999). Every magnet technology used in accelerators is represented in this compendium.
3. K. Buschow, *Permanent Magnet Materials and Their Application* (Trans Tech Publication, 1998). A technical resource for permanent magnets.
4. H. Brechna, *Superconducting Magnet Systems* (Springer, 1973). A somewhat dated book on superconducting magnet technology.
5. P. Luchini, *Undulators and Free-Electron Lasers* (Oxford University Press, 1990). There are many good bibliographic references included with this monograph.
6. Computer codes used for design of magnetic components in accelerators are found at the Los Alamos Accelerator Code group website, http://laacgl.lanl.gov.

Exercises

(6.1) Consider a quadrupole magnet designed only with current distributions.

    (a) Evaluate this case based on surface current distributions, relating the parameters a, $K_z(\phi)$, and B (the field gradient).

    (b) Construct a similar design using the superposition of four uniform current density cylinders, in analogy to Fig. 6.1.

    (c) Compare, for the case of small cylinder displacement, your results from part (b) to those of part (a).

(6.2) For the bifilar (double helix) helical undulator, which is shown in Fig. 2.15 (Exercise 2.16), the current density is

$$\vec{J} = I\delta(x - a\cos(k_u z))\delta(x - a\sin(k_u z))\hat{t} + \cdots$$
$$- I\delta(x + a\cos(k_u z))\delta(x + a\sin(k_u z))\hat{t},$$

where a is the radius of the helix, and $\hat{t}$ is the unit vector tangent to the positive winding.

(a) Evaluate the on-axis magnetic field associated with this current density, using Eq. (6.14).

(b) What is the purpose of the return winding?

(6.3) For a superconducting quadrupole magnet designed with surface current distributions only, as in Exercise 6.1, evaluate the peak magnetic pressure for a bore $a = 2$ cm and the peak field at the current surface of $B = 3$ T. Evaluate the direction and amplitude of the magnetic pressure at this surface.

(6.4) Consider a coil rotating in a magnetic field, as shown in Fig. 6.10.

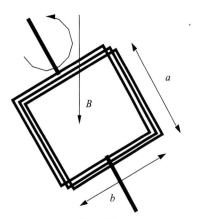

Fig. 6.10 Schematic of a rotating coil magnetic measurement device.

The coil, which has N turns, is rotated at frequency ω. This produces a voltage on the terminals of the coil by electromagnetic induction.

(a) Find this voltage. Write it explicitly in terms of the multipole moment strengths of the field.

(b) What is the effect of placing the coil at some position offset from the axis of the device? How could one determine if one has placed the coil correctly inside of a quadrupole?

(6.5) For a sextupole (six pole) of strength $b_3 = 10$ T/m^2 (see Eq. (2.39) for coefficient definition), and pole tip radius $a = 5$ cm, find the total current I through the windings around each pole.

(6.6) A version of an **H-magnet** (so termed because of the shape of the iron) is shown in Fig. 6.11.

(a) This magnet has a winding around each leg of the yoke. Does this configuration give twice the field as the C-magnet? Why or why not?

(b) What happens if one of the windings is turned off?

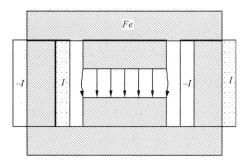

Fig. 6.11 Schematic view of H-magnet.

(6.7) Compare the total current (add all current sheet amplitudes) needed for exciting a symmetric Panofsky quadrupole to the current needed for exciting

(a) an iron-tipped quadrupole of the same aperture and strength, and

(b) a pure current-based quadrupole of the type discussed in Exercise 6.1.

(6.8) Consider a betatron, with magnetic field in the far interior of the magnet (where the field is nearly uniform) varying sinusoidally from approximately 0.1 to 1 T, at a cyclic frequency of 60 Hz, $\alpha_m = 0.1$ and $\mu/\mu_0 = 10^4$.

(a) Evaluate the power loss density associated with hysteresis $dP/dV|_{\text{hysteresis}}$.

(b) Assuming the magnet is made of iron, and that laminations have **not** been used, what is the ohmic power loss density $dP/dV|_{\text{ohmic}}$ associated with the induced eddy currents in the magnet interior? You may ignore the diamagnetic effect of the eddy currents in your answer.

(6.9) If one desires to keep the amplitude of the fifth harmonic of the vertical field in the symmetry plane of a Halbach undulator to 5 per cent of the amplitude of the fundamental harmonic, what must the ratio b/a be?

(6.10) For the Halbach undulator magnet analyzed above, find the field outside of the magnet ($|y| > b + a$). Find the form of the field in the limit $b \ll a$.

(6.11) Let us examine the field associated with a single permanent magnet (square in y and z, with dimension a, and very long in x), magnetized in the y direction.

(a) Calculate, using effective magnetic surface charges and Eq. (6.37), the potential ψ_H in all space.

(b) Calculate, using Amperian surface currents and the Biot–Savart law, the $\vec{B}$-field in all space, and compare that to the $\vec{B}$-field derived from the potential ψ_H found in part (a).

Accelerator technology II: waveguides and cavities

While the magnet technology introduced in Chapter 6 is primarily used to manipulate charged particle orbit directions, it is not central to the acceleration process. As discussed in Chapter 4, the most common technologies used in charged particle acceleration are based on (radio-frequency, or rf) electromagnetic waves. In this chapter, we discuss the principles behind the design and implementation of the most common rf devices, waveguides and resonant cavities. These objects are employed to confine and shape the electromagnetic fields, in order to allow power flow and acceleration. The analyses presented in this chapter are unified, in the sense that resonant fields in cavities and traveling waves in guides share some common wave descriptions, as well as useful notions of impedance and equivalence to simple circuit models. In addition to applications that pertain strictly in the context of rf accelerator technology, we also will explore a few topics along the way which help prepare for the discussion of photon beams in Chapter 8.

7.1 Electromagnetic waves in free space★

To set the stage for our discussion of electromagnetic waves in accelerators, waveguides, and lasers (in the following chapter), we first must review the characteristics of such waves in the absence of notable bulk material influences or boundary conditions. The wave equation governing electromagnetic fields in free space ($\varepsilon = \varepsilon_0, \mu = \mu_0$) is deduced from simultaneous use of Eqs (1.16)–(1.19), (1.21), and (1.22),

$$\left[\vec{\nabla}^2 - \frac{1}{c^2}\frac{\partial^2}{\partial t^2} \right] \left\{ \begin{matrix} \vec{E} \\ \vec{B} \end{matrix} \right\} = 0, \tag{7.1}$$

which is the same for all components of the fields. In the absence of the introduction of boundary conditions, these equations have simple, well-known solutions. The most commonly discussed of these are the **plane waves**, which are obtained by substitution into Eq. (7.1) of the traveling wave form $\vec{E}, \vec{B} \propto \exp[i(\vec{k}\cdot\vec{r}-\omega t)]$. Here $\vec{k} = k\hat{k}$ is the propagation vector, indicating both the wave number k and (unit) direction of propagation $\hat{k}$. This substitution yields the dispersion relation

$$\omega^2 = \vec{k}^2 c^2 = k^2 c^2, \quad \text{or simply } \omega = kc. \tag{7.2}$$

Waves in free space have equal **phase velocity** v_ϕ and **group velocity** v_g,

$$v_\phi = \frac{\omega}{k} = c \quad \text{and} \quad v_g = \frac{d\omega}{dk} = c. \tag{7.3}$$

The electric field, magnetic field, and propagation vectors are related by use of the curl relations

$$\hat{k} \times \vec{B} = \vec{E}/c, \quad \text{and} \quad \hat{k} \times \vec{E} = c\vec{B}. \tag{7.4}$$

In other words, the electric and magnetic fields are normal to each other and normal to the propagation vector. The electromagnetic field in this case is purely transverse, and this type of wave is termed **transverse electromagnetic** (TEM). The TEM case occurs naturally in the absence of imposed conducting boundary conditions and is thus important in free space (e.g. laser beams, as discussed in Chapter 8), but not in accelerators, where the waves are confined inside of conducting boundaries. For completeness, we note that TEM waves have two orthogonal (normal) possible polarizations that are transverse to the propagation vectors. When these two independent polarizations are both present in a TEM wave and are in phase with each other, the wave is **linearly polarized**, and when they are both present but not precisely in phase, the wave is **elliptically polarized**. When they are present in equal amplitudes and are $\pi/2$ out of phase the wave is **circularly polarized**.

The behavior described in this section is modified radically by the introduction of conducting boundaries. We begin the discussion of these modifications in the next section, where we examine the behavior of an electromagnetic wave in a strongly conducting medium.

7.2   Electromagnetic waves in conducting media★

In the remainder of this chapter, we will be concentrating on waves that are strongly confined by conducting materials that form accelerator cavity and waveguide walls. As a first approximation, one often assumes that both the electric field and magnetic field vanish inside of such conducting regions. This vanishing of fields is mathematically enforced in solving boundary value problems by insisting that the tangential component of the electric field as well as the normal component of the magnetic field must vanish at the interface between a conducting region and a non-conducting region. We shall see that this assumption is quite well justified, since the distance over which these fields penetrate highly conducting material is negligible for the wave frequencies (microwaves) of interest.

Let us examine the field associated with an electromagnetic disturbance in a conducting medium (with dielectric permittivity ε and $\mu = \mu_0$), with the conducting region of interest being directly adjacent to a free space region. We assume material has an ohmic conducting response, $\vec{J} = \sigma_c \vec{E}$. The field components are governed by the wave equation that describe electromagnetic fields in free space, written as

$$\left[\vec{\nabla}^2 - \mu_0 \sigma_c \frac{\partial}{\partial t} - \mu_0 \varepsilon \frac{\partial^2}{\partial t^2} \right] \left\{ \begin{matrix} \vec{E} \\ \vec{B} \end{matrix} \right\} = 0. \tag{7.5}$$

Substitution of $\vec{B} \propto \exp[\mathrm{i}(kz - \omega t)]$, a plane wave traveling in the z-direction as the solution (we assume ω is real), gives the dispersion relation

$$-k^2 + \mathrm{i}\omega\mu_0\sigma_c + \mu_0\varepsilon\omega^2 = 0. \tag{7.6}$$

In a highly conducting system, where the amplitude of the ohmic current density $\sigma_c \vec{E}$ greatly exceeds that of the displacement current density $\omega \varepsilon \vec{E}$, we have $\omega \varepsilon / \sigma_c \ll 1$. Thus, the dispersion relation can be approximately solved by ignoring the last term in Eq. (7.6), giving

$$k = \sqrt{\frac{\omega \mu_0 \sigma_c}{2}} (1 + i). \qquad (7.7)$$

The real and imaginary parts of the wave number given in Eq. (7.7) are equal. Strong damping of the wave occurs through ohmic power losses that are localized in a region of exponential field amplitude decay known as the **ohmic skin-depth**

$$\delta_s \equiv [\text{Im } k]^{-1} = \sqrt{\frac{2}{\omega \mu_0 \sigma_c}}. \qquad (7.8)$$

Thus, the transverse magnetic fields, and accompanying electric fields, that are driven by the existence of a tangential component of the field at the surface of a conductor (i.e. the plane $x = 0$, with the conductor occupying the upper half-space, $x > 0$) decay away from the boundary as $\exp[-\sqrt{\omega \mu_0 \sigma_c / 2} x] = \exp[-x/\delta_s]$.

In order to relate the power lost in the conducting medium to the value of the tangential field $\vec{B}_\parallel$ in the region across the boundary from the conductor, we note that in the limit that the skin depth is infinitesimally small, the surface current density that would be required to exclude the field from the conductor is $K_s = |\vec{H}_\parallel| = \mu_0 |\vec{B}_\parallel|$. In terms of bulk current density, the power loss (indicated by a negative sign) per unit volume in this system is simply

$$\frac{dP}{dV} = -\langle \vec{J} \cdot \vec{E} \rangle = -\sigma_c^{-1} \langle \vec{J}^2 \rangle = -\frac{J_0^2}{2\sigma_c} \exp[-2z/\delta_s], \qquad (7.9)$$

which can be integrated to give the power loss per unit surface area

$$\frac{dP}{dA} = \int_0^\infty \frac{dP}{dV} \, dz = -\frac{J_0^2 \delta_s}{4\sigma_c}. \qquad (7.10)$$

Since the peak bulk current density J_0 is related to the total surface current by $J_0 = K_s / \delta_s$, Eq. (7.10) can be written as

$$\frac{dP}{dA} = -\frac{K_s^2}{4\delta_s \sigma_c} = -\frac{K_s^2}{4} \sqrt{\frac{\omega \mu_0}{2\sigma_c}} = -\frac{K_s^2}{2} R_s, \qquad (7.11)$$

where we have defined the **surface resistance**,

$$R_s \equiv \frac{1}{2} \sqrt{\frac{\omega \mu_0}{2\sigma_c}}, \qquad (7.12)$$

which has units of ohm. The surface resistivity increases with frequency ω, because the square amplitude of the current density penetrating into the conducting medium increases as ω increases, more than offsetting the effect of the shorter skin depth, $\delta_s \propto \omega^{-1/2}$.

The last portion of Exercise 7.3 illustrates that the model we are presently employing for the conductivity has mathematical limitations. It also has physical limitations, in that the conduction electron response at high enough frequency does not obey Ohm's law, but rather displays **plasma** behavior. In simple circuit terms, the plasma response of the conduction electrons is said to be **reactive**, while the ohmic behavior is **resistive**. To illustrate these distinctions, we now can develop a rudimentary model for the conduction electron behavior. In conduction-electron **plasmas** (as gases of charged particles are called), the of motion for the electrons is written as

$$\frac{d\vec{v}}{dt} \cong \frac{\partial \vec{v}}{\partial t} = -\frac{e\vec{E}}{m_e} - v_c \vec{v}, \tag{7.13}$$

[1]This approximation for the microscopic behavior of conducting electrons was introduced first by P. Drude roughly 100 years ago and remains a common basis of waveconductor systems analysis.

where v_c is a collision frequency, assumed to be constant.[1] This model recovers the correct static flow limit, because for a constant applied field $\vec{E}$ the velocity of the conduction electrons is

$$\vec{v}_{stat} = -\frac{e\vec{E}}{m_e v_c}, \tag{7.14}$$

and the current density is

$$\vec{J} = -en_0\vec{v}_{stat} = \frac{e^2 n_0 \vec{E}}{m_e v_c}, \tag{7.15}$$

where n_0 is the density of the electrons. Thus, we have a simple microscopic model for the low frequency conductivity of a metal in terms of physical parameters,

$$\sigma_c = \frac{e^2 n_0}{m_e v_c}. \tag{7.16}$$

The collision frequency deduced from this relation, for copper ($n_0 \cong 8.5 \times 10^{22}$ cm^{-3}) is very high ($v_c \cong 4 \times 10^{13}$ s^{-1}), well above rf frequencies. Thus the relation given by Eq. (7.14) is an accurate predictor of the electron velocity for rf excitations.

The response of the system at very high (optical) frequency is more complicated. Writing the conduction current density in the linear approximation,

$$\frac{\partial \vec{J}}{\partial t} = -\frac{\partial}{\partial t} en\vec{v}_c \cong -en_0 \frac{\partial \vec{v}}{\partial t}$$

(assuming the electron density remains nearly constant $n \cong n_0$), we can write the full response of the electric field in the limit that collisions can be ignored as

$$\left[c^2 \vec{\nabla}^2 - \omega_p^2 - \frac{\partial^2}{\partial t^2} \right] \vec{E} = 0. \tag{7.17}$$

In deriving Eq. (7.17), we have assumed the dielectric polarization is negligible ($\varepsilon = \varepsilon_0$), and introduced the **plasma frequency** (cf. Ex. 2.14)

$$\omega_p^2 = \frac{e^2 n_0}{\varepsilon_0 m_e}. \tag{7.18}$$

Substituting $\vec{E} \propto \exp[i(kz - \omega t)]$, we obtain **plasma dispersion relation**,

$$\omega^2 = \omega_p^2 + k^2 c^2. \tag{7.19}$$

Equation (7.19) displays two terms with frequencies corresponding to free (plasma) and displacement currents, respectively. Equation (7.19) is one of the most common forms of dispersion relations found in nature—it will, for instance, be re-encountered in the next section in the context of waveguides.

The **cutoff** (plasma) frequency is defined to be the minimum frequency for which Eq. (7.19) has a real (propagating) solution, where $\omega = \omega_p$ and $k = 0$. For frequencies below cutoff, the wave is non-propagating, or **evanescent**, with a characteristic attenuation constant, $\kappa = \sqrt{\omega_p^2 - \omega^2}/c$. The maximum spatial attenuation rate occurs in the low frequency limit, $\kappa \cong k_p = \omega_p/c$, where k_p^{-1} is termed the **plasma skin depth**.

The power lost to walls is an inherent feature of all accelerators and waveguides, even accelerators based on superconducting cavities. In a superconductor exposed to a static voltage, current flows without resistance because a component of the material's conduction electron population forms Cooper pairs and does not undergo collisional losses during flow. Electrons that are not in a superconducting state do not flow after the applied voltage has been shielded by the motion of the superconducting pairs. With harmonically time-varying fields, however, the situation is different, as fields always penetrate a plasma skin depth into the material and excite the normal conducting electron population into motion. This normal conducting flow dissipates power, of course. A circuit analogy is that of the purely reactive LC circuit being driven in parallel with a resistor—the LC component does not dissipate power, but supports the voltage that drives current through the resistor, with resultant power loss. More circuit analogies used in describing microwave device physics are discussed in the following sections.

7.3   Electromagnetic waves in waveguides★

Free space propagation of electromagnetic waves at rf frequencies is inappropriate, due to extremely large diffraction effects as well as electromagnetic noise in the laboratory. Therefore, rf power and signals are transported from one point to another in accelerator laboratories using compact structures known as **waveguides**. In addition, traveling wave accelerators themselves can be viewed as a particular type of waveguide.

It is natural at this point to introduce the most common waveguide used—which many would not conceive of at first as a guide at all—the coaxial (concentric cylindrical metal surfaces) geometry, illustrated by the cross-sectional view shown in Fig. 7.1. The coaxial guide is ubiquitous in the laboratory and is most often encountered as a flexible cable. It is representative of a general class of waveguide structure, having two disjoint conductors where one may be entirely enclosed inside the other. We shall see that the waves supported by this types of guide have certain characteristics in common with the free space waves discussed above.

In order to analyze the possible forms of the fields, or **modes**, that may exist inside of a waveguide, we begin by assuming a solution to the wave equation in non-conducting media contained inside of a perfectly conducting guide,

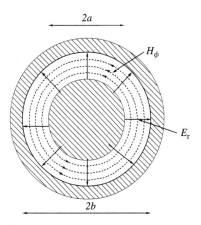

Fig. 7.1 Geometry of coaxial waveguide cutaway, with inner conductor of outer radius *a* and outer conductor of inner radius *b*. The electric and magnetic fields associated with electromagnetic waves are shown as radial (solid) field lines and azimuthal (dashed) field lines, respectively.

$$\left[\vec{\nabla}^2 - \mu\varepsilon \frac{\partial^2}{\partial t^2} \right] \begin{Bmatrix} \vec{E} \\ \vec{B} \end{Bmatrix} = 0, \qquad (7.20)$$

which is both harmonic in the waveguide propagation direction (i.e. normal to the cross-section shown in Fig. 7.1) and time. This assumption allows us to write $\vec{E} = \tilde{E}(\omega, \vec{r}_\perp) \exp[i(k_z z - \omega t)]$, where $\vec{r}_\perp = x\hat{x} + y\hat{y}$, and so any component of the field in Cartesian coordinates is of the form, for example, $E_x = f(\omega, x, y) \exp[i(k_z z - \omega t)]$. Substituting this form of solution into Eq. (7.20) gives the result,

$$\left[\vec{\nabla}_\perp^2 - k_z^2 + k_0^2\right] \left\{\begin{matrix}\tilde{E}\\\tilde{B}\end{matrix}\right\} = 0, \tag{7.21}$$

where we have defined $k_0 \equiv \omega\sqrt{\mu\varepsilon} = n_0\omega/c$ (see Eq. 8.4) as the wave-number which a plane wave propagating in such a medium would possess. The example of a plane wave, which clearly **cannot** be a solution to the wave equation in a region bounded transversely by conducting surfaces, is relevant here because substituting $k_z = k_0$ gives

$$\vec{\nabla}_\perp^2 \left\{\begin{matrix}\tilde{E}\\\tilde{B}\end{matrix}\right\} = 0. \tag{7.22}$$

The plane wave, in which there is no transverse variation of any field component, clearly and trivially satisfies Eq. (7.22). The classes of solution that are of interest here, however, are not so trivial. Clearly, one may state that Eq. (7.22) implies that in the case $k_z = k_0$, as solutions to the two-dimensional Laplace equation, the $\tilde{E}$ and $\tilde{B}$ fields have the same form as two-dimensional (purely transverse) **static** electric or magnetic fields.

The parameter $k_c \equiv \sqrt{k_z^2 - k_0^2}$, which measures the degree to which the wave conditions deviate from that assumed in Eq. (7.22), is termed the **cutoff wave number** for reasons which will become apparent below. Using the cutoff wave-number, we can rewrite the two-dimensional (transverse) Helmholtz equations in Eq. (7.21) as

$$\left[\vec{\nabla}_\perp^2 - k_c^2\right] \left\{\begin{matrix}\tilde{E}\\\tilde{B}\end{matrix}\right\} = 0. \tag{7.23}$$

The individual Maxwell equations (Eqs (1.16)–(1.19)) also become, under the current assumptions, and written in standard waveguide analysis form,

$$\tilde{H}_z = \frac{i}{k_z}\vec{\nabla}\cdot\tilde{H}_\perp, \tag{7.24}$$

$$\tilde{E}_z = \frac{i}{k_z}\vec{\nabla}\cdot\tilde{E}_\perp, \tag{7.25}$$

$$\tilde{E} = -\frac{Z_0}{ik_z}\vec{\nabla}\times\tilde{H}, \tag{7.26}$$

$$\tilde{H} = \frac{\vec{\nabla}\times\tilde{E}}{ik_z Z_0}, \tag{7.27}$$

where $Z_0 \equiv \sqrt{\mu/\varepsilon}$ is the **wave impedance** corresponding to the medium inside of the waveguide (recall that in plane waves the impedance is the of the ratio of electric and magnetic field amplitudes $\|\tilde{E}\|/\|\tilde{H}\|$). In many situations, the medium is free space, in which case $Z_0 = \sqrt{\mu_0/\varepsilon_0} \cong 377\,\Omega$. We note at this point that another, different definition of impedance, which arises in the context of waveguides, will be discussed in Section 7.5.

Equations (7.26) and (7.27) can be placed in similar form as Eqs (7.24) and (7.25) to explicitly show transverse and longitudinal components separately, by writing

$$\tilde{E}_\perp = -\mathrm{i}\frac{k_0}{k_c^2}\left[\hat{z}\times\vec{\nabla}_\perp(Z_0\tilde{H}_z) - \mathrm{i}\frac{k_z}{k_0}\vec{\nabla}_\perp(\tilde{E}_z)\right],\qquad(7.28)$$

and

$$Z_0\tilde{H}_\perp = -\mathrm{i}\frac{k_0}{k_c^2}\left[\hat{z}\times(\vec{\nabla}_\perp\tilde{E}_z) + \frac{k_0}{k_z}\vec{\nabla}_\perp(Z_0\tilde{H}_z)\right].\qquad(7.29)$$

It can be seen from Eqs (7.28) and (7.29) that, if one of the longitudinal (electric or magnetic) field components vanishes, one may solve for the remaining field components by initially finding the form of the non-vanishing longitudinal field. The two distinct cases we examine, corresponding to a vanishing longitudinal electric field $\tilde{E}_z$ and vanishing longitudinal magnetic field $\tilde{H}_z$, are termed transverse electric (TE) modes and transverse magnetic (TM) modes, respectively. There may also be cases where a superposition of TE and TM modes are needed in order to satisfy a particular boundary condition. These combined field patterns belong to what are termed the hybrid electromagnetic (HEM) modes.

There is obviously a third situation to consider in this analysis where both $\tilde{E}_z$ and $\tilde{H}_z$ vanish, that is termed a transverse electromagnetic (TEM) mode. In this case, taking the transverse Laplacian of Eqs (7.28) and (7.29) yields two equations that are consistent only if Eq. (7.22) holds and, in turn, implies that $k_c = 0$. Thus, we see that all TEM modes have zero cut-off frequency, the purely transverse fields obey Eq. (7.22), and the fields are of identical form to two-dimensional electrostatic or magnetostatic cases. For the example of Fig. (7.1), these are the familiar fields one calculates for infinitely long coaxial cylinders with uniform surface charge and current density on the conducting surfaces. The fields associated with these source densities are related through the wave impedance and the geometry of the guide. Note that the geometry needed to satisfy an electrostatic-like field form of the TEM mode must have separated conductors, much like the coaxial case shown in Fig. 7.1. This is because a single conductor would have a constant electrostatic potential, and any solution of Eq. (7.22) inside a two-dimensional region bounded by a constant potential is trivial (constant, with vanishing field). An even simpler TEM geometry than the coaxial case—although more challenging from the viewpoint of field calculations—would be that of two parallel conducting wires.

It can be seen that, in the TEM case, the wave dispersion relation reduces to the free-space expression. Thus, the phase velocity and group velocity are identical to those of plane waves in the medium contained between the conductors. If the medium inside of the guide itself is non-dispersive, pulsed wave-packets having large frequency spreads will experience negligible dispersion and distortion as they propagate (in what are commonly referred to as TEM transmission lines). In other words, any arbitrary time pulse can be synthesized from a Fourier sum of differing frequency (ω) components, and the pulse will propagate without distortion at a constant phase velocity ω/k. Because of this property, TEM transmission lines such as the coaxial waveguide are often used to transport electromagnetic signals in laboratory situations where time information is to be preserved.

The total instantaneous power transmitted in the TEM guide can be found by calculating the value of the Poynting vector $\vec{S} = \vec{E} \times \vec{H}$ (which measures electromagnetic power density per unit area), and integrating over the cross-section of the waveguide,

$$P = \iint \vec{S} \cdot d\vec{a} = \iint \vec{E} \times \vec{H} \cdot d\vec{a} = \frac{1}{Z_0} \iint \vec{E}^2 \, da$$

$$= v_\phi \iint \varepsilon \vec{E}^2 \, da = v_g \iint u_{\mathrm{EM}} \, da, \tag{7.30}$$

where u_{EM} is the electromagnetic energy density in the wave. Equation (7.30), which is valid for only for a TEM mode (where $v_\phi = v_g$), implies that the Poynting vector for this system is composed of the electromagnetic energy density multiplied by the group velocity of the wave, as one might have expected, just as in the case of free-space wave propagation. For the simple case of the dielectric-loaded coaxial waveguide, using the results of Exercise 7.7, we find that the total power transmitted down the waveguide is given in terms of the peak electric and magnetic fields inside the guide (on the inner conductor surface) by

$$P = \vec{E}_{\mathrm{max}}^2 \frac{2\pi a^2}{Z_0} \ln\left(\frac{b}{a}\right) = \vec{H}_{\mathrm{max}}^2 2\pi a^2 Z_0 \ln\left(\frac{b}{a}\right). \tag{7.31}$$

The transmission of electromagnetic power in waveguides comes at a price—the dissipation of power in the metallic walls of the guides.[2] As discussed in Section 7.2, this is a case corresponding to strong damping in ohmic media, and we may use the asymptotic analysis developed for this limit in calculating the power dissipation. The power lost in the walls is, as stated above, connected to the field (and power) values inside of the guide by noting that the surface current density is simply related to the tangential magnetic field just outside of the metal by $K_s = H_{||}$. Equation (7.11) for the dissipated power density can be then rewritten as

$$\frac{dP}{dA} = \frac{H_{||}^2}{2} R_s = \frac{H_{||}^2}{4} \sqrt{\frac{\omega \mu_0}{2\sigma_c}}. \tag{7.32}$$

[2]Power may also be dissipated in the media found in the interconductor region through the finite conductivity of the dielectric. This effect may be important in dielectric-loaded coaxial cables, but these are seldom used for high power transmission. High power, for driving accelerator cavities, is usually propagated in gas (e.g. SF_6) or evacuated TE or TM wave-guides, and for these cases the power loss in the region inside the guide is usually much smaller than that occurring in the walls.

Thus, for the case of the coaxial TEM waveguide, integrating over the azimuth of each cylindrical conducting surface, we obtain the power loss per unit length,

$$\frac{d}{dz} P(z) = -\frac{R_s}{2} \left[\oint H_\phi^2 \, d\phi \Big|_{\rho=a} + \oint H_\phi^2 \, d\phi \Big|_{\rho=b} \right] = -\pi a R_s H_{\mathrm{max}}^2 \left[1 + \frac{a}{b} \right]$$

$$= -\left[\frac{R_s}{Z_0} \cdot \frac{1 + (a/b)}{2a \ln(b/a)} \right] P(z)$$

$$\equiv -\eta_w P(z), \tag{7.33}$$

which we have written as a differential equation in z, the distance down the guide. Equation (7.33) has the familiar solution

$$P(z) = P(0) \exp(-\eta_w z). \tag{7.34}$$

The engineering units of the **attenuation constant** η_w are given the name **nepers per meter**. Note that the attenuation constant is frequency dependent, and higher frequency components (obtained by Fourier analysis of the time-dependence of the original waveform) of a propagating pulse in this system are more strongly attenuated than low frequency components. Thus, even though the TEM coaxial system is dispersionless, a waveform may distort during propagation due to frequency-dependent differential attenuation.

While the discussion of the TEM coaxial transmission line is very important for understanding general laboratory techniques, we have made it our first example here mainly to introduce the salient characteristics of waveguides. In the context of accelerators, however, TE and TM modes in waveguides are used to transport electromagnetic power and convert it to a useful configuration of acceleration fields. As before, we must decide on an initial geometry to provide a concrete, useful example of such waveguides. By far, the most common such waveguide geometry is the **rectangular guide**, shown in Fig. (7.2). Note that we have, in order to conserve symbols, recycled the letters a and b to indicate the width and height of the waveguide, respectively.

Because it is used to transport high powers, the interior of the rectangular waveguide is usually evacuated or, perhaps gas-loaded (e.g. using the high dielectric breakdown strength gas SF_6), and we can usually make the approximation $\varepsilon \cong \varepsilon_0$. In order to proceed with solving for the fields in the guide, we must first discuss boundary conditions. In the case of TM waves, where all fields can be derived by first solving for $\tilde{E}_z$, the boundary condition on the field of interest is obvious—$\tilde{E}_z = 0$ at the metallic surfaces. For TE waves, we solve initially for $\tilde{H}_z$, and the relevant boundary condition requires some thought. From Eq. (7.26), we have that

$$\tilde{E}_\perp = -\frac{Z_0}{ik_z}\vec{\nabla} \times \tilde{H}_z, \quad \text{or} \quad \tilde{E}_{x(y)} = \mp \frac{Z_0}{ik_z}\frac{\partial \tilde{H}_z}{\partial y(x)}, \qquad (7.35)$$

and for the tangential components of the electric field to vanish, the derivative with respect of surface-normal $\hat{n}$ of the H-field must also do so, $\partial \tilde{H}_z/\partial n = 0$.

Beginning with the TE case, we find by separation of variables that the solutions for the longitudinal magnetic field are of the form

$$\tilde{H}_z = H_0 \cos(k_x x)\cos(k_y y), \qquad (7.36)$$

where the boundary conditions are satisfied if

$$k_x = \frac{m\pi}{a}, \quad k_y = \frac{n\pi}{b}, \quad m, n = 0, 1, 2, 3, \ldots. \text{ not both} = 0. \quad (7.37)$$

The integers m and n are positive because negative integers are redundant, and only one integer can be equal to zero in a non-trivial solution. The different solutions, or **modes**, corresponding to different choice of integers, are indicated by subscripting the integers, writing TE$_{mn}$.

Using the transverse Helmholtz equation (Eq. 7.23) we can now relate the cutoff wave number to the transverse variation of $\tilde{H}_z$,

$$k_c^2 = k_x^2 + k_y^2 = \left(\frac{m\pi}{a}\right)^2 + \left(\frac{n\pi}{b}\right)^2. \qquad (7.38)$$

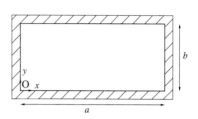

Fig. 7.2 Geometry of rectangular waveguide cutaway with length (in x) a and height (in y) b. The origin of the transverse coordinates is placed in the lower left-hand corner of the guide.

In order for a mode to propagate, its longitudinal wave number $k_z = \sqrt{k_0^2 - k_c^2}$ must be real, so we have the condition

$$k_0 = \frac{\omega}{c} > \sqrt{\left(\frac{m\pi}{a}\right)^2 + \left(\frac{n\pi}{b}\right)^2}. \tag{7.39}$$

Frequencies that are too low to satisfy this inequality do not propagate, but evanesce, decaying in characteristic length $\|k_z^{-1}\|$. The lowest frequency TE wave which can be propagated in a waveguide is, assuming $a > b$, the TE_{10} mode, where $k_0 > \pi/a$. This is the only possible propagating TE mode until the frequency satisfies $k_0 > \pi/b$, when the wave can also propagate in the TE_{01} mode. A more common way of describing the propagation condition uses frequency instead of wave number—the frequency ω of a propagating wave must be above the **cutoff frequency** $\omega_c = k_c c$

The analysis of TM modes proceeds in an analogous manner, using separation of variables and applying of the relevant boundary condition to obtain the solution for the longitudinal electric field,

$$\tilde{E}_z = E_0 \sin(k_x x) \sin(k_y y), \tag{7.40}$$

where the transverse wave-numbers are given by

$$k_x = \frac{m\pi}{a}, \quad k_y = \frac{n\pi}{b}, \quad m, n = 1, 2, 3, \dots. \tag{7.41}$$

In TM modes, there must be variation in both x and y, and neither m nor n can be zero. Therefore the lowest cutoff frequency (given also by Eq. 7.38) wave of this type is the TE_{11} mode, with $k_c = \sqrt{(\pi/a)^2 + (\pi/b)^2}$. Thus, we see that, for a band of frequencies given by

$$\frac{\pi}{a}c < \omega < \frac{\pi}{b}c \tag{7.42}$$

only the TE_{10} mode, and no TM modes, may propagate. In this condition, the wave/waveguide system is referred to a **single mode**, and the propagation characteristics of the wave are uniquely determined by the properties of the TE_{10} mode.

We now examine some properties of the TE and TM modes. The dispersion relation for the TE and TM modes is of the familiar form

$$\omega^2 = (k_z^2 + k_c^2)c^2, \tag{7.43}$$

which is identical to the plasma dispersion relation, with the substitution $k_c \Leftrightarrow k_p$. Thus the phase velocity is written

$$v_\phi = \frac{\omega}{k_z} = \frac{c}{\sqrt{1 - (\omega_c/\omega)^2}}, \tag{7.44}$$

and the group velocity is

$$v_g = \frac{d\omega}{dk_z} = c\sqrt{1 - (\omega_c/\omega)^2}. \tag{7.45}$$

While the group velocity of a propagating mode is strictly less than the speed of light, the phase velocity exceeds c. This implies that the wavelength of the wave in the guide λ_g is larger than the free-space wavelength λ_0,

$$\lambda_g = \frac{2\pi}{k_z} = \frac{\lambda_0}{\sqrt{1 - (\omega_c/\omega)^2}}. \tag{7.46}$$

It should be noted from Eqs (7.45) and (7.46) that, as the frequency approaches cutoff from above, the group velocity tends to zero and the guide wavelength becomes infinite.

In rf accelerators, we have seen that the electromagnetic field in the accelerating device is monochromatic and, therefore, the strong dispersion associated with rectangular waveguide modes which are used to fill the structures is not of concern. In fact, since the rectangular guide is a simple structure (both to understand and to build), it is ideal for handling the transmission of electromagnetic power from an rf source to the accelerator. In fact, single-mode TE_{10} operation in rectangular guide is most common method of such transmission. The instantaneous power transmitted in a TE_{10} mode is related to the transverse electric

$$\tilde{E}_\perp = i\frac{k_0 a}{\pi} Z_0 H_0 \sin\left(\frac{\pi x}{a}\right) \hat{y}, \tag{7.47}$$

and magnetic

$$\tilde{H}_\perp = iH_0 \frac{a k_0^2}{\pi k_z} \sin\left(\frac{\pi x}{a}\right) \hat{x}, \tag{7.48}$$

fields, by integrating the Poynting vector over the waveguide cross-section and then **averaging** in time. This exercise gives

$$\langle P \rangle = \frac{(k_0 a)^3 b}{(2\pi)^2 k_z} Z_0 H_0^2, \tag{7.49}$$

where we have written the result in terms of H_0 in order to connect the transmitted power to that dissipated in the surface of the guide. The average (in time) value of the dissipated power per unit length is obtained, by analogy, with Eqs (7.32) and (7.33),

$$\frac{d}{dz} P(z) = -R_s H_0^2 \left[\frac{a}{2} + b + \frac{(k_0 a)^2}{2\pi k_z} \right] \equiv -\eta_w P(z) \tag{7.50}$$

where we now have the attenuation constant,

$$\eta_w = \frac{R_s}{Z_0} \left[\frac{a}{2} + b + \frac{(k_0 a)^2}{2\pi k_z} \right] \frac{(2\pi)^2 k_z}{(k_0 a)^3 b}. \tag{7.51}$$

Note that, as the height of the guide b tends to zero, the attenuation is severe, because the transmitted power vanishes while the power dissipation (area over which the magnetic field is non-zero) remains finite in this limit. A similar phenomenon is seen when the inner and outer conductors are forced together in the coaxial TEM guide (cf. Eq. 7.33).

7.4   Resonant cavities

As we have seen in Chapter 4, electromagnetic fields for acceleration are supported by metallic boundaries that are termed **resonant cavities**. Such systems

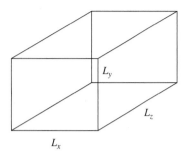

Fig. 7.3 Geometry of metallic boundary of a rectangular resonant electromagnetic cavity with length (in z) L_z, width (in x) L_x, and height (in y) L_y.

are termed resonant because the electromagnetic power is contained by the boundaries in such a way that: (i) many round-trips (at the speed of light) of a wave are allowed before the power is either dissipated in the walls or propagated out of the cavity; and (ii) the waves constructively interfere upon each round-trip. Furthermore, these cavity systems are divided into two categories standing and traveling wave devices. Perhaps the simplest way to proceed with the discussion of resonant cavities is to concentrate on the standing wave case beginning with the model problem of a single, isolated cavity or cell, which we do in this section. We will also extend the single-cell model to examine aspect of the role of the spatial harmonic description of electromagnetic waves in periodic systems. Then, in order to understand the temporal and frequency response of cavities, as well as cell-to-cell coupling in multiple-cell standing wave or traveling wave systems, we will first introduce equivalent circuit models for such electromagnetic structures in Section 7.5. At that point we will be able to draw strong connections between the standing wave and traveling wave systems.

To connect the discussion of cavities to the previous section on waveguides we introduce a simple example of a cell—the rectangular box, as shown in Fig. 7.3. The modes in this cavity can be thought of as derived from traveling wave solutions to a waveguide, having a rectangular cross-section of dimensions L_x and L_y, but with additional constraints on the allowed frequency due to the boundary condition imposed in z. These constraints give only discrete mode (**eigenmode**) frequencies. For a TM wave, which has a component of the electric field along the z direction and is therefore a candidate for a cavity mode that accelerates charged particles, the boundary condition due to the conducting walls normal to the z-direction requires that an integral number of half guide-wavelengths exist in the space of length L_z. Thus, the allowed frequency eigenvalues for TM cavity modes are given by

$$\omega_{lmn} = c\sqrt{\left(\frac{l\pi}{L_x}\right)^2 + \left(\frac{m\pi}{L_y}\right)^2 + \left(\frac{n\pi}{L_z}\right)^2}, \quad l, m = 1, 2, 3, \ldots;$$

$$n = 0, 1, 2, \ldots \text{ (TM modes)}.$$
$$(7.52)$$

Similar arguments can be made for the TE case with the result that

$$\omega_{lmn} = c\sqrt{\left(\frac{l\pi}{L_z}\right)^2 + \left(\frac{m\pi}{L_x}\right)^2 + \left(\frac{n\pi}{L_y}\right)^2},$$

$$l, m, n = 0, 1, 2, 3, \ldots \text{ (TE modes)},$$
$$(7.53)$$

where only one integer l, m, or n may vanish at a time.

The TM modes are obviously of primary importance to the subject of acceleration since TE modes have a vanishing longitudinal electric field E_z. The field E_z near the axis must be supported in these cavities by the entire TM mode, which contains a certain value of the electromagnetic energy. For example, with rectangular cavities in vacuum having only longitudinal electric fields ($n = 0$) and lowest transverse dependence ($l = m = 1$, the fundamental

TM mode), the time-averaged stored electromagnetic energy is given by

$$U_{EM} = \left\langle \iiint u_{EM}\, dV \right\rangle = \left\langle \iiint \frac{1}{2} \left(\varepsilon_0 \vec{E}^2 + \mu_0 \vec{H}^2 \right) dV \right\rangle$$

$$= \varepsilon_0 E_0^2 \frac{L_x L_y L_z}{16} \left[1 - \left(\frac{L_x}{L_y} + \frac{L_y}{L_x} \right)^{-2} \right] \tag{7.54}$$

where E_0 is the on-axis ($x = y = 0$) longitudinal electric field. In Eq. (7.54), we have explicitly used the cavity dimensions to parameterize the wave-numbers and frequency in order to emphasize the geometric dependences of the problem.

In order to obtain a picture of how efficient this cavity is at acceleration, the stored energy should be compared to the energy gain of a charged particle as it traverses the cavity in the z-direction. Assuming, as in Chapter 4 analyses, that the longitudinal velocity v of the particle is constant during this traversal, the maximum energy gain in the cavity is given by

$$\Delta U_{max} = qE_0 \int_{-L_z/2}^{L_z/2} \exp(i\omega t)\, dz = qvE_0 \int_{-L_z/2v}^{L_z/2v} \exp(i\omega t)\, dt$$

$$= 2qvE_0 \sin\left(\frac{\omega L_z}{2v} \right). \tag{7.55}$$

It is traditional to define, in the context of acceleration by a single cell cavity such as this, the transit time factor,

$$T_a \equiv \frac{\Delta U_{max}}{q \int E_z|_{t=\text{const}}\, dz} = \frac{\Delta U_{max}}{qE_0 L_z}, \tag{7.56}$$

which is the maximum energy gain normalized to the cavity "voltage",

$$V_c \equiv \int E_z|_{t=\text{const}}\, dz, \tag{7.57}$$

that has the simple value $E_0 L_z$ in the present example. For this case, we have

$$T_a = \frac{2v}{\omega L_z} \sin\left(\frac{\omega L_z}{2v} \right) = \sin c \left(\frac{\omega L_z}{2v} \right). \tag{7.58}$$

The transit time factor is strictly less than unity, as is illustrated by Eq. (7.58), and in fact vanishes for $\omega L_z/v = 2\pi$. In fact, for cavity lengths $\omega L_z/v > \pi$, one unavoidably has decelerating regions in the cavity and the maximum cavity length is chosen, in practice, to be $L_z \leq \pi v/\omega$ (a maximum rf phase advance of π during traversal). Note also that the definition of the rf voltage given in Section 4.7 in the context of "short" (no spatial field reversal) rf cavities can be written as $V_0 = T_a V_c$.

The maximum energy which can be given to a charged particle by acceleration in the cavity is, therefore, quoted as a fraction of the stored electromagnetic energy (for the minimum stored energy case $L_x = L_y$; cf. Ex. 7.11)

$$\frac{\Delta U_{max}}{U_{EM}} = \frac{16q\omega^2 T_a}{3\pi^2 c^2 \varepsilon_0 E_0}. \tag{7.59}$$

This energy gain arises at the expense of the stored energy, and Eq. (7.59) shows how much electromagnetic energy is extracted by the acceleration of beam particles. Equation (7.59) illustrates several phenomena related to this energy extraction. First, if we interpret the charge q as the total beam charge then we note that large beam charges may extract significant energy from the cavity. This effect is known as **beam loading**. Also, we can see that for a large accelerating field, the extracted energy is linear in E_0, while the stored energy is proportional to E_0^2. Thus, beam loading effects are mitigated for large E_0. Finally, we can see that for large frequency ω, the fractional extracted energy is large, because the volume of the cavity is small, and therefore so is the stored energy.

The existence of stored electromagnetic energy also implies that there is, as in the case of a waveguide, power loss to the walls of the cavity. For the present case, the power loss is given by

$$\langle P \rangle = \frac{1}{2} R_s \left\langle \iint H_\parallel^2 \, da \right\rangle$$

$$= \frac{1}{2} R_s (c\varepsilon_0 E_0)^2 \left[\frac{1}{2} L_x L_y + L_z (L_x + L_y) \right] \left[1 - \left(\frac{L_x}{L_y} + \frac{L_y}{L_x} \right)^{-2} \right] \quad (7.60)$$

The **quality factor Q** associated with resonant systems can be defined from Eqs (7.54) and (7.60) in the following way:

$$Q \equiv \frac{\omega U_{\text{EM}}}{\langle P \rangle} = \left[\frac{16\varepsilon_0 R_s c^2}{\omega} \left(\frac{1}{2L_z} + \frac{1}{L_x} + \frac{1}{L_y} \right) \right]^{-1}. \quad (7.61)$$

For the minimum stored energy case introduced in Exercise 7.11, Eq. (7.61) can be written as

$$Q = \left[\frac{16\varepsilon_0 R_s c}{\sqrt{2}\pi} \left(\frac{L_x}{2L_z} + 2 \right) \right]^{-1}. \quad (7.62)$$

Equation (7.62) indicates that the quality factor of a cavity does not have an explicit frequency dependence derived from the geometry of the cavity. The geometric factor in Eq. (7.62) can be eliminated by assuming a fixed relationship for L_x/L_z, for example one appropriate for maximum acceleration of a speed-of-light particle, $L_x = \sqrt{2}L_z$. Thus, the only scaling of Q with frequency arises from that embedded in the surface resistivity of the metal,

$$Q \propto R_s^{-1} \propto \omega^{-1/2}, \quad (7.63)$$

and Q decreases with increasing frequency as the inverse square-root of the frequency.

As illustrated by the example in Exercise 7.12, the Q is much greater than unity in practical rf cavities, taking values in the range of several thousand to tens of thousands. The use of a resonant cavity system to produce large electric fields for acceleration relies on such large Q—for a given instantaneous power the associated electric field is larger in a resonant cavity than in a waveguide by roughly a factor of $\sqrt{Q}$. This remarkable advantage is accompanied by a differing picture of the power flow in these two systems. In a waveguide, the power flows with only small losses along the propagation axis of the guide. In the resonant cavity, the power flows essentially only into the walls. This power must come from somewhere and, in fact, it is derived from a waveguide that is **externally coupled** to the cavity.

In practice, rf cavities for acceleration are not manufactured in rectangular cross-section geometry, but in circular cross-section,[3] shown in Fig. 7.4, which displays a "pill-box" shape. In order to see why this is so, we now analyze the mode structure in such a geometry. We begin by writing Eq. (7.1) for the longitudinal electric field in cylindrical coordinates, assuming axisymmetry

$$\left[\frac{1}{\rho}\frac{\partial}{\partial\rho}\left(\rho\frac{\partial}{\partial\rho}\right) + \frac{\partial^2}{\partial z^2} - \frac{1}{c^2}\frac{\partial^2}{\partial t^2}\right]E_z = 0. \qquad (7.64)$$

This equation is analyzed in the usual fashion, by separation of variables, where we write $E_z(\rho,z,t) = E_0\tilde{R}(\rho)Z(z)T(t)$, and the normalization of the functions is chosen so that E_0 is the maximum electric field. The substitutions of functional forms $\exp[i\omega t]$ in time, and $\cos[k_{z,n}z]$, with $k_{z,l} = n\pi z/L_z, n = 0, 1, 2 \ldots$ (satisfying the boundary conditions at $z = 0$ and $z = L_z$), reduce Eq. (7.64) to the radial equation

$$\left[\frac{1}{\rho}\frac{\partial}{\partial\rho}\left(\rho\frac{\partial}{\partial\rho}\right) - k_{z,n}^2 + \frac{\omega^2}{c^2}\right]\tilde{R} = 0. \qquad (7.65)$$

For $k_{z,n} < \omega/c$, Eq. (7.65) is the Bessel equation, while for $k_{z,n} > \omega/c$, it has the form of a modified Bessel equation, the solutions to which were encountered in Eq. (4.67). In the present case, we are interested in cavities where $n = 0$ and the solution is of standard Bessel form,

$$\tilde{R}(\rho) = \sum_{l=0}^{\infty} A_l J_0(k_{\rho,l}\rho). \qquad (7.66)$$

Each mode in the cylindrical cavity must, in order to satisfy the boundary condition at the outer wall, vanish at $\rho = R_c$. This condition, that successive zeroes of the Bessel function J_0 occur at $\rho = R_c$, determines the radial wavenumber, $k_{\rho,l}$. In the case of the fundamental mode, the first zero of J_0 occurs at approximately $k_{\rho,0} \cong 2.405/R_c$. Note that the dispersion relation implicit in Eq. (7.65) is $\omega_{l,n}^2 = [k_{\rho,l}^2 + k_{z,n}^2]c^2$, and so the fundamental resonant frequency is

$$\omega_{0,1} \cong \frac{2.405c}{R_c}. \qquad (7.67)$$

As with the rectangular cavity, the most efficient mode for acceleration is, in fact, the fundamental mode. In this case, the radial dependence of the azimuthal magnetic field is given by

$$H_\phi(\rho) = \frac{\omega\varepsilon_0}{k_{\rho,0}}E_0 J_1(k_{\rho,0}\rho) = c\varepsilon_0 E_0 J_1(k_{\rho,0}\rho). \qquad (7.68)$$

The stored electromagnetic energy in the fundamental mode is thus

$$U_{EM} = \frac{1}{4}\varepsilon_0 L_z E_0^2 \int_0^{R_c} \left[J_0^2(k_{\rho,0}\rho) + J_1^2(k_{\rho,0}\rho)\right]\rho\,d\rho$$

$$= \tfrac{1}{2}\varepsilon_0 L_z E_0^2 R_c^2 J_1^2(k_{\rho,0}R_c), \qquad (7.69)$$

[3]The choice of rectangular, as opposed to circular, cross-section waveguide is driven by the need to control the electric polarization of the TE mode that is propagated. In TM accelerating cavities, the electric polarization is longitudinal, and the choice of transverse cross-section does not affect it.

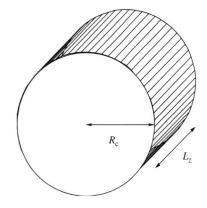

Fig. 7.4 Geometry of metallic boundary of a cylindrical ("pill-box") resonant electromagnetic cavity, with length (in z) L_z, and radius R_c.

and the average power dissipated in the cavity walls is

$$\langle P \rangle = \pi R_s (c\varepsilon_0 E_0)^2 \left[R_c L_z J_1^2(k_{\rho,0} R_c) + 2 \int_0^{R_c} J_1^2(k_{\rho,0}\rho)\rho \, d\rho \right] \tag{7.70}$$

$$= \pi R_s (c\varepsilon_0 E_0)^2 R_c J_1^2(k_{\rho,0} R_c)[L_z + R_c].$$

Combining Eqs (7.69) and (7.70), we have

$$Q \equiv \frac{\omega U_{\text{EM}}}{\langle P \rangle} = \frac{Z_0}{2R_s} \frac{2.405}{(R_c + L_z)} \tag{7.71}$$

for the quality factor of the fundamental mode.

It can be seen in the example of cylindrical symmetry that both the stored energy and the power loss for a given length of cavity are smaller than in the square cross-section case. This is fundamentally due to the fact that, while keeping the resonant frequency constant, a circular cross-section minimizes both the surface area of the cavity and its enclosed volume. Because of these attributes, essentially all accelerator cavities in use nominally cylindrically symmetric designs. This design imperative is derived from the high premium one must pay to obtain sufficient power from sources in the rf wavelength range. There is now much interest in using much shorter wavelength electromagnetic sources—lasers—to power accelerators. As the power available from such sources is relatively much higher, the resonant cavity geometry in an **optical acceler ator** may be chosen to have a slab-symmetric (Cartesian) symmetry, where one transverse dimension is very large compared to the other.

While the discussion of resonant electromagnetic cavities given above is sensible and useful from the viewpoint of frequency, energy and power analyses, the reader may at this point have significant worries about its application to acceleration. From the start, an eminently practical issue presents itself—the propagation of charged particles unimpeded by metal walls, which are clearly in place in Figs 7.3 and 7.4. The solution to this problem is obvious (and originally shown in Fig. 4.1): one must cut holes in the pill-box cavity near the axis to allow beam passage without material scattering. It turns out that the introduction of such holes, illustrated quantitatively in Fig. 7.5, also solves other issues concerning the form of the electromagnetic fields.

If one reviews the discussion of acceleration in periodic structures in Section 4.8, several apparent inconsistencies with respect to our electromagnetic analysis given above spring to mind. First, E_z is expanded in terms of the modified Bessel function I_0 in the acceleration analysis, while we have used the Bessel function J_0 in discussion of the pill-box cavity. Examination of Eq. (4.67) points out why this is so. In short, most of the spatial harmonics making up the expansion in Eq. (4.67) have phase velocity less than c because $k_{z,n}$ is large. In our single-cavity analysis, however, we have assumed no longitudinal variation of E_z, or equivalently $k_z = 0$. This is allowed only because we have assumed conducting boundary conditions at the upstream and downstream cavity walls. Thus, the second obvious conflict between this discussion and that of Section 4.8, arises—the total acceleration of an ultra-relativistic particle has $J_0(k_{\rho,0}\rho)$ dependence when these conducting walls are included, whereas our previous analysis concluded that such acceleration due to axisymmetric TM fields was nearly independent of ρ.

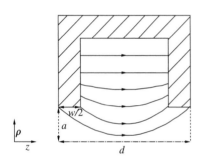

Fig. 7.5 Section of rotation of cylindrical electromagnetic cavity (hatching indicates metallic region), assumed as one period of a repetitive structure in z. Representative electric field lines are displayed. This structure can be described as a disk-loaded cylindrical waveguide.

Mathematically, the shift of using J_0 to I_0, which are simply related by the factor of i in their argument, can be motivated by anticipation of whether the spatial harmonics have $k_{\rho,n}^2$ that are positive or negative, respectively. In the second case, longitudinal phase velocities are physical, that is, less than c. The use of a series in I_0 to represent E_z also points to a physical difference in the two cases. The functional form of J_0 is appropriate for matching the outer wall boundary conditions, where E_z must vanish. On the other hand, the irises (holes in the disks) shown in Fig. 7.5 cause the field lines to be attracted, and E_z must be allowed to grow as a function of ρ in the region interior to the iris radius a. Thus, we see that, from the viewpoint of calculating the general rf characteristics of the cavity over the majority of its volume ($\rho > a$), the J_0 representation is convenient, but for the near axis fields, we should employ the series in I_0 given by Eq. (4.67).

Since we are ultimately concerned more with coupled-cell systems (e.g. Fig. 4.1) than isolated cavities, the introduction of a periodic geometry to explain the most relevant effects of holes on the fields does not cause much loss of generality. The qualitative discussion above can indeed also aid us in quantitative estimation of the effects of disk and iris shape on spatial harmonic content of the near-axis fields. To show this, consider the geometry shown in Fig. 7.5, which we take, for the sake of example, to be a period of a π-mode structure (this is in fact implied by the form of the field lines shown in the figure). In order to begin the calculation, we estimate the z-dependence of E_z on the line $\rho = a$. It clearly vanishes for the regions inside of the irises, and we then make the approximation that it is constant elsewhere,

$$\tilde{E}_z(z)|_{\rho=a} = \begin{cases} 0, & 0 < z < w/2, d - w/2 < z < d \\ E_0, & w/2 \leq z \leq d - w/2. \end{cases} \tag{7.72}$$

We can evaluate the spatial harmonic content of this field by Fourier analyzing Eq. (7.72) in z and equating equivalent terms in Eq. (4.67) to give

$$a_n \cong \frac{2(-1)^{(n-1)/2}}{n\pi I_0[k_{\rho,n}a]} \cos\left(\frac{n\pi w}{d}\right), \quad n \text{ odd}, \tag{7.73}$$

where $k_{z,n} = [2\pi(n + 1)]/d$, and $k_{\rho,n} = \sqrt{k_{n,z}^2 - (\omega/c)^2}$, and the even n coefficients vanish. Equation (7.73) indicates a general phenomenon—that the Fourier–Bessel amplitudes of the spatial harmonics decrease exponentially (the asymptotic form of I_0) in the limit of large n. This is due to large radial wave number of these harmonics and resulting poor coupling of the on-axis field to the boundary, which is located at large argument of the modified Bessel function, $k_{\rho,n}a \gg 1$.

7.5 Equivalent circuits for waveguides and cavities

That waveguides and cavities can be profitably described in terms of equivalent circuits is not immediately obvious to a student of physics, since circuits are composed of discrete ("lumped") elements with single characteristics such as resistance or inductance, while waveguides and cavities are clearly exten ded ("distributed circuit") electromagnetic objects with spatially varying fields The advantage in this conceptual change of emphasis is quite large, as we can examine these distributed systems using "short-hand" ideas such as imped ances, as well as notions concerning resonant circuits, which can be analyzed as simple damped oscillators. In order to motivate the assignment of circuit like parameters to devices containing electromagnetic waves, we first return to the simple illustrative case of TEM modes in a coaxial waveguide. We refer to the dielectric-loaded coaxial transmission line shown in Fig. 7.1, with the dielectric material in the region $a < r < b$ having permittivity $\varepsilon = \kappa_e \varepsilon_0$ and permeability μ_0, We have seen from Section 7.3 that the fields in such a system behave, as far as dependence on transverse coordinates, as if they were derived from static linear charge (electric) and current (magnetic) densities. Because of this, we may easily define two circuit parameters: the **capacitance per unit length**

$$C' \equiv \frac{\lambda}{V} = \frac{2\pi\varepsilon}{\ln(b/a)}, \tag{7.74}$$

and **inductance per unit length**

$$L' \equiv \frac{\Phi'}{I} = \frac{\mu_0}{2\pi} \ln(b/a), \tag{7.75}$$

where λ, V, I, and Φ', are the charge per unit length, voltage, current, and magnetic flux per unit length, respectively. An equivalent circuit schematic for this transmission line is shown in Fig. 7.6.

The **transmission line impedance** of this system can be defined in a natural way as the ratio of the instantaneous line voltage to the current

$$Z \equiv \frac{V}{I} = \frac{L'}{C'} \frac{\lambda}{\Phi'}. \tag{7.76}$$

Because the line charge and the current are trivially related in the TEM system

$$I = \lambda v_\phi = \frac{\lambda}{\sqrt{\mu_0\varepsilon}}, \tag{7.77}$$

the magnetic flux per unit length is simply

$$\Phi' = \frac{\mu_0 I}{2\pi} \ln(b/a) = \lambda \frac{Z_0}{2\pi} \ln(b/a). \tag{7.78}$$

Equation (7.76) can be thus be written as

$$Z = \frac{Z_0}{2\pi} \ln(b/a). \tag{7.79}$$

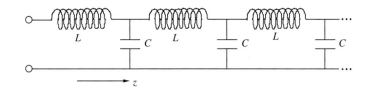

Fig. 7.6 Equivalent circuit of a transmission line such as that shown in Fig. 7.1, with capacitance and inductance per unit length defined as in Eqs (7.74) and (7.75).

The arguments leading to Eq. (7.79) are straightforward and unambiguous, because TEM line analyses are mathematically indistinguishable from simple electrostatics and magnetostatics. For other waveguide modes, these arguments need to be generalized by defining the line "voltage" V in a manner similar to that given in Eq. (7.57), and the current in a like fashion, as the maximum current passing a plane in z. By calculating the total charge per unit length on the waveguide walls and the maximum flux per unit length in the same fashion, the circuit parameters C' and L' can be defined in analogy to Eqs (7.74) and (7.75). For arbitrary transmission lines, we can use these parameters to generalize Eq. (7.79) to read

$$Z = \sqrt{\frac{L'}{C'}}. \tag{7.80}$$

The analysis of wave propagation in guides (and cavities, discussed below) in terms of equivalent circuit parameters is quite general and powerful, and is needed to understand phenomena such as impedance mismatch at discontinuities. In these cases, the concept of the impedance becomes not just a parameter determined locally, but is a complex (phasor) description of the wave propagation and reflection characteristics of the system. A reasonable development of such a description would easily double the length of this chapter, however. Since there are numerous existing expositions of the complex impedance of microwave circuits (defined roughly as circuits in which the size of the system is smaller than the wavelength of interest) available in the accelerator and electrical engineering literature, we forego a more detailed discussion in favor of referring the reader to the existing literature, and proceed in a less rigorous, heuristic manner.

The same considerations exist in identifying the equivalent circuit parameters that can be used in describing resonant cavities as we have encountered in waveguides. In the cavity case, however, it is much more important to include a resistive component to the model, as illustrated in Fig. 7.7(b). In this picture, the resonant cavity (the pill-box in Fig. 7.7(a)) is represented by a resonant RLC circuit. The other components of the circuit diagram in Fig. 7.7(b) correspond to the beam holes which allow the beam current (the current source) to exchange power with the cavity, and a coupling needed for external power input (the magnetic coupling). Note that in cavities we are now dealing with lumped circuit elements, which characterize the total value of R, L, and C of the cavity, rather than the distributed element description of waveguides.

We begin our equivalent circuit-based analysis by first concentrating on the RLC cavity circuit itself. A governing equation can be written for the current I_c in this cavity,

$$\frac{d^2 I_c}{dt^2} + \frac{R}{L}\frac{dI_c}{dt} + \frac{I_c}{LC} = 0, \tag{7.81}$$

which, in more standard form, is cast as

$$\frac{d^2 I_c}{dt^2} + \nu \frac{dI_c}{dt} + \omega_0^2 I_c = 0, \tag{7.82}$$

where we have defined $\omega_0 = (LC)^{-1/2}$ and $\nu \equiv R/L$.

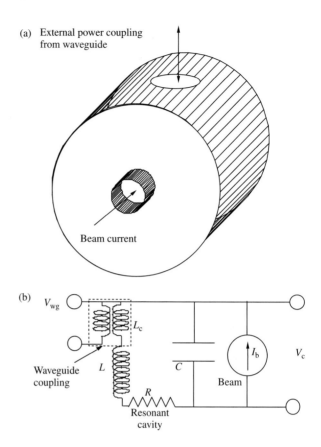

(a)   External power coupling
       from waveguide

(b)

V_{wg}

L_c

Waveguide
coupling

L

C

I_b

V_c

Beam

R

Resonant
cavity

Beam current

Fig. 7.7 (a) Physical picture of a resonant accelerator cavity, with external coupling
to the waveguide and to the beam current.
(b) Equivalent circuit of the resonant cavity system shown in (a), including external
coupling.

Equation (7.82) is describes a damped oscillator, with solutions of the form

$$I_c = \exp\left[-\frac{vt}{2}\right]\left\{A\cos\left[\sqrt{\omega_0^2 - \left(\frac{v}{2}\right)^2}\,t\right] + B\sin\left[\sqrt{\omega_0^2 - \left(\frac{v}{2}\right)^2}\,t\right]\right\}.$$

(7.83)

Assuming weak damping, $\omega_0 \gg v/2$, and an initial applied current I_{c0},
Eq. (7.83) can be approximated as

$$I_c(t) \cong I_{c0}\exp\left[-\frac{vt}{2}\right]\cos[\omega_0 t].$$

(7.84)

Since the stored energy in the *RLC* oscillator can be written in circuit terms as

$$U_c = L\|I_c\|^2 \propto \exp[-vt]\cos^2[\omega_0 t],$$

(7.85)

which secularly decays as $\exp[-vt]$. The average power lost due to this decay is

$$P_c = -\frac{d\langle U_c\rangle}{dt} = v\langle U_c\rangle,$$

(7.86)

and the quality factor can be written as

$$Q_0 = \frac{\omega_0\langle U_c\rangle}{P_c} = \frac{\omega_0}{v} = \frac{\omega_0 L}{R}.$$

(7.87)

Using Eq. (7.87), we may rewrite Eq. (7.84) as

$$I_c(t) \cong I_{c0} \exp\left[-\frac{\omega_0 t}{2Q_0}\right] \cos[\omega_0 t] \qquad (7.88)$$

and an averaged version of Eq. (7.85) as

$$U_c = U_{c0} \exp\left[-\frac{\omega_0 t}{Q_0}\right]. \qquad (7.89)$$

The current decays as $\exp[-\omega_0 t/2Q_0]$ while the stored energy decays as $\exp[-\omega_0 t/Q_0]$. These statements are quite general and apply to any weakly damped oscillator system.

A damped, undriven oscillator has limited practical interest since we would like to power these cavities to achieve an equilibrium voltage for accelerating particles. The cavity circuit shown in Fig. 7.7(b) allows for the RLC oscillator to be driven by an oscillating voltage (arising from rf power from a waveguide impinging on the external coupling iris shown in Fig. 7.7(a)). The cavity circuit equation becomes, upon inclusion of the external driving term,[4]

$$\frac{d^2 I_c}{dt^2} + \frac{R}{L}\frac{dI_c}{dt} + \frac{I_c}{LC} = \frac{1}{L}\frac{dV_{wg}}{dt}, \qquad (7.90)$$

or

$$\frac{d^2 I_c}{dt^2} + \frac{\omega_0}{Q_0}\frac{dI_c}{dt} + \omega_0^2 I_c = -\beta_c \frac{d^2 I_{wg}}{dt^2}. \qquad (7.91)$$

Here we have introduced the definition of the **coupling strength** $\beta_c \equiv L_c/L$. The analysis of this inhomogeneous equation representing a driven oscillator is usually begun in the frequency domain, where we assume a driving current of frequency ω, $I_{wg} = I_0 \exp(i\omega t)$, to give the response

$$I_c(\omega) = \frac{-\beta_c \omega^2 I_{wg}}{\omega^2 + i\omega(\omega_0/Q_0) - \omega_0^2} = \frac{-\omega^2 \beta_c I_{wg}}{\sqrt{(\omega^2 - \omega_0^2)^2 + (\omega\omega_0/Q_0)^2}} \exp(i\varphi_c),$$

$$= \frac{-Q_0 \beta_c I_{wg}}{\sqrt{Q_0^2(1 - (\omega_0/\omega)^2)^2 + (\omega_0/\omega)^2}} \exp(i\varphi_c), \qquad (7.92)$$

where the phase response of the cavity with respect to the driving current is given by

$$\tan(\varphi_c) = \frac{1}{Q_0}\left(\frac{\omega\omega_0}{\omega^2 - \omega_0^2}\right). \qquad (7.93)$$

In a damped, driven oscillator, the particular solution to this system, Eq. (7.92), is established after all of the transients (represented by the homogeneous solutions to Eq. (7.91)) have damped away. Note that the response is 90° out of phase with respect to the driving current at resonance ($\omega = \omega_0$). Also, the cavity current swings in phase from 0 to 180° with respect to the drive as the frequency is swept through resonance. In terms of amplitude, the cavity response at resonance ($\omega = \omega_0$) is Q_0 times larger than the driving current.

[4] We have used the approximation that the inductance of the coupling iris is much smaller than that of the bulk of the cavity, $L_c \ll L$. This implies that we do not include the possibility of power flow *out* of the coupling iris, and the only power loss in the cavity circuit arises from the wall resistance.

Equation (7.92) can be recast in **resonant approximation** by assuming $Q_0 \gg 1$, and expanding in the parameter $\Delta\omega \equiv \omega - \omega_0$

$$I_c(\Delta\omega) \cong \frac{\beta_c I_{wg}}{2(\Delta\omega/\omega_0) + (i/Q_0)} = \frac{\beta_c Q_0 I_{wg}}{\sqrt{(2Q_0(\Delta\omega/\omega_0))^2 + 1}} \exp(i\varphi_c). \quad (7.94)$$

The response of the cavity to an external driver is most often defined in terms of stored energy (or, equivalently, of power dissipated) in the cavity,

$$U_{EM}(\Delta\omega) \propto P(\Delta\omega) \propto \|I_c(\Delta\omega)\|^2 \propto \frac{Q_0\omega_0}{(2Q_0\Delta\omega)^2 + \omega_0^2}. \quad (7.95)$$

From Eq. (7.95), the width of the absorbed power spectrum at half maximum can be seen to be

$$\Delta\omega_{1/2} = \frac{\omega_0}{Q_0}. \quad (7.96)$$

This is an excellent operational way to define Q since the half-power points are an easy quantity to measure. Note that all of the ways in which Q has been defined thus far—stored energy-to-power ratio, decay time, and half-power width—are equivalent, because they are all based on the assumption of a weakly damped oscillator.

In this context, it is interesting to examine the difference between the cavity response given by Eq. (7.92) and the approximate resonant response given by Eq. (7.94). It can be seen from Fig. 7.8, which illustrates behavior of a resonator with $Q_0 = 5000$, that over the range where there is significant response ($\Delta\omega \approx \Delta\omega_{1/2}$), the two solutions are identical. In a system with many resonant modes (different values of ω_0), such as a cavity, it is operationally important that the resonant peaks corresponding to different resonances not overlap. This requirement may be stated as $|\omega_{0,j} - \omega_{0,i}| \gg \Delta\omega_{1/2,j} + \Delta\omega_{1/2,i}$ where the subscripts i and j indicate different, nearby resonances. The subject of **mode separation** requirements will be revisited in the following section.

In the time domain, one can solve Eq. (7.91) for the most commonly encountered case, where a constant power is turned on rather suddenly at $t = 0$

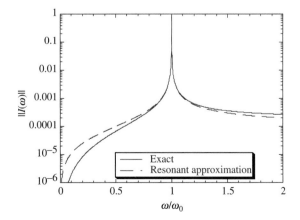

Fig. 7.8 Relative response of cavity system to current excitation as a function of ω, exact solution (Eq. (7.92)) and resonance response (Eq. (7.94)). The Q value is taken to be 5000 in this example.

o drive the cavity, $I_{wg} \propto \Theta(t - t_0)$. The full solution (homogeneous plus par-
cular, in correct combination to obey the initial conditions at $t = 0$) is, at
esonance,

$$I_c(t) \cong \frac{\beta_c Q I_{wg} \sin([\omega_0 t])}{\sqrt{2Q_0 \left(\frac{\Delta\omega}{\omega_0}\right)^2 + 1}} \left[1 - \exp\left(-\frac{\omega_0 t}{2Q_0}\right)\right]. \tag{7.97}$$

The current in the cavity asymptotically approaches a constant value, for times
much longer than the fill time $\tau_f \equiv 2Q_0/\omega_0$. Likewise, the stored energy in the
avity rises as

$$U_c(t) = U_{SS}\left[1 - \exp\left(-\frac{t}{\tau_f}\right)\right]^2, \tag{7.98}$$

where the subscript SS indicates a steady state value. When the exciting current
 turned off at $t = t_0$, rather than on, the current is

$$I_c(t) \cong I_c(t_0) \sin([\omega_0 t]) \exp\left(-\frac{(t - t_0)}{2\tau_f}\right), \tag{7.99}$$

nd the stored energy decays as

$$U_c(t) \cong U_c(t_0) \exp\left(-\frac{(t - t_0)}{\tau_f}\right). \tag{7.100}$$

Up until this point, because of the primary role the cavity current has in the
oupling model we are employing, we have emphasized the current over the
avity voltage. The two concepts are easily connected at resonance by

$$V_c \cong L\frac{dI_c}{dt} = i\omega_0 L I_c. \tag{7.101}$$

he reactive component of the cavity impedance Z of course vanishes at reson-
nce, as the voltage drops across the lumped inductor and capacitor are equal
nd opposite, and both are maximized when the current is changing most rap-
dly. When the cavity current itself is maximized, the impedance of the cavity
 purely resistive, and much smaller than that of either reactive elements. This
 of course the lumped cavity resistance

$$Z(\omega_0) = R = \frac{V_c}{I_c}. \tag{7.102}$$

he power loss associated with this real impedance is simply

$$P_c = \frac{\omega U_c}{Q_0} = \frac{L\|I_c\|^2}{\omega L/R} = \|I_c\|^2 R = \frac{\|V_c\|^2}{R} = \frac{V_c^2}{2R} = \frac{V_c^2}{Z_s}. \tag{7.103}$$

Here we have introduced the **shunt impedance** of the cavity, which can be seen
o be simply twice the lumped cavity resistance, $Z_s = 2R$. The shunt impedance
 related to the **beam impedance**, the ratio of the reactive voltage induced by
he passage of a beam to its current,[5]

$$Z_b(\omega_0) = \frac{\Delta V_c}{I(\omega_0)} = 2\sqrt{\frac{L}{C}}. \tag{7.104}$$

he beam impedance is a characterization of the reactive component of the
avity circuit and is employed to calculate the effects of beam loading. It is

[5] The current is that Fourier component $I(\omega)$ at the resonant frequency—the beam imped-
ance, like the shunt impedance, is a frequency domain concept. The reader may have noted that the TEM line impedance was not viewed in this way, but this is only because in that case the impedance is independent of frequency.

simply related to the resistive measure of the cavity, the shunt impedance, by $Z_b = Z_s/Q$. The factor of two found in Eq. (7.104) arises from consideration of the time-domain response of the system—most illustratively, a point charge only "feels" half of the field that it excites in the cavity mode. This is due to causality, since there is no induced field ahead of the point charge, so the factor of one-half can be considered as arising from averaging.

The circuit model of a cavity is, as we have seen, quite powerful and can even be extended to give a tool for predicting the effects of changing the cavity geometry. To see this, we note that the circuit model views the cavity essentially as an oscillator. As the action in an oscillator (see Section 1.3) is a constant, we may write the stored energy in the cavity as the product of a generalized action and the resonant frequency

$$U_{EM} = \iiint\limits_{V} u_{EM} \, dV = \iiint\limits_{V} \left[\tfrac{1}{2}\varepsilon_0 \vec{E}^2 + \tfrac{1}{2}\mu_0 \vec{H}^2\right] dV = J\omega_0. \quad (7.105)$$

Under small perturbations, the oscillator action is an invariant, and so we may write

$$J = \frac{U_{EM}}{\omega_0} = \text{constant}, \quad \text{so } \delta J = \frac{\delta U_{EM}}{U_{EM}} - \frac{\delta \omega_0}{\omega_0} = 0, \quad (7.106)$$

and

$$\frac{\delta \omega_0}{\omega_0} = \frac{\delta U_{EM}}{U_{EM}}. \quad (7.107)$$

Equation (7.107) is termed **Slater's perturbation theorem** and can be used to predict how geometry changes will affect the resonant frequency of a cavity or, conversely, to perturbatively measure the field profiles in a cavity. For small, local deformations of the resonant cavity geometry, we may rewrite Eq. (7.107) as

$$\frac{\delta \omega_0}{\omega_0} = \frac{\delta V_c}{U_{EM}} \left[\frac{1}{2}\varepsilon_0 \vec{E}^2 - \frac{1}{2}\mu_0 \vec{H}^2\right]. \quad (7.108)$$

The negative sign in front of the magnetic term (as opposed to the positive electric contribution) in Eq. (7.108), which is perhaps familiar to the reader from introductory magnetics, deserves some illustrative discussion. This discussion is best done by examples.

As an electric example, if a conducting bead of a small volume δV_c is placed on the axis of a cavity (on, e.g. a very thin filament), where there is only longitudinal electric field and no magnetic field, measurement of the change in the resonant frequency yields the electric field amplitude as

$$|E_z| = \sqrt{\frac{2U_{EM}}{\varepsilon_0 \delta V_c} \frac{\delta \omega_0}{\omega_0}}. \quad (7.109)$$

In the case of a conducting bead, the electric field is excluded from the bead volume (δV_c is negative) and so the resonant frequency change is negative. Another way of describing this effect is placing of a bead in a region with electric field increases the capacitance C, thus decreasing the resonant frequency.

To discuss magnetic perturbations, we examine the case of a perturbation of the outer cavity wall at a position where the electric field vanishes and the magnetic field is at a maximum. A re-entrant bump (again a negative δV_c) at such a position clearly decreases the cavity inductance L by displacing magnetic flux, and therefore raises the resonant frequency. The tuning of a resonant cavity during manufacture often proceeds by removing material from the outer wall to approach the desired resonant frequency from above. This is termed inductive tuning. An alternative method is to employ capacitive tuning, in which one removes material from the ends of the cavities near the axis (adjacent to the beam hole). Unfortunately, this prescription also changes the length of the cavity, and may therefore introduce unwanted changes in the acceleration properties of the system.

7.6 Cavity coupling to the waveguide and the beam

We are now in a position to complete our circuit description of the resonant accelerator cavity by examining the effects of external coupling—to the waveguide and the beam itself. The coupling of the cavity to the waveguide has not been included in Eq. (7.91), as if the waveguide circuit affects the cavity circuit, but not *vice versa*. The effect of this coupling is two-fold: power is lost from the cavity back into the guide and this power is radiated as a backward wave into the waveguide. The lost radiative power can be added to the power dissipated in the walls, and we can therefore define a coupling quality factor, Q_c, related to the entire loaded quality factor of the cavity, by

$$\frac{1}{Q_L} = \frac{1}{Q_0} + \frac{1}{Q_c}.$$
(7.110)

It can be seen from the discussion in Section 7.5 that the coupling quality factor is related to the external coupling strength and cavity (wall power dissipation-derived) quality factor by $Q_c = Q_0/\beta_c$.

The effect of the beam current on the cavity circuit is to excite voltage oscillations, an effect which can be included by use of the beam impedance. In describing both external coupling and beam effects mathematically, we rewrite Eq. (7.91) in terms of voltages,

$$\frac{d^2 V_c}{dt^2} + \frac{\omega_0}{Q_0} \frac{dV_c}{dt} + \omega_0^2 V_c = \beta_c \omega_0 \frac{d}{dt}(V_F - V_R) - \omega_0 Z_b \frac{dI_b}{dt}.$$
(7.111)

We have introduced here two waveguide voltages, one associated with the forward wave V_F (with longitudinal functional dependence $\exp[i(k_z z - \omega t)]$, z being distance along the waveguide) impinging on the cavity, and one radiated as a reflected wave V_R (of longitudinal dependence $\exp[i(k_z z + \omega t)]$) back into the waveguide. Further, continuity of fields requires that $V_c = V_F + V_R$, which allows us to recast Eq. (7.11) as

$$\frac{d^2 V_c}{dt^2} + \frac{\omega_0}{Q_L} \frac{dV_c}{dt} + \omega_0^2 V_c = \frac{2\omega_0}{Q_0} \frac{dV_F}{dt} + \omega_0 Z_b \frac{dI_b}{dt}.$$
(7.112)

Let us examine Eq. (7.112) first by concentrating on the role of external coupling term. In the absence of beam, and assuming again a forward wave which turns on suddenly, $V_F \propto \Theta(t - t_0)$, we have in analogy to Eq. (7.97), the solution (at resonance)

$$V_c(t) \cong \frac{2\beta_c}{1 + \beta_c} V_F \sin([\omega_0(t - t_0)]) \left[1 - \exp\left(-\frac{\omega_0}{2Q_L}(t - t_0)\right)\right]. \quad (7.113)$$

Note that the time response of the cavity is now controlled by the loaded Q_L, which is lower than the cavity Q_0, that is, the cavity responds more quickly to both turn-on and turn-off of the voltage, because of the added radiative path of energy exchange. For the condition $\beta_c = 1$, the cavity voltage asymptotically approaches the forward voltage, $\lim_{t \to \infty} V_c \cong V_F$, and the system is considered **impedance matched**, or **critically coupled**. In this case, the reverse voltage,

$$V_R \cong V_c - V_F \propto \frac{\beta_c - 1}{\beta_c + 1} V_F, \quad (7.114)$$

disappears. Thus, for $\beta_c = 1$, all of the incident rf power is absorbed in the cavity and, even though the time-response of the cavity is controlled by Q_L, the power and the cavity voltage are linked through the cavity Q_0 by Eq. (7.103). For the under-coupled case, $\beta_c < 1$, the reflected voltage asymptotically is $180°$ out of phase with the incident voltage, while in the over-coupled case, $\beta_c > 1$, it is in phase.

In general, the system is not perfectly impedance matched and its performance is characterized by the **voltage standing-wave ratio (VSWR)**,

$$\text{VSWR} \equiv \frac{\|V_F\| + \|V_R\|}{\|V_F\| - \|V_R\|} = \begin{cases} \beta_c, & \beta_c > 1 \\ \beta_c^{-1}, & \beta_c < 1. \end{cases} \quad (7.115)$$

The VSWR is a common measurement used to diagnose the degree of mismatch in a microwave circuit system. The matched condition, where all incident power is absorbed by the cavity and none is reflected, has the minimum achievable VSWR of 1. The minimum reflected power is only achieved after a time significantly longer than the **loaded fill time** $\tau_f = \omega_0/Q_L$. During this transient, the resonant cavity system is described as having a time-dependent impedance.

One may reasonably enquire into the physical basis of such a time-dependent system. At the start of the fill, nearly all of the incident power is reflected by the cavity (it behaves as a **short circuit**), while after many fill times, nearly all power is absorbed (the cavity is presents a **matched load**). The short circuit is straightforward to understand—the electromagnetic power impinges on a wall with only a small coupling hole, so nearly all power is reflected. The very small amount that couples into the cavity slowly excites the cavity into oscillation. After the cavity fills with stored energy, the wave reradiated out of the coupling hole back into the guide becomes non-negligible. Because it is of opposing phase to the reflected wave, it tends to cancel the reflected power. For $\beta_c = 1$, this cancellation is exact. Thus, a system with a serious impedance mismatch (the reflection from the coupling iris wall) can be effectively matched when another mismatch downstream (the back-radiated cavity power) cancels it. This is a common occurrence in microwave circuits in general, not just in resonant cavity systems.

Just as with the coupling to the waveguide, the effects of beam current on the cavity system are often not ignorable. A beam that is bunched at the resonant frequency of the cavity can give up power to cavity mode via beam loading. For example, in a klystron output cavity an intense bunched beam excites a TM mode and the beam-derived power built up is out-coupled (through some aperture like the slot shown in Fig. 7.7(a)) into a waveguide, where it is transported to the load (e.g. an accelerator cavity). In an accelerator cavity with an intense bunched beam, on the other hand, the voltage excited in the cavity by the incident waveguide power can be suppressed through the extraction of power by the beam. This effect is often used to produce a time-independent voltage amplitude $\|V_c\|$ in the cavity. According to Eq. (7.111), a solution of the form $V_c = \|V_c\| \exp(i\omega_0 t)$ is produced when the current satisfies

$$I_b(\omega_0) = \frac{V_F}{Z_b}\left(\frac{2}{Q_0} - \frac{V_c}{V_F Q_L}\right) = 2\frac{V_F}{Z_s}\left(2 - \frac{V_c}{V_F}\left(1 + \frac{Q_0}{Q_c}\right)\right). \qquad (7.116)$$

Equation (7.111) gives the physical condition that the power flow into the cavity during its (beam-off) fill is exactly cancelled by power extraction due to the beam acceleration, as illustrated in Fig. 7.9. The two cavity voltage amplitudes shown correspond to the cases of filling of a matched bare rf cavity, and one in which beam is turned on partially through the rf pulse, stopping the fill at a lower voltage, with an accompanying impedance mismatch. For heavily beam-loaded cases, where it is desired that a significant fraction of the rf power to be absorbed by the beam, the bare cavity is typically over-coupled, so that one may critically couple the system by turning on the beam at the correct time. In this case, as the cavity-plus-beam system absorbs higher power, it "looks" like a lower Q_0 cavity from the input viewpoint, and this more lossy cavity is matched to the coupling. With superconducting rf cavities, the cavity Q_0 is orders of magnitude larger than for normal conducting cavities (it may easily exceed 10^9), and the external coupling and beam loading are adjusted to give the system with a much lower Q_L, with nearly all of the electromagnetic power being absorbed by the beam after the steady state voltage is achieved.

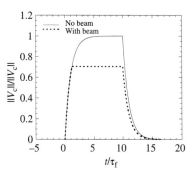

Fig. 7.9 Voltage amplitudes during cavity fill and emptying, for bare cavity (no beam) and beam-loaded cases. The rf power is constant during the time $t = 0, \ldots, 10\tau_f$, and in the beam-loaded case, beam is on during $t = 1.23\tau_f, \ldots, 10\tau_f$.

7.7 Intra-cavity (cell-to-cell) coupling

While in circular accelerators, rf resonators are encountered which have a simple single-cavity geometry, in linear accelerators the cavities consist of a number of sub-cavities coupled together. In this case, each sub-cavity is termed a cell, and the entire cavity is referred to as a multi-cell device. Multi-cell cavities are typically coupled to the waveguide power through one cell, and the other cells are then excited by cell-to-cell coupling. This coupling is often accomplished through the beam holes and such a near-axis effect is clearly capacitive (electric coupling), as is illustrated for a two-cell system in Fig. 7.10. In this system, the near-axis coupling is simply represented by a small capacitor linking the two resonant circuits corresponding to the cavities themselves.

The analysis of the two-cell resonant cavity system serves to illustrate many of the characteristics of multi-cell cavities in general. The differential equations describing the coupled circuits shown in Fig. 7.10(b) can be written as

$$\frac{d^2 I_1}{dt^2} + \omega_0^2(1 - \kappa_c)I_1 = -\kappa_c \omega_0^2 I_2 \qquad (7.117)$$

and

$$\frac{d^2 I_2}{dt^2} + \omega_0^2(1 - \kappa_c)I_2 = -\kappa_c \omega_0^2 I_1, \qquad (7.118)$$

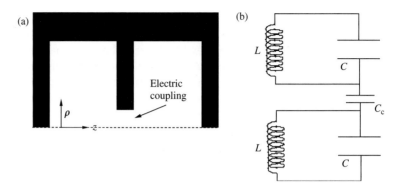

Fig. 7.10 (a) Physical picture of a two-cell resonant cavity, with near-axis cell-to-cell electric coupling, and with no external coupling. (b) Equivalent circuit for the system shown in (a), ignoring wall losses.

where $\kappa_c \equiv C/C_c$, and we have assumed that the two resonators have identical values of L, C, and thus the same resonant frequency. Note we have also ignored power dissipation, as it is not central to the present discussion. There are a number of ways to solve Eqs (7.117) and (7.118), but since we have already relied much on matrix and eigenvalue approaches, we adopt this method. Upon substitution of an $\exp(i\omega t)$ dependence for the currents (written as a vector $\mathbf{i} = (I_1, I_2)$), the determinant of the coefficient matrix in the resulting eigenvalue equation,

$$\begin{bmatrix} \omega_0^2(1 - \kappa_c) - \omega^2 & \kappa_c \omega_0^2 \\ \kappa_c \omega_0^2 & \omega_0^2(1 - \kappa_c) - \omega^2 \end{bmatrix} \mathbf{i} = 0, \tag{7.119}$$

when set to zero gives the secular equation

$$\omega^4 - 2\omega_0^2(1 - \kappa_c)\omega^2 + \omega_0^4(1 - 2\kappa_c) = 0. \tag{7.120}$$

The solutions for the frequencies (eigenvalues) of this system are

$$\omega = \omega_0 \quad \text{and} \quad \omega = \omega_0\sqrt{1 + 2\kappa_c}, \tag{7.121}$$

corresponding to the eigenvectors

$$\mathbf{i} = \frac{1}{\sqrt{2}}\begin{pmatrix} \pm 1 \\ 1 \end{pmatrix}, \tag{7.122}$$

respectively. The higher frequency solution corresponds to oscillations that are 180° out of phase in the two cavities—the π-mode ($\psi = \pi$) introduced first in Fig. 4.1. Likewise, the lower frequency solution corresponds to the zero-mode ($\psi = 0$, see Ex. 4.1), where the currents and fields are in phase with each other in the two cavities. The zero-mode frequency is the same as that for an isolated oscillator, $\omega = \omega_0$, because when this mode is established, the two oscillators are effectively decoupled, and the electric fields are essentially a simple superposition of the isolated oscillation cases. In the π-mode, however, a null is forced at the cavity interface in the middle of the beam hole (coupling iris). This induces a higher frequency because the mode is more strongly confined longitudinally, so k_z and thus $\omega = c\sqrt{\sum_i k_i^2}$ are increased.

The analysis of two-cell system, which is found in few applications such as radio-frequency electron guns (cf. Ex. 4.3), can be generalized to cover the many-cell (cell number $N_c \gg 1$) cavity structures that are typically used in linear accelerators. Ignoring for the moment the question of end-cell effects, the generalization of Eqs (7.117) and (7.118) for the current in the jth cell of the cavity is

$$\frac{d^2 I_j}{dt^2} + \omega_0^2(1 + 2\kappa_c)I_j = \kappa_c\omega_0^2[I_{j+1} + I_{j-1}], \tag{7.123}$$

with the coupling to cell number $j + 1$ and $j - 1$ are clearly displayed. Equation (7.123) could in principle be cast in matrix form, but this method does not easily yield a general, closed-form result.

A better approach to analyzing the N_c normal modes of the system described by Eq. (7.123) is inspired by noting that it is similar to the model one constructs to derive the equations of motion for waves on a discrete mass-loaded string. Since we are concerned now with a wave-like system, let us assume the solutions to Eq. (7.123) are of the form

$$I_j = a_j \exp(i\psi_j) \exp(i\omega_j t), \tag{7.124}$$

where the constants ω_j, ψ_j and a_j are real constants, the normal mode frequency and phase shift, and amplitude, respectively. Using Eq. (7.124), we have the recursion relation

$$-a_{j-1}\kappa_c\omega_0^2 + \left[\omega_0^2(1 + 2\kappa_c) - \omega^2\right]a_j - a_{j+1}\kappa_c\omega_0^2 = 0. \tag{7.125}$$

If we further assume that the phase shifts have the simple form $\psi_j = jb$, where b is a mode-dependent constant, we then have

$$-e^{-ib} + \left[\omega_0^2(1 + 2\kappa_c) - \omega^2\right] - e^{ib} = 0, \tag{7.126}$$

from which we can derive the dispersion relation giving the frequency dependence of b,

$$\omega^2 = \omega_0^2 + 2\kappa_c\omega_0^2[1 - \cos(b)] = \omega_0^2\left[1 + 4\kappa_c \sin^2\left(\frac{b}{2}\right)\right]. \tag{7.127}$$

The boundary conditions on the ending cells require that their current be at a local extremum (standing wave boundary), and so we can write

$$a_1 = 1, \quad \text{and} \quad a_{N_c} = \pm 1. \tag{7.128}$$

Equation (7.128) are satisfied if we take $N_c b = n\pi$, where $n = 0, 1, 2, \ldots,$ $N_c - 1$. We can thus choose N_c values of the constant b, corresponding to each of the normal modes,

$$b_n = \frac{n\pi}{N_c - 1}. \tag{7.129}$$

We can now identify the constants b_n as the Floquet phases (phase shift per cell) first mentioned in Chapter 4 during our introduction to electromagnetic waves in periodic systems. They may take on N_c values between 0 and π.

As the cavities are optimized for acceleration on only the fundamental spatial harmonic, the wave-number corresponding to each normal mode is $k_{z,n} = \psi_n/d$, and we may rewrite the dispersion relation in Eq. (7.127) as

$$\omega_n = \omega_0 \sqrt{1 + 4\kappa_c \sin^2\left(\frac{k_n d}{2}\right)} = \omega_0 \sqrt{1 + 4\kappa_c \sin^2\left(\frac{n\pi}{2(N_c - 1)}\right)}. \quad (7.130)$$

Thus, the minimum frequency supported by the multi-cell cavity system is again ω_0, and the maximum is $\omega = \omega_0\sqrt{1 + 4\kappa_c}$, corresponding to the zero and π modes, respectively. This range of frequencies is referred to as a **pass-band**, of which there are many, associated with the different resonances of the individual cells. It should be noted that this situation, where the frequencies rise as a function of the phase shift per cell, is predicated on the assumption that the parameter κ_c is positive. This may not always be the case, as when one couples the cells in a different way (e.g. magnetically), the sign of κ_c in Eq. (7.130) may change. In such a case, the dispersion relation indicates that the group velocity is in opposition to the phase velocity. The modes in the pass-band are then termed "backward waves."[6]

It is necessary that the normal mode one wishes to excite is not too near another within the pass-band (or, for that matter, in any other pass-band). The criterion for decoupling of each normal mode is that the separation between adjacent modes of interest is many times their width due to finite Q,

$$\|\omega_{0,n} - \omega_{0,n\pm1}\| \gg \frac{\omega_{0,n}}{Q_{L,n}} + \frac{\omega_{0,n\pm1}}{Q_{L,n\pm1}}. \quad (7.131)$$

When this set of criteria is violated, one can no longer guarantee that only the desired mode is excited by the application of rf power at frequency $\omega_{0,n}$. It can be seen that for modes with $n = 0$ and N_c (π-mode) the mode separation problem exerts its most stringent requirement. For the interesting case of the π-mode, in the limit of weak coupling $\kappa_c \ll 1$ and large N_c, we have approximately

$$\kappa_c \left(\frac{\pi}{2N_c}\right)^2 \gg \frac{1}{Q_L} \quad (7.132)$$

to ensure adequate mode separation. The narrow separation between modes at the edges of the pass-band is illustrated for a 12-cell structure in Fig. 7.11.

[6]Unfortunately, this term has already been used to designate the negative phase velocity component of a standing wave. Here we mean it to indicate a group velocity that is opposite to the phase velocity. This distinction becomes more clear when we discuss traveling waves.

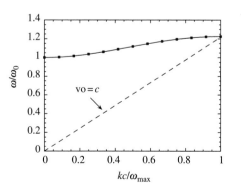

Fig. 7.11 Brillouin diagram of pass-band in 12-cell structure, with the π-mode chosen to have speed-of-light phase velocity, for $\kappa_c = 0.125$. Mode frequencies indicated by squares, phase velocity and group velocity curves also shown.

Up until this point we have assumed that the coupling between cells is weak, $\kappa_c \ll 1$. While this is often true for on-axis, electrically coupled cells, it may not be for other classes of cavities. A simple example of this is found in the zero mode-excited drift tube linac, shown in Exercise 4.1. In such an open, magnetically coupled structure, it seems almost inappropriate to refer to the cells individually, and the coupling (magnetic, cf. Ex. 7.24) is very large.

7.8   Traveling wave versus standing wave cavity structures

From the discussion in the previous section we can see that long, multi-cell rf cavity structures can support standing waves of certain forms. It should also be apparent that such a system can equally support traveling waves (cf. Eq. (4.2))— the standing wave solutions are just superpositions of forward and backward traveling waves. In fact, many of the parameters we are interested in from the viewpoint of traveling waves have already been uncovered by our standing wave analysis. In particular, since we have derived an approximate dispersion relation, we can obtain the phase and group velocities of the modes in a given structure with finite N_c,

$$v_{\phi,n} = \frac{\omega_n}{k_n} = \frac{\omega_0 \sqrt{1 + 4\kappa_c \sin^2\left(\frac{k_n d}{2}\right)}}{k_n}, \tag{7.133}$$

and

$$v_{g,n} = \frac{d\omega}{dk}\bigg|_{k=k_n} = 2\kappa_c \omega_0 d \frac{\sin\left(\frac{k_n d}{2}\right)\cos\left(\frac{k_n d}{2}\right)}{\sqrt{1 + 4\kappa_c \sin^2\left(\frac{k_n d}{2}\right)}}, \tag{7.134}$$

Assuming positive coupling, the phase velocity given by Eq. (7.133) decreases with increasing k_n (and ψ_n), and is chosen in the case of the operating mode to closely match the expected charged particle velocity in linear accelerator applications. On the other hand, the group velocity is maximized in band center, when $k_n \cong \pi/2d$, or $\psi_n \cong \pi/2$. These dependences are displayed as well for the 12-cell example structure in the Brillouin diagram (ω–k plot) of Fig. 7.11.

Because the π-mode typically has relatively high shunt impedance, it is often employed in standing wave structures, which do not rely on power flow along the structure. This reliance often imposes a limit on the number of cells, and thus the length, of a standing wave system due to the mode-separation considerations discussed above. The major difference between standing wave and traveling wave structures is that power indeed does flow (generally in the positive $z$direction, as indicated by Eq. (7.134)) along the structure in the traveling wave case. Because of this, in traveling wave structures, power is coupled in at one end of the structure and dumped at the other end into a load. In the interior of the structure, power is dissipated along its propagation via wall losses, as in the waveguide[7] case discussed in Section 7.3. The power dissipation is governed by an equation,

[7]Traveling wave accelerator sections are also sometimes referred to as "waveguides" by rf engineers.

$$\frac{d}{dz}P(z) = \frac{d}{dz}v_g(z)u_{EM}(z) \tag{7.135}$$

which has similar form to Eq. (7.50), but with the additional possibility that the group velocity may change (through, e.g. beam iris modifications) as the wave propagates through the structure. One may attempt to increase the shunt impedance per unit length of the device,

$$Z_s' = \frac{dP/dz}{\langle E_0^2 \rangle}, \tag{7.136}$$

where the amplitude of the accelerating wave E_0 is held constant by such changes, to produce a **constant gradient structure**. If for simplicity of design the traveling wave accelerator cavity is purely periodic, however, it is termed a **constant impedance structure**.

In a constant impedance structure, the power loss per unit length is strictly proportional to the instantaneous power flow,

$$\frac{d}{dz}P(z) = -\frac{\omega_0}{2Qv_g}P(z) \equiv -\eta_{TW}P(z), \tag{7.137}$$

where $\eta_{TW} = (\tau_f v_g)^{-1}$. The solution to Eq. (7.137) gives the familiar exponential decay.

$$P(z) = P(0)\exp(-\eta_{TW}z). \tag{7.138}$$

This power profile yields an acceleration amplitude profile,

$$E_0(z) = E_0(0)\exp\left(-\frac{\eta_{TW}z}{2}\right), \tag{7.139}$$

which can be integrated over the length of the structure L_s to give the maximum acceleration possible,

$$\Delta U = q\int_0^z E_0(\tilde{z})\,d\tilde{z} = \frac{2qE_0(0)}{\eta_{TW}}\left[1 - \exp\left(-\frac{\eta_{TW}L_s}{2}\right)\right]. \tag{7.140}$$

It can be seen that after a length L_s more than, say, twice the power attenuation length η_{TW}^{-1}, here is little additional acceleration, The attenuation length is therefore the approximate practical limit on the length of a traveling wave structure. In any case, the acceleration in a constant impedance structure is strictly limited by $2qE_0(0)\eta_{TW}^{-1}$.

The finite group velocity of the wave also means that the fill time of a traveling wave accelerator differs from its standing wave counterpart, in that power must flow for a time

$$T_f = \frac{L_s}{v_g} \tag{7.141}$$

to completely fill the structure. Defining the total attenuation in the structure to be $\alpha_{TW} \equiv \eta_{TW}L_s$, we can link this fill time to that introduced in Eqs (7.98)–(7.100) by

$$T_f = \alpha_{TW}\frac{Q_0}{\omega_0} = \alpha_{TW}\tau_f. \tag{7.142}$$

Since $\alpha_{TW} \equiv \eta_{TW}L_s$ is of order unity, the two definitions of the fill time are numerically not too different in practice, even though they indicate very different underlying physical processes.

Another way in which traveling wave structures differ from their standing wave counterparts is that there is no reflected wave inherent in the sudden application of rf power to the traveling wave system. This is because the input coupling cell (the first cell) acts as an impedance matching network, in which waves reflected from the high impedance interior of the structure are cancelled by a reflection at the start of the input cell. Thus, the input cell can be considered a type of standing wave resonator.

The shunt impedance per unit length is also related to the beam impedance per unit length, as in the case of a single cell cavity, by

$$Z_b' = \frac{Z_s'}{Q} \left[1 - \frac{v_g}{v_b} \right] \cong \frac{Z_s'}{Q} \left[1 - \frac{v_g}{c} \right]. \tag{7.143}$$

The factor $1 - (v_g/v_b)$ accounts for the fact that the electromagnetic wave that is excited in the cavity structure by the beam is not trapped in a non-propagating mode, as in the single cavity, but is continually "catching up" to the beam. Equation (7.143) is equally valid for both traveling and standing wave accelerator structures.

Traveling wave accelerators in general have complicated field patterns, due to the fact that disk-loaded geometries are not simple boundary conditions. This complexity is reflected by the need to use many spatial harmonics in the field description. On the other hand, there exists a simple geometry that allows a simple mode, with appropriate phase velocity, supported in a cylindrical metallic tube that has dielectric (with a beam hole) inserted in it. This geometry can be analyzed exactly, with a single spatial harmonic. The details of this system are explored further in Exercise 7.25.

7.9 Electromagnetic computational design tools*

Even though there are exact analytical approaches to understanding waveguide and cavity systems, as we have spent some considerable effort introducing above, there are many reasons why these cannot be used alone to guide the design and production of accelerator cavities. Two salient features of these cavities prevent the purely analytical approach. First, even for nominally cylindrically symmetric cavities, exact analysis of relevant geometries (e.g. beam holes) is not tractable. Three-dimensional features (coupling slots, etc.) make this electromagnetic analysis even more challenging. When one also considers that cavity systems are resonant, and therefore need to be tuned very accurately in frequency, the need for more tools is easily understood.

This situation, which is similar to, but even more difficult than, that encountered in magnet design, gives rise to reliance on measurements and sophisticated computer modeling. Much of the detailed analysis and design work on electromagnetic cavities is performed by use of two and three-dimensional design codes, computer programs that solve the Maxwell equations (or their equivalent, e.g. Eq. (7.1), or in the frequency domain, Eq. (7.23)) on a spatial discrete mesh. The actual details of the calculational approaches are quite intricate, and thus these algorithms are beyond the scope of the present discussion. We therefore discuss here briefly only the uses one typically makes of these codes.

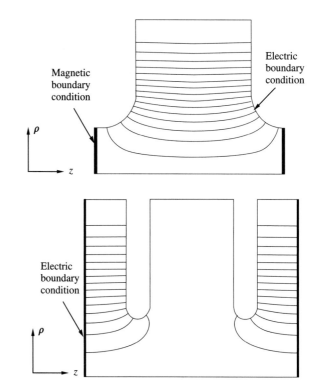

Fig. 7.12 Contours of constant enclosed magnetic flux in a cell of a π-mode structure, from SUPERFISH simulation. Magnetic boundary conditions applied in beam apertures.

Fig. 7.13 Contours of constant enclosed magnetic flux in a minimal period (1 + 2/2 cells) of a (π/2)-mode structure, from SUPERFISH simulation. Electric boundary conditions applied at ends, which are midplanes of the cells.

[8] The SUPERFISH graphical output shows the contours of constant azimuthal magnetic flux. These lines are parallel to the electric field lines.

The first electromagnetic design codes were two-dimensional (axisymmetric) and are typified by SUPERFISH, which solves the Helmholtz equation given a particular geometry and appropriate boundary conditions. The boundary conditions employed are perfectly conducting—electric boundary—on a metal surface, and electric or magnetic (electric field parallel to the surface) boundary conditions at appropriate longitudinal positions. The graphical output obtained from SUPERFISH simulation of a single cell of a π-mode structure, shown in Fig. 7.12, illustrates the electric field[8] behavior at electric and magnetic boundaries. Note that the field is parallel to the boundary in the beam holes, as the electric field is a null at the upstream and downstream edges of the π-mode cell. This behavior can be contrasted with that shown in Fig. 7.13, which displays the fields in 1 + 2/2 cells (midplane-to-midplane of the upstream and downstream adjacent cells) of a (π/2)-mode structure, and where electric boundary conditions have been chosen at the upstream and downstream edges.

Note that in Fig. 7.12 the upstream and downstream halves of the cell are calculationally redundant, as they have reflection symmetry. Thus one could obtain the same information with half of the computational grid, by applying an electric boundary condition at the mid-plane of the π-mode cell. Likewise, the upstream and downstream halves of the computational grid are redundant in Fig. 7.13, and the same information could have been obtained by applying a magnetic boundary condition at the grid mid-plane.

One of the most important features of cavity behavior one obtains from SUPERFISH simulations is, not too surprisingly for a frequency-domain code, the resonant frequencies of the cavity modes. One also obtains the fields, from which calculation of the tuning behavior of boundary modifications from Slater's theorem and the power dissipation on the walls (from knowledge of H_ϕ)

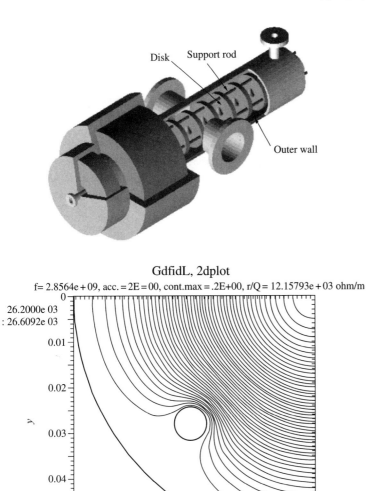

Disk Support rod

Outer wall

Fig. 7.14 Rendered cutaway drawing of a plane-wave transformer (PWT) photoinjector structure (at the UCLA PEGASUS laboratory), along with focusing solenoids (to the left). Note that there is an open region around the disks inside of the rf structure and that the disks are supported by four longitudinally-oriented rods.

GdfidL, 2dplot

f= 2.8564e + 09, acc. = 2E = 00, cont.max = .2E+00, r/Q = 12.15793e + 03 ohm/m

26.2000e 03
: 26.6092e 03

Fig. 7.15 Contours of constant enclosed magnetic flux at mid-plane of PWT cell, from a three-dimensional frequency domain simulation performed by the code GdfidL.

can be derived. In addition, the accelerating fields and their spatial harmonic content of the may be deduced, thus also allowing calculation of the shunt impedance. Field information is also typically exported to other simulation programs that model the beam dynamics. This process of field exporting is also used with the magnetic simulations discussed in Chapter 6.

The recent explosive growth in available computing power has popularized the use of both time-domain codes and three-dimensional modeling tools. Time-domain codes, which must follow the fields over many cycles to be of use, can be used for wave propagation and coupling analyses. In addition, many of these codes allow the introduction of beam charge and current, to model wake-fields and there effects on beam propagation.

A graphical output from a three-dimensional electromagnetic frequency domain calculation (using the code GdfidL) is shown in Fig. 7.15, which models the inherently three-dimensional plane-wave transformer (PWT) accelerator shown in Fig. 7.14. This structure has a gap between the outer wall and the disks, effectively providing an enormous magnetic coupling between cells. The disks must be supported, however, and this is accomplished by the use of four rods running longitudinally along the structure. The distortion in the fields introduced by these rods can be seen in Fig. 7.15, which displays the minimal computational space around the azimuth needed to account for the four-fold symmetry of the structure.

Three-dimensional electromagnetic simulation tools are increasingly used for designing and modeling accelerator cavities and related systems. Intricate structures such as the PWT, shown in Fig. 7.14, have many mechanical (e.g. vibrations and electromagnetic stresses) and thermal (from high levels of rf power dissipation in modern, high power structures) issues that need to be addressed during the design and fabrication process. These problems are also examined through computational modeling, and thus electromagnetic design codes are increasingly being integrated with not only beam dynamics simulations, but with mechanical/thermal simulations as well.

7.10   Summary and suggested reading

This chapter has surveyed aspects of electromagnetic fields inside of regions bounded by conductors, in scenarios of relevance to accelerator technology. This discussion was based on general considerations of electromagnetic wave propagation in vacuum, dielectrics, and conductors. After review of the propagation modes—TE, TM, and TEM modes—and their characteristics in waveguides, we have built our discussion of resonant cavity modes upon relevant concepts in waveguide theory. A detailed examination has been undertaken to elucidate the forms of fields in coupled cavity linacs, devices that are also profitably viewed as TM waveguides in their own right. This discussion clarifies the simple view of these traveling wave systems first introduced in Chapter 4.

Equivalent circuit models, and several variants on the concept of impedance have been introduced in the context of waveguides. These methods have been found to be quite powerful when extended to the study of resonant cavities. In particular, one may to understand the frequency and time response of resonant, coupled systems such as rf cavity cells using circuit models. Quality (Q) factors have been introduced that parameterize these responses, and aid in understanding power handling in cavity–waveguide–beam systems. Examining long cavity systems with coupled-oscillator theory also has allowed us to classify the modes in traveling and standing wave linacs according to the phase shift between adjacent cells.

Additionally, we have exploited our view of rf cavities as generalized oscillators to introduce the Slater perturbation theorem, a powerful tool for tuning and measuring cavity characteristics. The analytical and perturbative approaches described in this chapter form the basis of understanding and classifying the behavior of electromagnetic cavities and waveguides as they are implemented in accelerators. In practice, one must often complement these tools with sophisticated computational modeling, as has been reviewed in the final section of this chapter.

The details of rf or microwave engineering that the reader may find relevant or helpful extend well beyond the treatments presented in this chapter

In addition to the emphases we have placed here on fields and equivalent circuit models, descriptions of microwave systems in more general engineering texts will emphasize tools based on complex impedances. The following references are recommended:

1. R. Wangsness, *Electromametic Fields* (Wiley, 1986). An overview of waveguide and resonant cavity theory is found here, as in other electromagnetism texts listed in Chapter 1.
2. T. P. Wangler, *Principles of RF Linear Accelerators* (Wiley, 1993). This text reviews aspects of electromagnetic cavities in linear accelerators.
3. R. E. Collin, *Foundations for Microwave Engineering* (McGraw-Hill, 1966). This is a very good text to start with when approaching the field of rf engineering.
4. R. E. Collin, *Field Theory of Guided Waves* (IEEE Press, 1991). A more advanced book by Collin.
5. C. G. Montgomery, R. H. Dicke, and E. M. Purcell, *Principles of Microwave Circuits* (McGraw-Hill, 1948), MIT Radiation Laboratory Series. The MIT series contains information which is both historical (it documents the science behind the US radar development program in Second World War) and extremely useful.
6. J.C. Slater, *Microwave Electronics* (D. Van Nostrand, 1950).
7. D. H. Whittum, *Introduction to Electrodynamics for Microwave Linear Accelerators*, SLAC-PUB-7802 (Stanford Linear Accelerator Center Report, April 1998). A detailed and useful digestion of much of the material in the above four references, placed in the context of accelerators.
8. H. Padamsee, J. Knobloch, and T. Hays, *RF Superconductivity for Accelerators* (Wiley, 1998). A modern treatment of RF engineering and physics with superconducting materials.
9. Computer codes used for design of radio-frequency components in accelerators are found at the Los Alamos Accelerator Code group website, http://laacgl.lanl.gov.
10. Information on the three-dimensional electromagnetic simulation program GdfidL can be found at http://www.gdfidl.de/.
11. R.B. Neal, editor, *The Stanford Two-Mile Accelerator* (W.A. Benjamin, 1968). Details of such critical technologies as klystrons and linear accelerator cavities that eventually underpinned the Stanford Linear Accelerator Center's 50 GeV linac can be found in this book.

Exercises

(7.1) The mean average of the Poynting vector $\vec{S} = \vec{E} \times \vec{H}$ represents an average power density flow in electricity and magnetism. Consider the case of a plane wave and show that this power density is $|\langle \vec{S} \rangle| = \langle u_{\text{EM}} \rangle v_{\text{g}}$, where $\langle u_{\text{EM}} \rangle$ is the average value of the electromagnetic energy density, $u_{\text{EM}} = \frac{1}{2}(\vec{E} \cdot \vec{D} + \vec{H} \cdot \vec{B})$.

(7.2) Write Eq. (7.1) in spherical polar coordinates. Show that the components of a **spherical wave** of the form

$$\vec{E} \propto \exp[ik(r - ct)]/r$$

are solutions to this wave equation and that $\hat{r} \times \vec{E} = 0$ (the wave polarization is transverse). The spherical wave is analogous to the plane wave in that the phase fronts are uniform (the field does not vary in amplitude along $r - ct = $ const. surfaces). Waves related to this type of solution will be explored in detail in Chapter 8.

(7.3) Because it is commonly available, relatively inexpensive, and an excellent conductor, copper is used extensively in waveguides and cavities. The conductivity of copper is $\sigma_c = 5.96 \times 10^7 \, \Omega^{-1} \, m^{-1}$.

    (a) For a common S-band frequency, $f_{rf} = 2.856 \, \text{Ghz}$. What is the skin depth and surface resistivity of copper at this frequency? What is the ratio $\omega\varepsilon/\sigma_c$ for this case?

    (b) When applying high power, long wavelength (far infrared) lasers such as the CO_2 laser ($\lambda = 10.6 \, \mu m$), copper mirrors are often used due to their low absorption losses and ability to withstand high field. What is the skin depth and surface resistivity of copper at this wavelength? What is the ratio $\omega\varepsilon/\sigma_c$ for this case?

    (c) At what wavelength of an electromagnetic (light) wave is $\omega\varepsilon/\sigma_c = 1$?

(7.4) Show that the form of the dispersion relation given in Eq. (7.19) arises from the condition

$$v_\phi v_g = c^2.$$

(7.5) By multiplying Eq. (7.19) by $\hbar^2$, convert it into a momentum–energy relation. By examining the constant term in this relation, determine the **effective rest mass** of a photon in a plasma.

(7.6) In copper, the conduction electron density is $n_0 \cong 8.5 \times 10^{22} \, cm^{-3}$.

    (a) What is the plasma frequency of this metal?

    (b) What is the collision frequency in copper, as determined from the conductivity (see Ex. 7.3)?

    (c) At what frequency is $k_p\delta_s = 1$? In this region of frequencies, a dispersion analysis which includes both collisions and plasma effects must be used.

(7.7) For the dielectric-loaded coaxial waveguide (transmission line) shown in Fig. 7.1, assuming the dielectric in the material in the region $a < \rho < b$ has permittivity $\varepsilon = \kappa_e\varepsilon_0$ and permeability μ_0,

    (a) Calculate the electric field associated with a uniform pulse with line charge of linear density λ_e on the inner conductor $\rho = a$. In order to exclude the electric field in the outer conductor, what must the linear charge density on the surface $\rho = b$? What is the voltage associated with this electric field?

    (b) Noting that the velocity of the surface charge density supporting this TEM mode must be the phase velocity, $v_\phi = (\mu\varepsilon)^{-1/2}$, calculate the resulting

surface currents on the inner and outer conductors and the associated magnetic field.

    (c) Calculate the wave impedance of this transmission line mode using the definition $Z_0 \equiv \|\vec{E}\|/\|\vec{H}\|$. Is it given by $\sqrt{\mu/\varepsilon}$ as expected?

(7.8) A commonly encountered dielectric-loaded ($\kappa_e = 2.25$) coaxial transmission line is has inner radius of $a = 2 \, mm$ and $b = 5 \, mm$, respectively. If the conductors are made of copper, what is the power attenuation of a 500 MHz (cyclic frequency) signal over 100 m of coaxial line? What is the voltage attenuation?

(7.9) An alternative way of viewing power flow in electromagnetic waves is to write, as in Exercise 7.1, the power flow density $|\langle \vec{S}\rangle| = dP/dA = \langle u_{EM}\rangle v_g$. Show that this expression holds for TE and TM modes in rectangular waveguides.

(7.10) Calculate the attenuation per unit length η_w for electromagnetic power flow in both TE and TM modes in rectangular waveguides. Discuss its behavior as the frequency of the wave approaches cutoff from above.

(7.11) Show that for a fixed resonant frequency, the stored energy in a given fundamental TM rectangular cavity is minimized by the choice $L_x = L_y$ ($L_z = $ constant, given by resonant velocity of particle).

(7.12) Consider now TM modes in rectangular cavity for arbitrary mode numbers l, m, and n, with fixed geometry $L_x = L_y = \sqrt{2}L_z$.

    (a) Write the electric and magnetic fields associated with these modes.

    (b) Calculate the stored energy in the cavity.

    (c) Calculate the energy loss in the cavity.

    (d) Find the quality factor Q_0 of these modes. For what mode is the Q_0 maximized?

    (e) For a commonly encountered rf frequency, 3 GHz, what is this Q_0?

(7.13) For a copper-walled pill-box cavity operated at $f = 3 \, \text{GHz}$ and with length appropriate for a speed-of-light particle to traverse the cavity in one-half of an rf wavelength, what is Q?

(7.14) As was discussed first in Section 4.8, the electromagnetic fields in a cavity may be expanded in a series of multipole with azimuthal dependence $\exp(in\phi)$.

    (a) By including the azimuthal derivatives in the Laplacian operator in Eq. (7.1), generalize Eq. (7.64) to include the description of modes with $m \neq 0$.

    (b) With separation of variables, develop a full multipole description of the longitudinal electric field in TM modes supported by the cylindrical cavity o

Fig. 7.4. What is the resonant frequency of these modes as a function of mode numbers l, m, and n?

(c) Why does the expression given in Eq. (4.74) not contain Bessel functions?

7.15) It was seen in Chapter 4 that only the fundamental (synchronous) spatial harmonic has the correct phase velocity to impart net acceleration to a charged particle of approximately equal longitudinal velocity. Therefore, it is desirable to minimize the contributions to the field profile due to the other (non-synchronous) spatial harmonics. In the example given in Fig. 7.5 (a π-mode structure), what is the condition that the next highest amplitude ($n = 3$) harmonic coefficient vanish?

7.16) According to our analysis, there are solutions with longitudinal variation ($n \neq 0$) supported by single cell cavities. Instead of multi-cell cavities, why can we not simply build a long cavity with $n \neq 0$ to achieve a wave with an ultra-relativistic phase velocity?

7.17) The TEM coaxial cable ubiquitously found in physics labs has a phase velocity of $0.66c$ and an impedance of $50\,\Omega$. What is the ratio of b/a in this case?

7.18) In a single cavity there are many resonances, as discussed in Section 7.4. Since they have generally distinct resonant frequencies ω_0, they must have differing values of lumped capacitance C and inductance L. Explain why this is in physical terms.

7.19) Consider now the off-resonant time-domain behavior of the driven resonant cavity. Assume an exciting current with frequency $\omega \neq \omega_0$, turned on instantaneously to $t = t_0$, so that we may write $I_{wg} = I_0\Theta(t - t_0)\exp[i\omega t]$.

(a) Using a particular solution proportional to $\exp[i\omega t]$ and a homogeneous solution proportional to $\exp[i\omega_0 t]$, solve the Eq. (7.91) in the time-domain with correct initial conditions applied.

(b) Calculate the stored energy in the cavity from this solution (hint: remember in a complex analysis you need to write $I_c^2 = \|I_c\|^2 \equiv I_c I_c^*$). Note that the energy has an oscillatory component—what is its frequency?

7.20) For the copper-walled pill-box cavity of Exercise 7.13,

(a) calculate the shunt impedance, and

(b) find the beam impedance.

7.21) One reason that the VSWR is traditionally quoted as a measured of the fractional of power traveling in the reflected direction is because it is easy to measure, simply by measuring the local electromagnetic power as a function of distance down the guide z.

(a) Power detectors give a measure of the mean-square voltage $V_{rms}^2(z) = \langle V^2(z, t)\rangle$. Write this quantity in terms of $\|V_F\|$ and $\|V_R\|$.

(b) Find the minimum and maximum values of $V_{rms}^2(z)$ in terms of $\|V_F\|$ and $\|V_R\|$. Write the VSWR in terms of these values.

(7.22) Derive the conditions for matching the rf power in a heavily beam loaded case where the power flowing into the beam is 10 times that being absorbed into the walls. When (in terms of fill times) should the beam be turned on after the start of the rf pulse to achieve this match?

(7.23) Magnetic coupling between cells is accomplished by cutting an off-axis slot, as shown in Fig. 7.16 above.

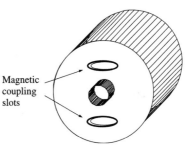

Fig. 7.16 Schematic of magnetic coupling slots at the end of pill-box cavity.

(a) Using a circuit diagram based on Figs 7.7(b) and 7.10(b), draw an equivalent circuit to model magnetic coupling between two cells.

(b) Analyze this system and comment on any differences between it and the electric coupling analyzed in this section.

(7.24) Consider a 12-cell resonant rf structure (with a passband illustrated in Fig. 7.11) operated in the π-mode at $f = \omega_0/2\pi = 2.998$ GHz. The loaded Q of the structure is known to be 6000 for this and nearby modes. If we require the mode separation between the π-mode and the $10\pi/11$-mode is at least five times the resonance bandwidth $\omega_{1/2}$, what is the minimum allowable coupling constant κ_c?

(7.25) In recent years, an old idea has been resurrected, that of using a cylindrical piece of dielectric (of permittivity $\varepsilon > \varepsilon_0$) with a beam hole cut in it, and placed inside of a cylindrical metal shell, as an accelerating structure (shown in Fig. 7.17). The accelerating mode in such a device is, or course, a TM mode. Assuming it has, in the vacuum region $\rho \leq a$, speed-of-light phase velocity, the longitudinal electric field is $E_z = E_0\cos[k_0(z - ct)]$, where $k_0 c = \omega$. In this region there is no transverse dependence, but inside of the dielectric $a < \rho \leq b$, there is a radial variation, $E_z = A\cos[k_0(z - ct)]\sin[\sqrt{(\varepsilon/\varepsilon_0) - 1}k_0(b - \rho)]$, where we have obeyed the metal boundary condition at $\rho = b$ as well as the dispersion relation in the dielectric.

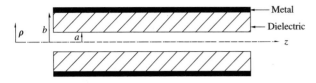

Fig. 7.17 Cross-section of dielectric-loaded traveling wave accelerator structure.

(a) Find the form of the other field components E_ρ and H_ϕ.

(b) Find the constant A by employing the correct boundary conditions on the electric field at $\rho = a$.

(c) Write the dispersion relation for these waves and the group velocity.

(d) Assuming $\rho = a$ mm, what is the outer radius b needed for the wave of cyclic frequency 10 GHz to have the correct phase velocity?

(e) Find the time-averaged electromagnetic energy density in this type of system.

(f) Find the Poynting vector in this type of system.

(g) From parts (e) and (f) deduce what the group velocity must be. Does this agree with the results of part (c)?

(h) Assuming the metal shell at $\rho = b$ is made of copper, find the form of the power dissipation per unit length.

(i) What is the shunt impedance per unit length of the 10 GHz example of part (d)? What is the beam impedance?

(7.26) What is the minimum number of cells needed to model a $(2\pi/3)$-mode structure in SUPERFISH? This is a common choice for traveling wave accelerators. The SUPERFISH calculation is apparently standing wave in nature. How do you derive traveling wave fields from the standing wave solutions?

(7.27) What is the lowest order rf multipole (cf. Eq. 4.74) beyond $m = 0$ that one would expect from the calculation in Fig. 7.13?

Photon beams

This chapter is intended to expand and generalize the concepts of classical charged particle beams introduced in the previous chapters, to allow description of photon beams, systems that often demand a quantum mechanical or wave analysis. The straightforward borrowing of analytical methods associated with quantum mechanics in nominally classical systems (such as particle beams), and vice versa, will be emphasized in this chapter. The considerable mathematical apparatus developed by the physics community during the quantum revolution has without doubt raised the level of theoretical sophistication in classical physics in general, and charged particle beam physics in particular. For example, the development of matrix treatments in charged particle optics discussed in Chapter 3 borrows some basic notions from matrix methods used in quantum mechanics. The interplay between methods developed for analysis of classical mechanics problems and quantum systems such as photon beams is bi-directional, as stated above, and we will see that our classical analyses can be used to describe photon beam (or **paraxial wave**) propagation. Again, the most relevant example is the use of matrices in analysis of beam transport. Using this method, one may deduce the evolution of the wave field distribution (light intensity) in photon beams from single particle (or **ray** in light optics) trajectory analysis, in a similar way as we found the evolution of the classical distribution function in particle beams in Chapter 5.

In this chapter, we will, by comparing the evolution of classical particle beams and photon beams using the same methods, explore the physical connections between the paraxial wave intensity profile (which is simply related to the photon's quantum probability function) and the classical distribution function. We will also exploit the single particle (ray) stability analysis of Chapter 3, as well as some of the notions concerning power and energy in confined electromagnetic systems introduced in Chapter 7, to understand the design of **optical resonators**. We shall see that these resonators, which are used to store photons through many round trips inside of the device, have direct and interesting analogies to both rf cavity resonators and to transverse dynamics in circular accelerators. Although mechanisms for analyzing paraxial waves show marked similarities to classical charged particle beams, our discussion of the transformations of each type of beam will also highlight the differences in the two systems. In particular, the phenomenon of wave diffraction, an effect that has no classical analogue, is an essential feature in coherent photon beams.

Once the basic methods for understanding photon optics and paraxial waves have been introduced, a relatively brief discussion of lasers is undertaken. These devices are central to the field of photon beams because the laser is the most common, powerful tool for creating coherent photon beams.

The last sections of this chapter are concerned with some of the most basic implications of the radiative electromagnetic process known as **synchroton radiation** have on accelerator performance. This effect is present in all bending systems based on magnetic fields, and has interesting implications for the motion of charged particles in storage rings. The subject of synchrotron radiation leads naturally to a final subject for this chapter—creation of coherent synchrotron light by electron beams in undulator magnets, or the **free-electron laser**.

8.1   Photons

The point at which students begin to learn light optics is most often the introduction of the light ray. The ray can be considered a classical approximation to the trajectory of a given "particle" of the light—the photon. For many systems in which wave phenomena such as diffraction are not important, the ray approximation is quite adequate in predicting characteristics of optical transport. However, the light ray is not strictly a classical object, in either its birth or its application. The light ray was originally derived by a semi-classical argument familiarly known as **Huygens' construction**. Even though it was not based on a rigorous wave theory, Huygens' construction anticipated the prediction of the wave behavior of light later made possible by Maxwell's theory of electromagnetism. As for the application of ray optics, we shall see that simple knowledge of the wave's initial conditions combined with the predictions of ray optics are enough to allow prediction of photon beam evolution, even for wave-dominated behavior.

If we can ignore the wave nature of photons in a given physical system (how to quantitatively justify this assumption is discussed below), then we can see that the trajectories of light rays are strictly analogous to classical charged particle trajectories. Furthermore, light beams will display the same behavior, in this approximation, as the beams made up of classical particle distributions discussed in Chapter 5. In this case, we may term the distribution of trajectories a **bundle** of rays. In fact, the wave nature of light is nearly ignorable in many common applications of paraxial optics—cameras, telescopes, optics of the human eye, etc. In all of these applications, one is dealing with **incoherent** light, in which the phases of the waves corresponding to individual photon fields are not correlated to each other. This is obviously the situation one encounters when the waves have a large spread in frequency, commonly referred to as **white light** in the above applications. For systems with narrow spectral width, however, the photon waves may have a strong correlation to each other, and the system is termed partially or wholly **coherent**. With the invention of various types of lasers[1] in the last 40 years, however, it has become commonplace both inside and outside of physics laboratories to encounter coherent light sources. Thus, methods have been introduced in recent years to describe the optics of coherent light beams, which in the context of electromagnetic wave theory are often referred to as **paraxial waves**.

We now examine the general characteristics of beams of photons, whether they are coherent or incoherent. We must begin, as in the case of charged particle beams, with the single particle dynamics. Photons, viewed as massless

[1] Note that the sources of electromagnetic waves that drive rf accelerator cavities are also examples of coherent wave sources.

"particles" of light, have energy

$$U = \hbar\omega, \tag{8.1}$$

where ω is the angular frequency of the light wave, and $\hbar = h/2\pi$, with h being Planck's constant. This energy is generally thought of as being constant, since standard optical transformations do not affect the frequency of the wave.[2] The amplitude of the photon's momentum, on the other hand, is given by

$$p = \hbar k, \tag{8.2}$$

where $k = 2\pi/\lambda$ is the wave number of the photon.

The photon momentum and energy are related by the dispersion relation,

$$k = \frac{\omega}{c} n(\omega). \tag{8.3}$$

In Eq. (8.3), the index of refraction of the medium $n(\omega)$ is defined as a given function of a particular frequency. This index, of course, gives the phase velocity of the light wave,

$$v_\phi = \frac{\omega}{k} = \frac{c}{n(\omega)}. \tag{8.4}$$

Simple inspection of Eq. (8.4), with the phase velocity given by the ratio the ratio U/p, indicates that it is not (necessarily) a particle velocity, a conclusion which we may anticipate from our discussion of special relativity in Chapter 1, but is of the form pc^2/U in free space. The photon particle velocity—the velocity associated with energy flow of the photon field—is, of course, the group velocity introduced in Chapter 7,

$$v_g = \frac{d\omega}{dk} = \frac{c}{n(\omega) + \omega n'(\omega)}. \tag{8.5}$$

Equation (8.5) indicates that for a non-dispersive medium ($n(\omega) = n$ is constant, independent of ω), or in free space, the phase velocity and group velocity are identical. For the remainder of the present section we will assume this case, as it simplifies the discussion somewhat.

As a beam of photons encounters regions of changing index n during its propagation (lenses, prisms, etc.), the photon may change its momentum and thus its phase and group velocity. Like we have seen in Chapter 3, when distance along a transport line z is considered to be the independent variable, the equations of motion for a ray depend on the velocity of ray. Therefore, this change in momentum will affect our analysis of ray optics.

It is useful at this point to recall Snell's law of refraction at a planar boundary between two media,

$$n_1 \sin(\theta_1) = n_2 \sin(\theta_2), \tag{8.6}$$

where $n_{1,2}$ and $\theta_{1,2}$ are the indices or refraction and the angles measured from interface normal in the two media, respectively. In terms of momentum conservation, Eq. (8.6) clearly indicates that the momentum component transverse to the interface is conserved, while Eqs (8.2) and (8.3) indicate that the total momentum changes when $n_1 \neq n_2$, and thus the longitudinal momentum must change as well. Thus, some longitudinal momentum is exchanged with the

[2]The frequency of light can change if the medium it is propagating in changes its electric polarizability or magnetic permeability as a function of time. This is rare, occurring, for example in a plasma ionizing under the influence of an intense laser beam, in which case the laser photons may change frequency as the plasma index of refraction changes in time.

medium at a boundary. This momentum change may be accompanied, in the (classical or quantum) wave picture, by a partial reflection of electromagnetic power.

It can be seen that the energy of a monochromatic (single frequency) beam of N photons is given by

$$U_{\text{beam}} = N\hbar\omega, \tag{8.7}$$

which is independent of the medium in which the photons propagate. Equation (8.7) has an analogue in charged particle beams, $U_{\text{beam}} = NU = N\gamma m_0 c^2$. In order to increase the energy of a charged particle beam, one increases the energy per particle $\gamma m_0 c^2$ while keeping the number of particles N constant. In photon beams, however, increasing the number of particles (photons) N while keeping the energy per particle $\hbar\omega$ constant increases the beam energy. We shall return to this point later when we discuss lasers.

The photon also has the property of transverse polarization of its electric field. Polarization is critical in practice, as many devices for manipulating photons are polarization sensitive—for example, many laboratory lasers use polarization to accomplish switching (spatial redirection) of photon pulses. In this chapter, we will not dwell heavily on these practical concerns. It is interesting to note, however, that in addition to the massless attribute of the photon, the importance of polarization in light propagation is one of the key characteristics that set photon optics apart from charged particle optics.

8.2   Ray optics

The single particle dynamics of photons are referred to as ray optics, and are easily treated using the paraxial ray matrix formalism introduced in Chapter 3. We begin this discussion by rewriting Snell's law (Eq. (8.6)) for the horizontal projection of the transverse (paraxial) angle θ_x at a boundary normal to the design trajectory approximately as

$$n_1 x_1' = n_2 x_2'. \tag{8.8}$$

In order to keep the area preserving property of the ray transformations in trace space, we normalize the ray's (horizontal) trace space vector as follows,

$$\vec{x}_r \equiv \begin{pmatrix} x \\ nx' \end{pmatrix}. \tag{8.9}$$

At a planar boundary between media the transformation of this vector is simply the identity transformation, while a drift transformation matrix is given by

$$\vec{x}_r(z) = \mathbf{M}_d \cdot \vec{x}_r(z_0), \quad \text{with } \mathbf{M}_d = \begin{bmatrix} 1 & \frac{z-z_0}{n} \\ 0 & 1 \end{bmatrix}. \tag{8.10}$$

For symmetric lenses made up of solid media, the lens boundary is spherical, with radius of curvature R, as illustrated in Fig. 8.1. For paraxial angles ($x' = \sin(\theta_x) \cong \tan(\theta_x) \cong \theta_x$) and small offsets ($x \ll R$), the angle of the boundary's

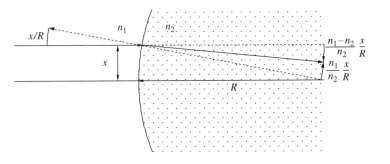

Fig. 8.1 Angular transformation of rays at spherical interface between two media.

interface normal changes linearly with offset as x/R, and the transformation of the ray angle is given by

$$\Delta x' = \frac{n_1 - n_2}{n_2} \frac{x}{R}.$$
(8.11)

The matrix transformation across the spherical interface is, therefore,

$$\mathbf{M}_{\text{sph}} = \begin{bmatrix} 1 & 0 \\ \frac{n_1-n_2}{R} & 1 \end{bmatrix}.$$
(8.12)

Note that the sign convention adopted here has assigned a positive radius of curvature as corresponding to the center of the radius lying in the downstream medium (2). If the center of curvature lies in the upstream medium (1), then one must change the sign of R in Eqs (8.11) and (8.12).

For the simple case of a double convex lens of index n, length L, and upstream (downstream) radii of curvature of $R_1(R_2)$, placed in air or vacuum ($n_1 \cong 1$), the total matrix transformation is obtained by multiplying the interface and drift matrices, to find

$$\mathbf{M}_{\text{T}} = \begin{bmatrix} 1 & 0 \\ \frac{n-1}{R_2} & 1 \end{bmatrix} \cdot \begin{bmatrix} 1 & \frac{L}{n} \\ 0 & 1 \end{bmatrix} \cdot \begin{bmatrix} 1 & 0 \\ \frac{1-n}{R_1} & 1 \end{bmatrix}$$

$$= \begin{bmatrix} 1 + \frac{L}{R_1}\frac{1-n}{n} & \frac{L}{n} \\ \left(\frac{n-1}{R_2}\right)\left(1 + \frac{L}{R_1}\right)\frac{1-n}{n} + \frac{1-n}{R_1} & 1 + \frac{L}{R_2}\frac{n-1}{n} \end{bmatrix}.$$
(8.13)

The effective thin-lens focal length can be found in the limit that $L/R_{1,2} \ll 1$, in which case we recover the lens-maker's formula for determining the focal length of a lens,

$$\frac{1}{f} \cong \mathbf{M}_{\text{T}21} = \left(\frac{n-1}{R_2}\right)\left(1 + \frac{L}{R_1}\frac{1-n}{n}\right) + \frac{1-n}{R_1} \cong (1-n)\left[\frac{1}{R_1} - \frac{1}{R_2}\right].$$
(8.14)

For spherical lens transport systems, the ray transport in the other transverse (i.e. vertical) dimension is identical. Sometimes, however, one wishes to focus the rays only in one transverse dimension, in which case a cylindrical lens may be employed. The cylindrical lens is constructed of cylindrical, not spherical, surfaces, and has no focal effect in the transverse dimension parallel to the cylinder axes.

One finds additional complications in the matrix description when using mirrors. A simple back-reflection (180°) by a planar mirror requires a flip in the

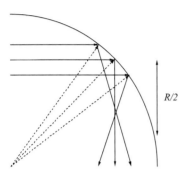

Fig. 8.2 Diagram of parallel incident rays in the *p*-plane focused by a spherical mirror oriented at $\theta_M = 45°$.

propagation direction, which because we utilize a right-handed coordinate system to describe the ray position, requires a reflection transformation ($\mathbf{M} = -\mathbf{I}$) on one of the transverse trace space vectors. For a mirror positioned with its normal at arbitrary angle θ_M with respect to the propagation axis, the axis about which the mirror is rotated is termed the *s*-direction while the other transverse dimension is called the *p*-direction. For $\theta_M \neq 0$, it is obvious that the reflection transformation must be assigned to the *p*-direction. For many optical transport systems, the reflections are all in one plane, and the *s*-and *p*-designations do not change, just as the *x*-direction is defined to be in the bending plane of a circular accelerator. If a reflection out of the nominal *p*-plane is encountered, the *s*-and *p*-designations (the *x* and *y* directions in the transport matrices) must be exchanged.

Spherical mirrors are also often employed to focus photon beams. The angular kick received by a ray of radial offset *r* at a mirror with a spherical surface of radius *R* oriented at 180° with respect to the incident propagation direction is simply $\Delta\theta = -2r/R$, with the sign of *R* taken as before. The projection of this angle onto the two transverse Cartesian dimensions is straightforward, for example, $\Delta x' = 2x/R$, implying a focal length $f = R/2$. On the other hand, if the mirror is rotated with respect to the initial propagation axis by an angle θ_M, the focal effects in the *s*- and *p*-directions are different. In Fig. 8.2, the focusing of rays offset in the *p*-direction is shown. In this case, while the angle of the surface normal relative to that found at the mirror center encountered by the ray is only a function of distance *x* from the mirror center *r*. This distance is a function of θ_M for a fixed value of *x*. Thus, the matrix for rays offset in the *p*-direction, including reflection flip of the coordinate system, is

$$\mathbf{M}_p = \begin{bmatrix} 1 & 0 \\ -\frac{2}{R\cos(\theta_M)} & 1 \end{bmatrix} \cdot \begin{bmatrix} -1 & 0 \\ 0 & -1 \end{bmatrix} = -\begin{bmatrix} 1 & 0 \\ -\frac{2}{R\cos(\theta_M)} & 1 \end{bmatrix}. \tag{8.15}$$

The focal length for rays offset in the *p*-direction is shorter by the factor $\cos(\theta_M)$ than the case of normal incidence, as is illustrated geometrically in Fig. 8.2.

For rays offset in the *s*-direction, the focusing effect is also changed compared to the normally oriented mirror. The matrix corresponding to the transformation of rays offset in the *s*-direction by reflection from a spherical mirror is given by

$$\mathbf{M}_s = \begin{bmatrix} 1 & 0 \\ -\frac{2\cos(\theta_M)}{R} & 1 \end{bmatrix}. \tag{8.16}$$

In applications where one wishes both to use a tilted focusing mirror and to keep the *s*- and *p*-offset focal lengths the same, an off-axis paraboloid mirror often replaces the spherical mirror. In this device, the curvature of the mirror surface in the *p*-plane is smaller than the curvature in the *s*-plane by the factor $\cos^2(\theta_M)$.

The focusing mirror and lens have been in use in light optics for over 400 years. In more modern times, a new method of focusing light over very long distances has emerged—the use of fiber optics. While some optical fibers are analogous to very narrow wave guides, their inherent strong dispersion makes them less useful than gradient index fibers for many applications. In a gradient index system, the index of refraction is a function of distance from fiber center ρ. The lowest non-trivial symmetric radial dependence is quadratic,

$$n(\rho) = n_0 - \tfrac{1}{2}n_2\rho^2 + \cdots , \tag{8.17}$$

where we have chosen the sign of the quadratic coefficient n_2 to correspond to a stable guiding system, as will be clear from the following discussion.

The evolution of off-axis rays in a quadratic gradient index fiber can be understood by again invoking Snell's law, with the interface between two distinct index media now generalized to a continuously varying index medium. The direction of index variation is radial in this system, so the angle θ is measured relative to the radial normal, as shown in Fig. 8.3. On the other hand, we are more interested in the paraxial radial angle $\rho' \cong \psi$, which we employ to rewrite Eq. (8.6) as

Fig. 8.3 Angular conventions for gradient index Fiber system.

$$n(\rho) \sin(\theta) \cong (n_0 - \tfrac{1}{2}n_2\rho^2)\cos(\rho') = \text{constant}. \tag{8.18}$$

The constant found in Eq. (8.18) can be recast by absorbing all additive constants on the right-hand side, and expanding the expression to lowest order in $r' \cong \psi$, to obtain

$$\frac{1}{2}n_0\left[(\rho')^2 + \frac{n_2}{n_0}\rho^2\right] = \text{constant}. \tag{8.19}$$

Equation (8.19) has the familiar form of an energy, or equivalently a Hamiltonian, for the simple harmonic oscillator system. This system, therefore, is equivalently described by the equation of motion

$$\rho'' + \kappa_f^2\rho = 0, \quad \text{where } \kappa_f^2 = \frac{n_2}{n_0}. \tag{8.20}$$

The general solution to a ray system governed by Eq. (8.20) was explored in Chapter 3, and the matrix corresponding to the solution was found to be

$$\mathbf{M}_{\text{fiber}} = \begin{bmatrix} \cos(\kappa_f(z - z_0)) & \frac{1}{n_0\kappa_f}\sin(\kappa_f(z - z_0)) \\ -n_0\kappa_f \sin(\kappa_f(z - z_0)) & \cos(\kappa_f(z - z_0)) \end{bmatrix}, \tag{8.21}$$

where we have reverted to the convention $\rho' \Rightarrow n\rho'$ in order to have consistency between Eqs (8.10), (8.12), (8.15), (8.16), and (8.21). If a medium is encountered where the quadratic coefficient is positive, the gradient index system is defocusing, and the matrix is analogous to Eq. (3.17).

8.3 Optical resonators

The optical resonator is in many ways similar to the resonant electromagnetic cavities used in rf acceleration introduced in Chapters 4 and 7. It is a device that uses boundaries to confine electromagnetic waves, yet in contrast to the resonant rf cavity, it has an open geometry, with no significant material on the sides (transverse to photon propagation direction) of the confined region. In addition, in the case of accelerator cavities, a boundary is encountered at least every one-half of a free-space radiation wavelength. In optical resonators, on the other hand, because the wavelength of the confined electromagnetic waves is so small, the resonator often contains a very large number of wavelengths along the propagation direction. We have spoken here in general terms of boundaries—this is another point of contrast between rf and optical resonators, as rf resonators are nearly always constructed from metal boundaries, while the mirrors used to define optical resonators often utilize layered dielectrics (Bragg reflectors) instead of metallic surfaces.

The optical resonator, examples of which are shown in Fig. 8.4, can be viewed from both the electromagnetic wave and the ray points of view. Electromagnetically, it can be seen that an integer number of half-wavelength of light must fit between adjacent mirror surfaces in order for a resonance to be achieved. This means that the axial mode separation, $\Delta f_{\mathrm{mode}} = c/L \cong f/N_\lambda$, where N_λ is the number of wavelengths contained in the resonator of round-trip length L, is very small. Many mode frequencies can be supported by the optical cavity for a typical electromagnetic source contained in the cavity, which often has a bandwidth well in excess of the axial mode separation. Thus, most laser oscillators (optical resonators with an internal active or lasing medium, as shown in Fig. 8.5) are multi-mode devices, and the exact geometry of the resonator does not dictate the precise mode of operation, unlike the rf cavity case. Instead, the excited mode frequencies are dictated by the characteristics of the lasing medium, and by filters (such as etalons, flat slabs of dielectric which pass certain resonant wavelengths well) that are placed inside of the optical cavity. While we are not yet in a position to discuss in great detail the physical role of the lasing medium in such a resonator, we can note that laser oscillation in the resonator represents a power balance between cavity losses, and the power in the photons derived from gain in the lasing medium.

Power losses in optical resonators arise, as illustrated in Fig. 8.5, from resistive, transmissive, and diffractive effects. The transmissive effects are due to imperfect mirror reflection, and control the frequency response of the resonator system (see Exercise 8.7). They are somewhat analogous to external coupling in rf cavities, and are introduced in a controlled way in order to allow power to be output from a laser oscillator. Resistive losses, which can arise in intracavity media such as the gain medium, as well as in mirrors, are analogous to wall losses in accelerator cavities. Finally, diffractive losses are due to the finite transverse size of the confining mirrors, which allow a portion of the confined power to escape around the outside of the mirrors. The closest analogy to this effect in rf cavities is the power loss from radiation through coupling holes back into the waveguide.

In such an optical resonator, the Q value can be defined as in Eq. (7.61),

$$Q = \frac{\omega U}{P} = \frac{\omega L}{c\delta_c} = \frac{2\pi N_\lambda}{\delta_c}, \tag{8.22}$$

where δ_c the fractional power loss suffered by the photon field per cavity round trip traversal. It can be seen that even if these losses are very high (δ_c of a few per cent is common), the Q of an optical resonator is very high (typically 10^5–10^6), and cavity filling and emptying takes very many optical periods. This high Q, which is surprising by normal conducting rf cavity standards, is due to the large number of stored wavelengths N_λ (the cavity is heavily overmoded in the longitudinal dimension), in combination with localized losses, mainly at the mirrors. On the other hand, because of their complete mode confinement characteristics, rf cavities experience loss over the entire length of the device

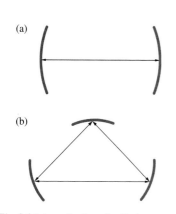

(a)

(b)

Fig. 8.4 Schematic view of optical resonators with (a) linear and (b) ring geometries.

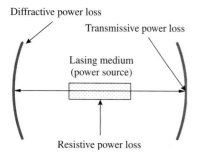

Diffractive power loss

Transmissive power loss

Lasing medium (power source)

Resistive power loss

Fig. 8.5 Schematic view of a laser oscillator within an optical resonator. The optical power source is from laser gain, and power losses arise from resistive, transmissive, and diffractive contributions.

It should also be noted that the resonance width $\Delta f_{res} \cong f/Q$ of a given mode is smaller than the mode separation by the factor $\delta_c/2\pi$. Thus, the axial modes are typically well separated from each other in an optical resonator.

We now move on to discuss the characteristics of optical resonators from the ray viewpoint. The ray, seen as a trajectory of a point-like object, is mathematically equivalent to a charged particle trajectory. Further, the resonator systems shown in Fig. 8.6 are periodic, and so the stability of the ray's transport inside of the resonator can be analyzed exactly as the periodic focusing lattice for charged particles was analyzed in Chapter 3. In this way, we can see that the ray optics of an optical resonator are equivalent to those of a circular accelerator.

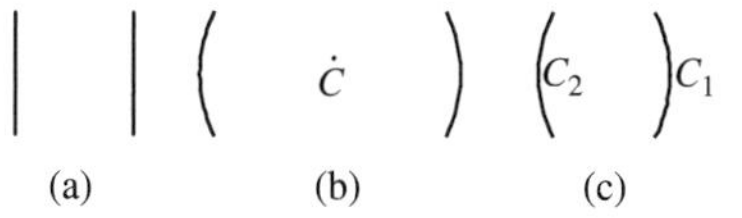

Some commonly encountered optical resonator configurations are shown in Fig. 8.6. These geometries are characterized in matrix-based stability analysis by the phase advance per period μ, given by $\cos \mu = 1/2 \, \mathrm{Tr}(\mathbf{M_T})$. It can be seen that the resonators shown in Fig. 8.6(a) and (b) represent the limits of stability for symmetric mirrors. In Fig. 8.6(a), the plane parallel resonator shown has no focusing at all, and $\mu = 0$. For the concentric resonator displayed in Fig. 8.6(b), the center of both mirrors radii of curvature is coincident, and one can easily show[3] that $\mu = \pi$, and the focusing is as strong as it can become while giving a stable system. In the center of the stable region we find that the confocal resonator geometry shown in Fig. 8.6(c) places the mirrors one radius of curvature apart, a focus + drift system with $f/L_d = R/2L_d = 1/2$, and $\cos \mu = 0$, or $\mu = \pi/2$.

(a)   (b)   (c)

Fig. 8.6 Common optical resonator geometries: (a) plane parallel, (b) concentric, and (c) confocal resonators.

[3]Note that the shortest periodicity defined by the symmetric mirror geometries in Fig. 8.6 is one-half of a round trip.

8.4   Gaussian light beams

The equivalence of the optics of non-diffractive photon beams, which can be thought of as bundles of rays, to particle beams, is intuitively obvious. What is not so obvious is that Gaussian light beams, which are the lowest-order mode of a localized (paraxial) electromagnetic wave, are also equivalent to particle beams of non-vanishing emittance. The elliptically shaped (or round, in the simplest case) intensity profile of a Gaussian light beam arises as solution to the paraxial wave equation for the transverse electromagnetic field components,

$$E_i, B_i = u(x, y, z) \exp\left[ik\left(z - \frac{ct}{n_0}\right)\right]. \tag{8.23}$$

The purpose of introducing the assumed field form in Eq. (8.23), which is to be substituted into a simplified version of the wave equation, is is to remove the fast variations in the field through use of the exponential, leaving an expression for the smoothly varying amplitude function $u(x, y, z)$. In the paraxial limit, the transverse second derivatives of $u(x, y, z)$ are much larger than the longitudinal second derivative, since the fast longitudinal oscillations in the field have been removed through use of Eq. (8.23). With the further approximation of slowly varying phase, $k|\partial_z u| \gg |\partial_z^2 u|$, and slowly varying envelope, $|\vec{\nabla}_\perp^2 u| \gg |\partial_z^2 u|$, a simplified version of the wave equation, the paraxial wave equation, is obtained,

$$\left[\vec{\nabla}_\perp^2 + 2ik\frac{\partial}{\partial z}\right] u(x, y, z) = 0. \tag{8.24}$$

The rigorous solution of Eq. (8.24) is a bit involved, and so we begin the analysis with a heuristic discussion of the solutions, which connects naturally to the ray optics analysis discussed above.

In the world of lasers, in order to formulate the transport of paraxial waves (coherent photon beams) using transport matrices, it is traditional to introduce the complex radius of curvature q. This quantity is defined as

$$\frac{1}{q(z)} = \frac{1}{R(z)} + i\frac{\lambda}{\pi n_0 w^2(z)}, \qquad (8.25)$$

where R is the radius of curvature of the phase fronts in the paraxial wave, λ is the free-space wavelength, n_0 is the on-axis index of refraction, and w is a measure of the transverse width of the wave. The motivations for this definition are not clear at this point, but should become apparent in the following discussion.

The imaginary part of the complex radius of curvature can be thought of as a mathematical artifice that allows the well-known spherical wave solution of the electromagnetic wave equation (cf. Ex. 7.2),

$$E_\perp \cong \frac{\exp(ikR)}{R} \cong \frac{1}{R}\exp\left(ikR\left(1 + \frac{x^2 + y^2}{2R^2}\right)\right), \qquad (8.26)$$

to have a term in the exponent that converts to a paraxial ($x, y \ll z \approx R$) Gaussian in the transverse coordinates. This can be seen upon substitution of $q^{-1}(z)$ from Eq. (8.25) into Eq. (8.26), and use of Eq. (8.23), giving

$$u(x, y, z) \cong \left[\frac{1}{R(z)} + i\frac{\lambda}{\pi n_0 w^2(z)}\right]\exp\left(ik\frac{x^2 + y^2}{2R(z)}\right)\exp\left(-\frac{x^2 + y^2}{w^2(z)}\right)$$

$$\cong \left[\frac{1}{R^2(z)} + \left(\frac{\lambda}{\pi n_0 w^2(z)}\right)^2\right]^{1/2}\exp[i\varphi_g(z)]$$

$$\times \exp\left(ik\frac{x^2 + y^2}{2R(z)}\right)\exp\left(-\frac{x^2 + y^2}{w^2(z)}\right). \qquad (8.27)$$

The scalar quantity u can be seen to be composed of three factors—a slowly varying phase and normalization factor ($q^{-1}(z)$), a complex exponential that displays the curvature of the wave-fronts, and a Gaussian envelope in x and y.

The propagation of the complex radius of curvature can be physically motivated by extension of the analogy with the real radius of curvature in spherical waves. Using this **ansatz**, we expect that in a lens-free drift, the real part of the complex radius of curvature is linearly proportional to the distance from the source point,

$$q(z) = q(z_0) + z - z_0. \qquad (8.28)$$

If the distance $z - z_0$ is negative, the real part of the radius of curvature is negative, and the waves are converging towards the source point at z_0. The imaginary part of q is constant in a drift space, and q is purely imaginary at $z = z_0$.

In a true spherical wave at the source point, the radius of curvature tends toward zero. However, with an imaginary component to the radius of curvature in the paraxial wave, this is no longer the case. Using Eq. (8.28), we can write he propagation of the real and imaginary parts of $q^{-1}(z)$ as follows:

$$\frac{1}{q(z)} = \frac{q_{Re} - iq_{Im}}{q_{Re}^2 + q_{Im}^2} = \frac{z - z_0 - i\lambda/\pi n_0 w^2(z_0)}{(z - z_0)^2 + (\lambda/\pi n_0 w^2(z_0))^2}. \tag{8.29}$$

Here the constant coefficient of the imaginary component of q is

$$q_{Im} \equiv Z_R = \frac{\pi n_0 w^2(z_0)}{\lambda} = \frac{\pi n_0 w_0^2}{\lambda}. \tag{8.30}$$

This coefficient is termed the **Rayleigh length**, Z_R, and is a function of the beam width w_0 at $z = z_0$, as well as the photon wavelength. Thus, the real component f $q^{-1}(z)$, which is identified as the inverse of the radius of curvature, gives

$$R(z) = \frac{q_{Re}^2 + q_{Im}^2}{q_{Re}} = \frac{Z_R^2}{z - z_0} + z - z_0. \tag{8.31}$$

he radius of curvature for this paraxial wave is infinite at both $z = z_0$ and $= \infty$. The imaginary component of $q^{-1}(z)$ is, on the other hand, associated vith the beam width,

$$w^2(z) = -\frac{\lambda}{\pi n_0(1/q)_{Im}} = \frac{\lambda}{\pi n_0}\frac{q_{Re}^2 + q_{Im}^2}{q_{Im}} = w_0^2\left[1 + \frac{(z - z_0)^2}{Z_R^2}\right]. \tag{8.32}$$

he square of the beam width varies quadratically away from its minimum value t the waist, with characteristic length Z_R. This behavior is exactly analogous o that found in charged particle beams, as can be seen from by comparing q. (8.32) to Eqs (5.48) and (5.49).

This analogy can be explored further by comparing the distribution function f a Gaussian charged particle beam to the Gaussian paraxial wave **intensity**. he photon beam probability density n_γ is simply proportional to the intensity I, nd thus the square of the electric field amplitude, as

$$I(x, y, z) = n_\gamma(x, y, z)\hbar kc^2 = \frac{\varepsilon_0}{2}n_0 cu^2(x, y, z). \tag{8.33}$$

hus, the photon density distribution associated with a Gaussian paraxial wave as the functional dependence in one transverse coordinate of

$$n_\gamma(x, z) \propto u^2(x, z) \propto \exp\left(-\frac{2x^2}{w^2(z)}\right), \tag{8.34}$$

nd we can identify the width parameter as being equal to twice the rms size, $\prime = 2\sigma$. With this identification, we note that the Rayleigh length is analogous o the minimum β-function

$$Z_R \Leftrightarrow \beta_x^*. \tag{8.35}$$

his analogy then suggests that we identify the **effective emittance of the** photon as

$$\varepsilon_x = \frac{\lambda}{4\pi n_0}. \tag{8.36}$$

he physical interpretation of this emittance is purely quantum mechanical— : simply represents the minimum uncertainty in phase space (x, p_x), as first formulated by Heisenberg,

$$\sigma_x \sigma_{p_x} \geq \frac{\hbar}{2} \quad \text{or} \quad \varepsilon_x = \sigma_x \sigma_{x'} \cong \frac{\sigma_x \sigma_{p_x}}{\hbar k} \geq \frac{\lambda}{4\pi n_0}. \tag{8.37}$$

Equation (8.37) shows that the Gaussian paraxial wave is the minimum uncertainty solution to Eq. (8.24). This fundamental quantum mechanical result should be familiar to those who have studied the Schrödinger equation and its solutions.

The quantum mechanical probability density of Eq. (8.34), which is proportional to the light intensity, gives a density profile analogous to the commonly encountered classical Gaussian density profile. However, the quantum mechanical distribution is a property of each photon, not a merely a property of the distribution as an ensemble—therefore the emittance given by Eq. (8.36) might be termed a type of inherent (single particle) emittance. This quantum mechanical emittance is also present in charged particle beams, as $p = h/\lambda$, and the minimum uncertainty in the "inherent normalized emittance" is simply $\sigma_x \sigma_{p_x}/m_0 c = \hbar/2 m_0 c = \lambda_c/2$, or one-half of the particle's Compton wavelength. This normalized emittance is at largest 1.93×10^{-13} m in the case of electrons, which is over three orders of magnitude smaller than has been achieved in any macroscopic beam. Nevertheless, when one considers, for example, the type of image formed in an electron microscope, this emittance represents a fundamental effect constraining resolution limits.

As we have seen, the analogy between classical and quantum (paraxial wave) Gaussian distributions is striking, and it is interesting to press it further, to ask what is the classical meaning of the radius of curvature in the paraxial wave case. The radius of curvature, as illustrated by Fig. 8.7 can be seen to be related to the local angle of propagation (phase front normal, or Huygen's ray direction), and the offset position as

$$\theta = \frac{\rho}{R}, \quad \rho = \sqrt{x^2 + y^2}, \quad \text{or} \quad x' \cong x/R, \tag{8.38}$$

if we consider the offset and the angle in only one Cartesian dimension. The angle x' defined in this way has an analogue in classical Gaussian (bi-Gaussian phase space) charged particle beams, in that it is the mean angle at a given point in x,

$$\langle x'(x) \rangle \equiv \int x' f(x, x') \, dx'. \tag{8.39}$$

This mean angle is the direction of the current density vector component at a given offset, $J_x(x)$, in a charged particle beam.

The identification of the direction of the phase front normal also suggests the correct algorithm for transporting the complex radius of curvature in an arbitrary optical system. This algorithm is obtained by generalizing the rule introduced in Eq. (8.28), through substitution of $q = R$ in Eq. (8.38),

$$q(z) = \frac{x(z)}{x'(z)} = \frac{\mathbf{M}_{11}x(z_0) + \mathbf{M}_{12}x'(z_0)}{\mathbf{M}_{21}x(z_0) + \mathbf{M}_{22}x'(z_0)} = \frac{\mathbf{M}_{11}q(z_0) + \mathbf{M}_{12}}{\mathbf{M}_{21}q(z_0) + \mathbf{M}_{22}}, \tag{8.40}$$

where $\mathbf{M}_{ij}$ are elements of the transport matrix giving the correct transformation of the ray vector from z_0 to z. Equation (8.40), which correctly

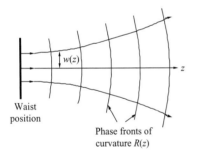

Waist position

Phase fronts of curvature $R(z)$

Fig. 8.7 Ray propagation and optical phase fronts in paraxial Gaussian wave near waist.

describes the evolution of the complex radius of curvature, is termed the ABCD-transformation rule in paraxial light optics. This is because in light optics the standard notation for representing the transport matrix is

$$\mathbf{M} \equiv \begin{bmatrix} A & B \\ C & D \end{bmatrix}, \quad \text{so} \quad q(z) = \frac{Aq(z_0) + B}{Cq(z_0) + D}. \tag{8.41}$$

As we have seen, it is possible to identify most aspects of Gaussian paraxial waves with some analogous attribute of a classical Gaussian charged particle beam. In particular, the description of beam envelope distributions are identical, and so, in principle, one may apply the methods of Twiss parameters, envelope equations, and σ-matrix to determine the paraxial wave envelope w, simply by identifying the emittance of the wave as $\lambda/4\pi n_0$. Conversely, one may just as easily borrow the complex radius of curvature formalism in order to examine the envelope behavior of a charged particle beam, if one assigns it an "effective wavelength" proportional to the emittance as $4\pi n_0 \varepsilon$.

One of the ways in which the complex radius of curvature approach to envelope descriptions is superior to the methods introduced in Chapter 5 for charged particle beams is in the geometrical understanding of simple optical resonators. For instance, since one knows the radius of curvature of the mirrors, one also knows, in the case of linear resonators, that the radii of curvature of the phase fronts at the mirrors must be identical to that of the mirrors. This condition in required because we have required a periodic solution to the paraxial wave equation in the resonator. Thus, by use of Eq. (8.31), one knows the distance from the mirror to the waist as a function of the Rayleigh length. As one has two equations (corresponding to measuring distance from either mirror) one may determine the Rayleigh length by requiring the waists be coincidental, and trivially construct the periodic paraxial wave solution inside of the resonator.

There are of course aspects of light beams that do not have classical analogues. An example of this is the phase factor $\varphi_g(z)$ found in Eq. (8.27), which is termed the Guoy phase shift. This phase shift is a correction to the paraxial approximation that assumes that the wave's z-dependence is given purely by $\exp[ikz]$ (Eq. (8.23)). The correction is necessary because this lowest-order dependence is, in fact, true only for plane waves. For waves of a finite extent, as the beam becomes smaller the transverse wave number associated with the wave $k_\perp$ is approximately $2/w$ (the transverse curvature of the wave amplitude) and the longitudinal frequency must become smaller, as indicated by use of the dispersion relation, $k_z = \sqrt{k^2 - k_\perp^2} \cong k - k_\perp^2/2k$, where $k = n_0\omega/c$. A spatial frequency error of this sort gives a monotonic phase shift, $\varphi_g(z) = \int_{-\infty}^{z} \Delta k_z \, d\tilde{z}$ (where $\Delta k_z \equiv k - k_z$), or

$$\varphi_g(z) \approx \int_0^z \frac{2 \, d\tilde{z}}{kw^2(\tilde{z})} = \frac{2}{kw_0^2} \int_0^z \frac{d\tilde{z}}{1 + \left(\dfrac{\tilde{z}}{Z_R}\right)^2} = \frac{2Z_R}{kw_0^2} \arctan \frac{\tilde{z}}{Z_R}$$

$$= \arctan \frac{\tilde{z}}{Z_R}, \tag{8.42}$$

where the integration is taken to be relative to a waist at $z_0 = 0$. This physical interpretation of the Guoy phase shift is in complete agreement with the definition implicit in Eq. (8.27),

$$\tan(\varphi_g(z)) \equiv \frac{Im(1/q)}{Re(1/q)} = \frac{\lambda}{\pi n_0 w^2(z) R(z)}$$

$$= \frac{Z_R}{\left[1 + \left(\frac{z}{Z_R}\right)^2\right]\left[\frac{Z_R^2}{z} + z\right]} = \frac{z}{Z_R}, \tag{8.43}$$

where we have again taken the waist to be at $z_0 = 0$.

Not every light beam, of course, is a coherent, Gaussian wave. The Gaussian beam is simply a lowest-order mode of the paraxial wave equation. There are higher-order modes to be considered as well, and they are discussed in the following section. In addition, many sources of light are incoherent, behaving much more like a classical collection of rays, and less like a paraxial wave. The case where this is so can now be clearly quantified: the spread in angles at the source multiplied by the source size must be much greater than the photon emittance, $\Delta\theta \cdot \sigma \gg \lambda/4\pi$. Incoherent light beams may arise from collimated thermal sources, that is, apertured beams existing inside of a telescope or camera observing a thermal source (a black-body radiator such as the sun, or its reflected light). Incoherence may also arise from the passage of coherent light beams through a diffusing medium, or diffuse reflection from a very rough surface.

8.5   Hermite–Gaussian beams

The general solution to the paraxial wave equation (Eq. (8.24)) is similar to the one encountered in the quantum mechanical simple harmonic oscillator. This is not too surprising considering that the fundamental solution is a Gaussian, just as in the quantum oscillator case. As Eq. (8.24) is separable in transverse coordinates, the form of its nth freely propagating (no gradient in index) solution is given in separable form $u(x, y, z) = u_x(x, z)u_y(y, z)$ by

$$u_{x,n}(x, z) = \left(\frac{2}{\pi}\right)^{1/4} \left(\frac{1}{2^n n! w_{x,0}}\right)^{1/2} \left(\frac{q_{x,0}}{q_x(z)}\right)^{1/2} \left(\frac{q_{x,0} q_x^*(z)}{q_{x,0}^* q_x(z)}\right)^{1/2}$$

$$\times H_n\left(\frac{\sqrt{2}x}{w_x(z)}\right) \exp\left[-i\frac{kx^2}{2q_x(z)}\right], \tag{8.44}$$

with a similar expression in the y coordinate. Note that a complex radius of curvature is now defined in both transverse dimensions

$$\frac{1}{q_{x,y}(z)} = \frac{1}{R_{x,y}(z)} + i\frac{\lambda}{\pi n_0 w_{x,y}^2(z)}, \tag{8.45}$$

and that the beam sizes and real radii of curvature of the phase fronts may be different in x and y. These complex radii of curvature are propagated along the beamline using the transport matrices for x and y, respectively, by use of Eq. (8.40). A more physically transparent version of Eq. (8.44) can also be written, therefore, as

$$u_{x,n}(x,z) = \left(\frac{2}{\pi}\right)^{1/4} \left(\frac{\exp[i(2n+1)\varphi_g(z)]}{2^n n! w_x(z)}\right)^{1/2} H_n\left(\frac{\sqrt{2}x}{w_x(z)}\right)$$

$$\times \exp\left[-\frac{x^2}{w_x^2(z)} - i\frac{kx^2}{2R_x(z)}\right], \qquad (8.46)$$

with an analogous equation for the wave profile in the y coordinate.

The functions $H_n(\chi)$ are the (real argument) Hermite polynomials of order n, where

$$H_0(\chi) = 1, \quad H_1(\chi) = \chi, \quad H_2(\chi) = 2\chi^2 - 2, \qquad (8.47)$$

and the higher-order polynomials are found by the **recursion relation**,

$$H_{n+1}(\chi) = 2\chi H_n(\chi) - 2n H_{n-1}(\chi). \qquad (8.48)$$

The Hermite–Gaussian functions given by Eq. (8.44) are orthonormal,

$$\int_{-\infty}^{\infty} u_n^*(x,z) u_m(x,z)\, dx = \delta_{nm}, \qquad (8.49)$$

where the orthonormality condition in Eq. (8.48) does not depend on z. This orthonormal **basis** can thus be used to represent an arbitrary paraxial wave

$$\upsilon(x,y,z) = \sum_{n=1}^{\infty}\sum_{m=1}^{\infty} c_{nm} u_{x,n}(x,z) u_{y,m}(y,z), \qquad (8.50)$$

where the coefficients c_{nm} are found by integral projection onto the basis functions,

$$c_{nm} = \int_{-\infty}^{\infty}\int_{-\infty}^{\infty} \upsilon(x,y,z) u_n^*(x,z) u_m^*(y,z)\, dx\, dy. \qquad (8.51)$$

The representation (Eq. (8.50)) of an arbitrary paraxial wave in terms of the basis functions given in Eq. (8.44) is well defined, but not unique. It depends explicitly on the choice of fundamental Gaussian widths w_{x0} and w_{y0}, which can be chosen at will. Thus, one may have an infinite number of representations of this form, with no clear way of choosing among them, unless one is dealing with a pure bi-Gaussian wave profile. In this case, one can obviously make the choice of w_{x0} and w_{y0} such that $c_{00} = 1$, and all other c_{nm} vanish. Failing the existence of this trivial case, or others in which only one Hermite–Gaussian mode exists, perhaps the most appealing way of choosing w_{x0} and w_{y0} is to require that the sum $\sum_{m,n} \|c_{nm}\|^2$ (the square of the "vector length" in a given representation) be minimized. This prescription can be easily implemented computationally using a finite, but large, number of basis functions.

The lowest-order modes allowed in the two-dimensional expansion of Eq. (8.50) are illustrated in Fig. 8.8. The symmetric Gaussian mode is the "ideal" beam, as it displays the best diffractive behavior. Other higher-order modes may appear in commonly encountered laboratory situations, for example, in laser oscillators that have asymmetries due to mirror misalignment.

The result of Exercise 8.11 points out a striking difference of classical distributions from quantum-mechanical waves. When one blocks the outer edges of

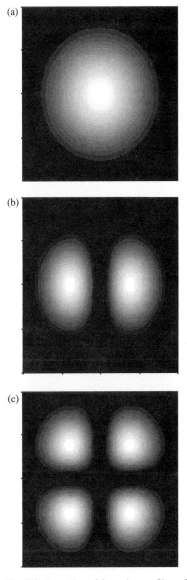

Fig. 8.8 Examples of intensity profiles of Hermite–Gaussian beams: (a) lowest-order bi-Gaussian beam ($m = n = 0$); (b) beam that is lowest-order symmetric in y and lowest-order antisymmetric in x ($m = 1, n = 0$) and (c) beam lowest-order antisymmetric in x and y ($m = n = 1$).

a classical beam, the subsequent profile produced is narrower than that with no obstacle present. On the other hand, for a coherent paraxial wave, the existence of blocking boundaries causes enhanced diffraction, and the beam is larger far from the obstacle. This can be qualitatively understood in terms of emittances. When one blocks the spatial edges of a classical beam with an uncorrelated trace space, the emittance is reduced, while the angular distribution of particles is unchanged. In the case of the analogous coherent paraxial wave, however, the effective emittance of the beam is not reduced (in fact, it is always increased, see Exercise 8.12), and so the angular spread of the wave is increased when the edges of a coherent light beam are removed.

As can be seen from Exercises 8.11 and 8.12, once a coherent light beam's profile ceases to be Gaussian by the cutting, or aperturing, of its outer edges, it becomes in some respects ill-behaved. The fast oscillations of the collimated profile in the asymptotic limit are commonly known as diffraction patterns, and in many applications (e.g. where uniform illumination of an object is desired) are to be avoided. A method for removal of these patterns in practice is known as relay imaging.

For example, relay imaging can be used to take a strongly apertured cylindrically symmetric light beam that is nearly uniform after passing an obstructing iris, and reproduce its nearly uniform profile at some point downstream of the iris. This process can be accomplished simply by placing the obstructing iris at a waist, and then imaging the obstruction itself. One can see that this prescription indeed works by examining the transport of a complex radius of curvature from the collimation point to a downstream point where $\mathbf{M}_{21} = \mathbf{M}_{12} = 0$, so that

$$\frac{1}{q(z)} = \frac{\mathbf{M}_{22}}{\mathbf{M}_{11}} \frac{1}{q(z_0)}. \tag{8.52}$$

Since we have assumed that $1/q(z_0)$ is purely imaginary (waist condition), the quantity $1/q(z)$ will also have no real component, and is therefore also at a waist. Further, Eq. (8.45) indicates that this waist is magnified by the factor $\sqrt{\mathbf{M}_{11}/\mathbf{M}_{22}}$. The assumed condition $\mathbf{M}_{12} = 0$ is obvious in our assumptions leading to Eq. (8.52), as it implies that a given ray's final offset is independent of initial angle (point-to-point focusing). The condition $\mathbf{M}_{21} = 0$ is less intuitive, as it means that the final angle is independent of the initial offset, which can be thought of as the analogue of the point-to-point condition for angles. In this way, we can see that the waist-to-waist transformation of Eq. (8.52) implies two conditions—the beam's spatial profile is reproduced (up to a magnification factor), and the radius of curvature of the paraxial wave must become infinite (the mean ray angle vanishes at all offsets).

Thus, the relaying of an image allows the transformation of one paraxial wave profile into another of nearly identical form. On the other hand, we have seen that the profile changes somewhat radically as the beam exits a collimating aperture and eventually takes on an asymptotic form, which expands in a self-similar fashion (the angular profile discussed in Exercise 8.11 is nearly unchanging after $\Delta z \gg z_R$). Examination of Eq. (8.46) informs us what is taking place— only the relative phases of the Hermite–Gaussian components change during the region $\Delta z \leq z_R$ near the waist, as the Guoy phase of each component is related to the fundamental Gaussian phase by $\varphi_{g,n}(z)(x,z) = (2n + 1)\varphi_g(z)$. After the phase shift $\Delta\varphi_g(z) \to \pi/2$ ($\Delta\varphi_{g,n}(z) \to \pi/2 + n\pi$), the relative phases of all components do not change, and the profile becomes self-similar in its expansion. In fact, the transformation of a beam either to or from a focus is equivalent to the **Fourier transformation** of the wave's transverse profile, as is discussed in Exercise 8.17.

As can be seen by the results of Exercise 8.14, the larger n Hermite–Gaussian components of the wave can be viewed as having a larger extent in k_x space. The physical way to see this is that a larger k_x content to a wave implies faster spatial changes in the wave, which is clearly true of the Hermite–Gaussian wave profiles as n is increased. The insight that the transverse profile of a paraxial wave can be easily converted to its Fourier transform by simply focusing the light has inspired the technique of **spatial filtering**. This technique commonly uses the optics shown in Exercise 8.14(a), with an aperture placed between the two lenses, at a distance of one focal length away from each lens. The aperture then blocks the larger k_x components of the original profile ($k_x > w_{x.0}^{-1}$), allowing a wave profile at the exit that more closely approximates a fundamental Gaussian wave.

8.6 Production of coherent radiation: lasers

The subjects of coherent radiation production in general, and the detailed physical principles of the laser in particular, are well beyond the scope of this text. However, in this section, we will give a short introduction to the subject of coherent radiation production by lasers in order to illustrate some physical principles that fall most neatly within the context of the rest of this book.

At the most fundamental level, electromagnetic radiation is produced by the acceleration of charged particles. In the non-relativistic case, the classical formulation of radiation production yields an emitted power from a free particle under acceleration due to a Lorentz force $\vec{F}$ that is given by

$$P = \frac{q^2\vec{F}^2}{6\pi\varepsilon_0 m_0^2 c^3}. \tag{8.53}$$

This radiation is emitted over a range of frequencies (and therefore photon energies) specified by the details of the applied forces.

On the other hand, laboratory sources of coherent electromagnetic radiation most commonly use quantum mechanical systems—atomic, molecular, or solid-state systems. In such cases the allowable emitted photon energies are restricted to be near to the differences in energy levels, $\hbar\omega_{ij} = \Delta E_{ij} = E_i - E_j$, between the states of the atomic electrons, electronic bands in semiconductors, or of molecular vibrations and rotations. This type of quantum level system is illustrated schematically in Fig. 8.9. Quantum mechanics dictates that transitions between levels proceeds both **spontaneously** at a certain mean rate, and at an enhanced rate when the photon density (wave intensity) is non-zero. The enhancement of the transition rate in the presence of non-vanishing wave intensity is termed **stimulated emission or absorption**.

Einstein originally deduced the relationship between the rates of spontaneous emission and stimulated emission/absorption in the context of black body radiation. In his study, he noted that the total rate of downward transitions $\dot{N}_{ij}$ between two quantum levels (i and j) is given by contributions from

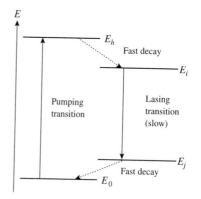

Fig. 8.9 Energy level diagram for typical four-level quantum system.

expected spontaneous rates (A) and stimulated rates ($BI(\omega_{ij})$), where $I(\omega_{ij})$ is the electromagnetic spectral energy density at the transition frequency, as

$$\frac{dN_{ij}}{dt} = A_{ij}N_i + B_{ij}I(\omega_{ij})N_i. \qquad (8.54)$$

Conversely, as there is no spontaneous upward transition, the upward transition rate between the two levels is only due to stimulated absorption,

$$\frac{dN_{ji}}{dt} = B_{ji}I(\omega_{ij})N_j. \qquad (8.55)$$

The coefficients B_{ji} and B_{ij} are known from time-dependent Schrödinger perturbation theory to be equal, as they are proportional in the same way to the same transition matrix element. In addition, Boltzmann statistics dictate that in equilibrium, the ratio of the populations in this two-level system must be

$$\frac{N_i}{N_j} = \exp\left[-\frac{\Delta E_{ij}}{k_B T}\right] = \exp\left[-\frac{\hbar\omega_{ij}}{k_B T}\right]. \qquad (8.56)$$

In equilibrium, the upward and downward transition rates must be equal, giving

$$I(\omega_{ij}) = \frac{A_{ij}/B_{ij}}{\exp[\hbar\omega_{ij}/k_B T] - 1}, \qquad (8.57)$$

which is identical to Planck's black-body radiation law, if the identification is made

$$\frac{A_{ij}}{B_{ij}} = \frac{\hbar\omega_{ij}^3}{\pi^2 c^3}. \qquad (8.58)$$

When stimulated transitions dominate over spontaneous transitions (this is generally *not* a case of black-body mediated equilibrium), we have $I(\omega_{ij}) \gg A_{ij}/B_{ij} = \hbar\omega_{ij}^3/2\pi c^3$, and the wave is strong, with large energy density (or large number of photons in the relevant volume—a cubic wavelength—around the quantum level system). Furthermore, if the population density of the states is not in thermal equilibrium, $N_i > N_j$, more photons are emitted than absorbed by the population until the populations are nearly equalized. In a laser (light amplification by stimulated emission of radiation) just such a system is utilized—the population of a state higher in energy is made, by a variety of techniques, to be larger than that of a state lower in energy. When this population inversion occurs and a wave of the correct photon frequency ω_{ij} is introduced, photon emission is stimulated, with the emitted photons added in a definite phase relationship to the original wave. In this way, the wave intensity may grow in the lasing medium, which contains the correct population-inverted material.

A common method of achieving population inversion, the four-level quantum system, is diagrammed in Fig. 8.9. In this scheme, the system is initially excited from the ground state, where the overwhelming majority of the population resides in equilibrium, by a strong pump wave of frequency $\omega_{h0} = (E_h - E_0)/\hbar$. Many of the levels in the population then decay quickly (large A_{hi}) down to state i, which does not easily decay to state j by spontaneous emission. Once the system is in state j, however, it decays very quickly back to the ground state. These conditions guarantee population inversion between states i and j if one has a strong enough pump wave. Thus the lasing is allowed at the frequency

$\omega_{h0} < \omega_{ij}$. It may be noted that the inherent frequency uncertainty in this system due to the spontaneous emission lifetime $\tau_{ij} = A_{ij}^{-1}$ leads to a finite line width $\Delta\omega = A_{ij}$. This spread in allowed frequencies can be understood mathematically as being analogous to the spread in frequency response in a resonant electromagnetic cavity, as discussed in Chapter. In the cavity case, the "lifetime" is due to decay of the trapped electromagnetic field via wall resistivity.

Unfortunately, the overview of lasing mechanisms given in preceding paragraphs cannot be easily expanded—the subject is simply too vast. Interested readers are referred to other books that specialize in such a discussion listed in Section 8.10. Instead, we will present a heuristic discussion of wave propagation in a lasing, or gain, medium, that is based on the discussion of waves in a medium with finite conductivity, as introduced in Chapter 7.

In the present case, however, we will assume a generalized conductivity, defined as usual by $\vec{J}_e = \sigma_c \vec{E}$, but where we now allow the possibility that the constant σ_c may take on positive or negative values. The wave equation, is therefore, identical to the one discussed in Section 7.2, with the same dispersion relation (Eq. (7.6)). However, in this case it is evaluated in a regime where the amplitude of the displacement current density $\omega\varepsilon\vec{E}$ is much larger than the ohmic current density $\sigma_c\vec{E}$, or $\omega\varepsilon/\sigma_c \gg 1$. For the assumed conditions, we can write the approximate solution for the wavenumber k as

$$
\begin{aligned}
k &= \sqrt{\mu_0\varepsilon\omega^2 - i\omega\mu_0\sigma_c} \\
&\cong \frac{\omega c}{n_0}\left[1 - i\frac{\sigma_c}{2\omega\varepsilon}\right] \equiv k_0 - i\frac{\Gamma}{2},
\end{aligned}
\tag{8.59}
$$

where $n_0 = (\mu_0\varepsilon)^{-1/2}$. Thus, the wavenumber has a small imaginary component ($\Gamma \ll 2k_0$) and the spatial dependence of the wave amplitude can be written as

$$
\vec{B} \propto \exp\left[ik_0 z - \frac{\Gamma}{2}z\right],
\tag{8.60}
$$

which damps in z if $\Gamma > 0$ and yields exponential gain if $\Gamma < 0$.

The idea that the generalized conductivity could be considered to be negative deserves further comment. In order to evaluate this possibility, we examine the power dissipation by the wave as it propagates in an ohmic medium, where $\Gamma > 0$. In this case, the intensity of the wave has the spatial dependence

$$
I \propto \|\vec{B}\|^2 \propto \exp[-\Gamma z],
\tag{8.61}
$$

which can also be viewed as the solution to the differential equation

$$
\frac{dI}{dz} + \Gamma I = 0.
\tag{8.62}
$$

The presence of the classical intensity, or equivalently, the quantum mechanical photon density, stimulates absorption of photons at a rate proportional to the intensity itself. The quantity Γ can be expressed in terms of microscopic physical attributes of the material, in particular, the power loss due to electron collisions within the medium.

Conversely, if we posit that $\Gamma < 0$, we have an exponential gain per unit length $|\Gamma|$ of the intensity, or photon density. The microscopic "picture" we must develop in order to explain this case can be expressed in semi-classical terms as follows: each photon may "collide" with a net inverted population density $n_{inv} = n_i - n_j = (N_i - N_j)/V$ (which can be positive or negative) in the medium, and each member of this population has a characteristic "size" or interaction **cross-section** Σ_c which is proportional to the quantum mechanical transition probability (B_{ij}). The mean free path for the photons in such a system is thus $l_f = |n_{inv}\Sigma_c|^{-1}$, and at each interaction the photon is either absorbed or replicated. In this simple picture, the gain or damping constant is given by $|\Gamma| = |n_{inv}|\Sigma_c$ and can be seen to be the inverse of the mean free path for a stimulated emission ($\Gamma > 0$) or absorption ($\Gamma < 0$) collision.

In a standard laser physics textbook treatment of gain, the idea of a generalized conductivity is not typically employed. Instead, one introduces a complex atomic or molecular **polarizability**, which is mathematically equivalent to our generalized conductivity, but implies a different physical picture. It is, in fact, closer to the oscillator picture of an rf cavity, as discussed in Chapter 7, in that a classical oscillator may be excited into a state where much energy is stored, but if the exciting force is reversed in phase by 180° the power will efficiently flow **out** of the oscillator. It should be emphasized that neither a generalized conductivity nor an atomic polarizability model gives a clear quantum mechanical picture of the microscopic physics of laser gain. We have introduced generalized conductivity in this discussion in order to make a pedagogical connection with the ohmic power dissipation encountered in Chapter 7.

Lasers are devices based on the gain media discussed above. In a continuous wave (cw) **laser oscillator**, the gain medium, located inside of an optical resonator, is kept at a roughly constant level of population inversion. In such a system, the round trip gain of the photon field is $\exp(\Gamma L_m)$, where L_m is length of the medium encountered in a round trip through the resonator. For cw operation, the power circulating inside of the resonator must be constant, and so fractional the power lost per round trip must be equal to the fractional power added per round trip,

$$\exp(\Gamma L_m) - 1 = \delta_c. \tag{8.63}$$

In a laser oscillator both the round trip gain and power loss are typically much smaller than unity, and as a result this relation can be approximated as

$$\Gamma L_m \cong \delta_c. \tag{8.64}$$

Therefore, the laser oscillator can provide a steady supply of coherent light flux through out-coupling from the cavity mirrors (or even around the outer edge of the mirrors in an unstable resonator). While the powers associated with typical laser oscillator outputs are modest (on the order of a few watts for commonly encountered laboratory lasers), it should be noted that the power circulating inside of the cavities is a factor of δ_c^{-1} times larger than the power lost from the cavity.

Ultra-high power laser applications require a device in which the single pass gain $\exp(\Gamma L_m)$ exceeds unity by a large factor (many orders of magnitude in some cases). This is termed a **high gain laser amplifier**, and plays a key role in physics research in the area of laser–matter interactions. In order to extract the most power from the gain medium in this case, one attempts to fill the medium up to its boundary—typically a cylindrical rod. In such a scenario, the output power will mimic the geometry of the medium, and a wave transverse profile with a sharp radial "edge" at the output results. In order to take this wave and efficiently transport it to a downstream lasing medium rod (for further amplification in a multistage amplifier), the relay-imaging configuration of Exercise 8.14(b) is often used. At the end of the chain of amplifiers, spatial filtering may be applied to "clean up" the wave and to create a well-behaved Gaussian beam.

In a high-gain laser system, the wave may **saturate** the amplifier's gain medium, in effect degrading the gain until the population levels equalize and the gain vanishes. The **rate equations** (ignoring spontaneous emission) for the population densities in this process can be written approximately as

$$\frac{dn_i}{dt} = B_{ij}I(\omega_{ij})(n_j - n_i) \quad \text{and} \quad \frac{dn_j}{dt} = B_{ij}I(\omega_{ij})(n_i - n_j), \qquad (8.65)$$

with the equation for the wave intensity given by

$$\frac{d}{dt}I(\omega_{ij}) = \hbar\omega(n_i - n_j)B_{ij}I(\omega_{ij}), \qquad (8.66)$$

and where we have ignored the possibility of pumping the upper state during the time of interest. The system described in Eqs (8.65) and (8.66) is a bit oversimplified, because the variables $I(\omega_{ij})$, n_i, and n_j, in addition to their time dependence, are also functions of spatial variables, in particular the propagation direction z.

The purpose of this section, and of this chapter for that matter, is to point out the most obvious connections between photon beams and our previous discussions of particle beams, their optics, and technical underpinnings. The topics not approached in this cursory section are incredibly numerous—polarization control, mode locking, Q-switching, nonlinear dielectric materials for harmonic light generation, to mention a few. The interested reader who has need of a more detailed treatment is again referred to some excellent books listed in Section 8.10.

8.7   Electromagnetic radiation from relativistic charged particles

The generalization of the theory of electromagnetic radiation from a charged particle in a magnetic field to include relativistic effects is, like many phenomena in electromagnetism, natural and yet fraught with subtlety. The photons emitted by relativistic charged particles display a rich structure in both angular and wavelength spectrum. One of the most striking of these spectral characteristics is that the typical angle of emission, which is centered along the instantaneous direction of the particle motion, is localized in a band of width approximately γ^{-1}. This angular spread can be quite narrow for ultra-relativistic electron beams.

Most other aspects of the emitted radiation need not be discussed in detail. For the purpose of this section, we simply state that the power emitted due to the transverse Lorentz force $\vec{F}_\perp$ is

$$P = \frac{\gamma^2 q^2 \vec{F}_\perp^2}{6\pi\varepsilon_0 m_0^3 c^3}. \tag{8.67}$$

Equation (8.67) indicates that there is now an energy dependence—or more explicitly and importantly, a γ-dependence—of the synchroton radiation power loss.[4] Because of this γ-dependence, synchrotron radiation plays a critical role in the physics and design of electron (or positron) storage rings, synchrotrons that store particles either for high-energy collisions (electron–positron colliders) or for dedicated synchrotron radiation sources used for X-ray beam production.

A significant amount of power can be lost by the beam particles in such a ring. If we hold the average magnetic field strength constant $\bar{B}$ and allow the energy to vary (as is the case for the operation of such a machine with a finite spread in particle energies), the emitting synchrotron radiation power has the following useful form:

$$P = \frac{\gamma^2 q^4 v^2 B^2}{6\pi\varepsilon_0 m_0^2 c^3} = \frac{2}{3}\gamma^2 r_c m_0 c^2 \left(\frac{qv\bar{B}}{m_0 c^2}\right)^2, \tag{8.68}$$

where r_c is the classical particle radius (see Eq. (1.23)). Equation (8.68) gives an average energy loss per revolution in a circular accelerator

$$\Delta U = -\frac{4\pi R}{3c}\gamma^2 r_c m_0 c^2 \left(\frac{qv\bar{B}}{m_0 c^2}\right)^2 = -\frac{4\pi}{3}r_c qc\bar{B}\beta^2\gamma^3. \tag{8.69}$$

This finite energy loss per turn must, in a steady state (e.g. in a storage ring), be compensated by the acceleration due to rf voltage. Thus, the synchronous (stationary) phase of a design particle in such a non-accelerating storage ring is no longer zero, but is defined by

$$qV_0 \sin(\phi_0) = \frac{4\pi}{3}r_c qc\bar{B}\beta^2\gamma^3. \tag{8.70}$$

In electron (or positron) circular machines, the Lorentz factor γ is quite large for all synchrotron rings presently in operation, and this phase shift due to radiative energy loss is non-negligible. On the other hand, in all present heavy particle machines this effect can be ignored, because for a given beam energy γ (and r_c) is smaller by the ratio of the electron-to-heavy particle mass.

The limits of design in electron storage rings due to synchrotron radiation power loss can be seen by writing Eq. (8.67) in terms of an average radius of curvature in a circular accelerator,

$$P = \frac{2r_c c(\beta\gamma)^4}{3R^2}m_0 c^2. \tag{8.71}$$

For a given radius of curvature the electron beam power due to synchrotron radiation scales as γ^4. This energy dependence is quite dramatic, and has positively motivated beam physicists to build synchrotron light sources, devices that can produce copious amounts of radiation in directed beams, at wavelengths down to the X-ray regions. Unfortunately, the details of the wavelength and angular spectra due to synchrotron (in simple bends, or in undulator magnets) radiation

[4] Synchrotron radiation refers to electromagnetic radiation emitted due to acceleration of a relativistic charged particle in a magnetic field. In the non-relativistic limit, this type of radiative emission is more commonly referred to as cyclotron radiation. Note that these fundamental processes—like betatron oscillations—were named after the accelerators in which they were originally observed.

are beyond the scope or our present treatment and will be taken up in the next volume of this text.

If the radius of curvature is not allowed to exceed a certain size (set by, e.g. geographical or financial considerations), Eq. (8.71) has negative implications, in that so much synchrotron radiation is emitted that the rf system needed to restore this energy loss becomes unmanageably large. This situation has been encountered already in the world of high-energy colliders—the largest electron–positron collider, LEP, as is illustrated by Exercise 8.21.

8.8 Damping due to synchrotron radiation*

Energy loss due to synchrotron radiation not only causes a shift in the stationary phase of the design particle in a synchrotron, but it also leads to phase space damping. This is because the energy loss per turn is larger for particles with relative energy offset $\delta U / U_0$; for small offsets it has linear dependence

$$\Delta U \cong -\frac{4\pi R}{3c} \gamma_0^2 r_c m_0 c^2 \left(\frac{q v_0 \bar{B}}{m_0 c^2}\right)^2 \left[1 + 2\frac{\delta U}{U_0}\right]. \qquad (8.72)$$

A force that is proportional to the energy (or momentum) offset produces damping in the motion. To illustrate this, we generalize the small amplitude equation of motion, Eq. (4.48), near φ_0 including the force on the design particle and the force linear in energy offset as given by Eq. (8.72),

$$\frac{d^2}{dt^2}(\delta U) + \frac{\omega_c q V_0 \sin\varphi_0}{2\pi U_0} \frac{d}{dt}(\delta U) + \frac{|\eta_\tau| h\omega_c^2 q V_0 \cos\varphi_0}{2\pi \beta_0^2 U_0} \delta U = 0, \qquad (8.73)$$

or in more standard form,

$$\frac{d^2}{dt^2}(\delta U) + 2\nu_d \frac{d}{dt}(\delta U) + \omega_s^2 \delta U = 0. \qquad (8.74)$$

Equations (8.73) and (8.74) ignore coupling between horizontal and longitudinal motion that arises because energy loss to radiation occurs in bend magnets, where the dispersion η_x is non-vanishing. Thus, a particle that loses energy by radiation in a bend immediately finds itself referenced to a different central horizontal betatron orbit for betatron oscillations, or vice versa. In this way, betatron and synchrotron motion are coupled, and horizontal damping is coupled to longitudinal damping in a more complicated way than we indicate here. As a result, Eq. (8.74) is only approximate and illustrative.

In Eq. (8.74), we have used the synchrotron frequency

$$\omega_s = \sqrt{\frac{|\eta_\tau| h q V_0 \cos\varphi_0}{2\pi \beta_0^2 U_0}} \omega_c, \qquad (8.75)$$

and introduced the damping coefficient,

$$\nu_d = \frac{\beta_0^2 \omega_s^2}{|\eta_\tau| h\omega_c} \tan\varphi_0 = \frac{2r_c qc\bar{B}}{3m_0 c^2}(\beta_0\gamma_0)^2 \omega_c, \qquad (8.76)$$

where we have employed the stationary phase φ_0 as defined by Eq. (8.70).

Assuming, as is always true for existing storage rings, that $\nu_d \ll \omega_s$, the solution to Eq. (8.74) can be written in the form

$$\delta U \cong \delta U_0 \cos(\omega_s t + \theta_s) \exp(-t/t_d), \qquad (8.77)$$

where the **damping time** $t_d \equiv \nu_d^{-1}$, and θ_s is an arbitrary phase introduced to handle all initial conditions. From this we see that the synchroton motion is harmonic with slowly decaying (characteristic exponential decay time t_d) amplitude. Eventually, all synchrotron motion must cease, and no matter what the initial conditions are, the particles end up at the longitudinal phase space (fixed) point $\delta U = 0$, $\varphi = \varphi_0$. However, the derivation Equation (8.77) assumes a classical synchrotron radiation process. In the classical picture, the power loss is continuous, but in the valid quantum picture the power emission takes place in discrete energy jumps associated with individual photons. Thus, a radiating particle does not converge to the classical fixed point, but jumps around it in a stochastic manner due to the discrete statistical nature of emission. This effect leads to a minimum finite final state emittance of a beam damped by synchrotron radiation.

In a synchrotron, the damping of the longitudinal phase space trajectories is also accompanied, as mentioned above, by damping of the transverse motion. This can be seen schematically through the diagram shown in Fig. 8.10. The initial state (Fig. 8.10(a)) shows a large transverse momentum p_x. The process of emitting a photon of momentum $\hbar k$ induces a recoil of essentially equal and opposite charged particle momentum, as indicated in Fig. 8.10(b). Finally, Fig. 8.10(c) illustrates that the reacceleration of the charge particle restores the radiated longitudinal momentum but not the component of momentum transverse to the motion, in this case p_x. Thus, the radiation process can also damp the transverse motion of a particle circulating in a synchrotron.

Because the radiated component of the synchrotron radiation is mainly longitudinal, the quantum noise inherent in longitudinal phase space damping is generally much larger than that which affects the transverse motion directly. One expects that the final transverse normalized emittance at equilibrium in the transverse phase planes would be much smaller than that found in the longitudinal phase plane. We have already noted that is dispersion leads to a coupling of the longitudinal and horizontal phase planes. Therefore, in the absence of x–y phase plane coupling, one finds that the equilibrium vertical emittance due to radiation damping effects in synchrotrons is very small, while the horizontal emittance is much larger due to its coupling to the higher-temperature longitudinal phase space. The relative damping rates and equilibrium values of the horizontal and longitudinal emittances are determined by the dispersion function around the ring. All of the damping rates are related, in that they obey a sum rule—the connection between horizontal and longitudinal damping is obvious from our discussion above, while the interplay between horizontal and vertical is simply dictated by lattice choice.

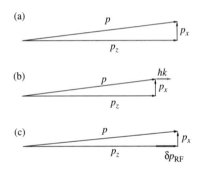

Fig. 8.10 Schematic of momentum vector evolution during transverse phase space damping in a synchrotron: (a) initial particle state; (b) state after photon emission (photon momentum $\hbar k$); and (c) state after reacceleration to original p_z by rf system's addition of δp_{RF}.

8.9   Undulator radiation and the free-electron laser*

As we have mentioned above, free electrons can be used to create coherent radiation. Many longer wavelength (e.g. radio-frequency) devices that

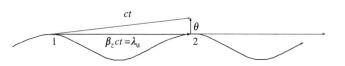

Fig. 8.11 Schematic of constructive interference effects in undulator radiation. The solid undulating line indicates the electron orbit due to the magnetic field, and the thick shaded lines show the constructively interfering photons' small angle emission direction.

create high-power fluxes, such as klystrons and magnetrons, utilize free, moderately energetic ($<500\,\text{keV}$) electron beams. These devices often employ resonant cavities in order to produce narrow frequency-spread output radiation (as in the laser) and are, in fact, used to power accelerator cavities.

On the other hand, of the more promising new devices for creating very short wavelength electromagnetic radiation, the free-electron laser (FEL), employs relativistic electron beams as its source of energy. Because quantum-based systems using bound electrons have limitations on both intensity (due to material breakdown) and minimum wavelength (due to available transition energies), the FEL is viewed as perhaps the most likely path to development of lasers in the deep ultraviolet to X-ray range.

In order to understand the basic physical mechanisms upon which the FEL is based, we must first begin with **undulator radiation,** which is generated by an electron that performs small amplitude oscillations in a magnetic undulator. The total power emitted by an electron as it oscillates in the undulator (see discussion in Section 2.8) can be deduced from the synchrotron radiation formula given by Eq. (8.67). However, the wavelength spectrum of the undulator radiation is quite different than the broad-band radiation characteristic of simple bend magnet-excited synchrotron radiation. In the case of undulator radiation, the spectrum collapses to narrow bands about a certain **undulator radiation frequency** and its harmonics.

The cause of this frequency selectivity in the radiation process can be understood by examination of Fig. 8.11. Consider the emission of light of a given frequency $\omega_r = k_r c$ at a finite, yet small, angle θ from the electron at point 1. In order for a wave-front emitted at point 2 to be at the correct phase to reinforce the wave-front emitted at point 1, the following relationship must exist:

$$k_r c (\cos \theta - \beta_z) t = 2\pi. \tag{8.78}$$

Equation (8.78) indicates that the radiation's phase front outruns the electron by one wavelength $\lambda_r = 2\pi / k_r$ every undulator wavelength λ_u. For relativistic longitudinal velocities, small wiggle motion and $\theta \ll 1$, Eq. (8.78) can be expressed as,

$$\lambda_r \cong \frac{\lambda_u}{2\gamma^2} (1 + a_u^2 + (\gamma \theta)^2). \tag{8.79}$$

Here we have introduced the undulator strength parameter a_u, which measures the degree to which the wiggle motion "slows down" the longitudinal motion. For a planar undulator of peak field B_0, we can see from averaging Eq. (2.71) that

$$a_u \cong \frac{q B_0}{\sqrt{2} k_u m_0 c}. \tag{8.80}$$

From Eq. (8.80), we deduce that a_u is, in fact, the rms transverse momentum associated with the wiggle motion in the undulator if measured in units of $m_0 c$. Thus, it is a measure of how relativistic the transverse oscillation appears in the Lorentz frame moving at the average longitudinal velocity of the electron. Note that in a helical undulator, the transverse momentum has constant amplitude, and the factor of $\sqrt{2}$ in the denominator of Eq. (8.80) is replaced by 1.

Equation (8.79) indicates that the undulator radiation spectrum has a shift towards longer wavelengths for off-axis ($\theta \neq 0$) emission. However, as was stated above, most of the radiation intensity is located in angles $\theta < \gamma^{-1}$ and as such this effect, which may be termed **off-axis Doppler shift**, does not introduce significant problems when one wishes to create nearly monochromatic photon beams from undulator radiation.

Harmonics in the radiation appear due the fact that, for $a_u \cong 1$ or larger the transverse oscillation is relativistic in the beam frame, and therefore certainly not simple harmonic. The resulting nonlinear motion contains Fourier components that have odd harmonics of the fundamental frequency λ_u. Thus odd harmonics of the fundamental radiation wavelength (Eq. (8.79)) are introduced into the emission spectrum. The strength of the harmonics is obviously a function of a_u, and when this parameter is much larger than unity, one expects large powers to be emitted in the higher harmonics. Note that such harmonics appear for planar undulators, but not for helical undulators, where the transverse momentum is constant, and the problem of relativistic oscillations is avoided.

The physical picture of the how coherent light is amplified by a beam of electrons in an FEL is shown in Fig. 8.12. The beam of electrons is initially unbunched, having a spatial density distribution that is essentially uniform in z, as in Fig. 8.12(a). The interaction of the wiggling electrons with a light wave of correct wavelength $\lambda_r \cong [\lambda_u/2\gamma^2](1 + a_u^2)$ then causes the beam to bunch at a spatial periodicity equal to this wavelength. As the beam becomes increasingly well **microbunched** by this effect (Fig. 8.12(b)), the undulator radiation is enhanced because the phase relationship of the light emitted within a microbunch is constructive. One can say that it resembles a "longitudinal grating" that is in relativistic motion. In this scenario, the radiated electric field scales as N_μ, the number of particles in the microbunch, and the radiation intensity scales as N_μ^2. This enhancement of radiated power at the appropriate wavelength λ_r obviously serves to strengthen the wave that caused the bunching interaction, and so a positive feedback system (or instability) is established. This feedback mechanism leads to exponential gain behavior, just as in a standard quantum level system laser.

The discussion in the previous paragraph has an obvious deficiency—we have not yet described the interaction between the wiggling electrons and the coherent light wave, which we must do to understand the physical mechanism that causes microbunching. In order to describe this interaction, we introduce the one-dimensional (we assume pure wiggle motion in x, no motion in y) near-axis Hamiltonian for an electron in a planar undulator with a plane polarized copropagating electromagnetic wave of frequency ω_r and electric field strength E_r

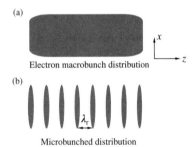

(a)

Electron macrobunch distribution

(b)

Microbunched distribution

Fig. 8.12 Representation of (a) initial and (b) final states of an electron beam's spatial distribution in an FEL. When the beam is fully microbunched at the radiation wavelength λ_r, the FEL radiation output is saturated.

$$H = \sqrt{q^2 c^2 \left[\frac{B_0}{k_u} \cos(k_u z) + \frac{E_r}{\omega_r} \cos(k_r z - \omega_r t) \right]^2 + p_{c,z}^2 c^2 + (m_0 c^2)^2}.$$

$$(8.81)$$

Assuming that the longitudinal momentum is much larger than the transverse momentum, we can approximate Eq. (8.81) as

$$H = \gamma_0 m_0 c^2 \left[1 + \frac{1}{2} \left(\frac{q}{\gamma_0 m_0 c} \right)^2 [A_u \cos(k_u z) + A_r \cos(k_r z - \omega_r t)]^2 \right].$$

(8.82)

Here we have written the transverse momentum in terms of the vector potential amplitudes $A_u \equiv B_0/k_u$, and $A_r \equiv E_r/\omega_r$ of the undulator and radiation fields, respectively. We are explicitly assuming $p_z \gg qA_u, qA_r$, and can therefore expand Eq. (8.82) as

$$H = \gamma m_0 c^2 \left[1 + \frac{1}{2} \left(\frac{q}{\gamma m_0 c} \right)^2 A_u^2 \cos^2(k_u z) + A_r^2 \cos(k_r z - \omega_r t) \right.$$

$$\left. + \cdots \frac{A_u A_r}{2} [\cos((k_r + k_u)z - \omega_r t) + \cos((k_r - k_u)z - \omega_r t)] \right]. \quad (8.83)$$

The squaring of the transverse vector potential terms produces four wave-like components. The first has zero phase velocity and cannot interact resonantly with a relativistic electron,[5] nor can the last, which has super-luminal phase velocity. The term that is quadratic in A_r has phase velocity equal to the speed of light, so if it were large (as in the case of violent acceleration in a linac), it would be non-negligible. In fact, because the frequency of the light is so high (recall that $\omega_r \propto \gamma^2$), we have an additional hierarchical relationship, $A_u \gg A_r$. This disparity in transverse momenta implies two important points: that the wiggle motion can be deduced by only looking at the undulator vector potential (as was done in Section 2.8), and that we may ignore the effects of the term proportional to A_r^2.

Thus, we are left with one resonant term in the Hamiltonian given in Eq. (8.83), and we can proceed to recast it in averaged form, ignoring the fast oscillating, constant, and ignorably small terms

$$\langle H \rangle \cong \gamma m_0 c^2 + \frac{q^2 A_u A_r}{4\gamma m_0} \cos[(k_r + k_u)z - \omega_r t]. \quad (8.84)$$

In this Hamiltonian, a particle that is resonant with the traveling wave has longitudinal velocity

$$v_z = v_\phi = \frac{\omega_r}{k_r + k_u}. \quad (8.85)$$

We note that the resonant term in the Hamiltonian in Eq. (8.84) is second order in the field strengths, and is the main potential term that survives when averaging over a period. It is therefore, another example of a ponderomotive potential, and is sometimes referred to as a **ponderomotive bucket**. The use of the term "bucket" is not a coincidence, as one can easily see that the form of Eq. (8.84) is formally identical to that of Eq. (4.17) if we perform a canonical transformation to the coordinate $\zeta = z - v_\phi t$. Since this is an example of a "gentle" interaction,

[5] This term gives the slowing down of the electron's motion in z due to finite wiggle amplitude when one correctly leaves in the formal dependence of the Hamiltonian on $p_{x,c}$.

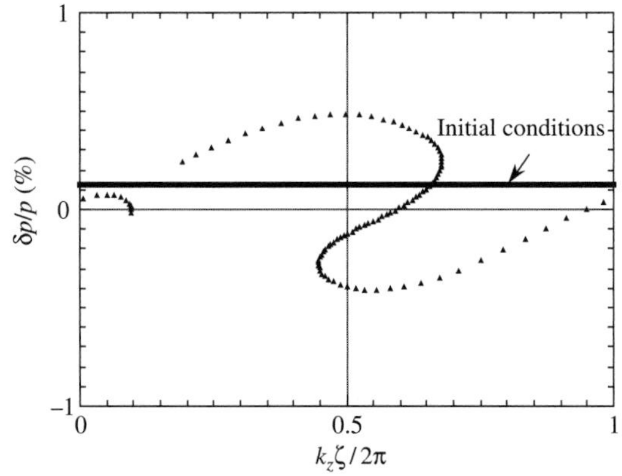

Fig. 8.13 Initial and final longitudinal phase space distributions for electrons in an FEL oscillator. The distribution initially is at an energy above resonance, and undergoes approximately one-half of a small-amplitude synchrotron oscillation to give up net energy to amplify the radiation wave.

we can expand the Hamiltonian about the resonant momentum, to obtain, in analogy to Eq. (4.27),

$$\tilde{H}(\zeta, \delta p_\zeta) \cong \frac{\delta p^2}{2\gamma_0^3 m_0} + \frac{q^2 A_u A_r}{4\gamma_0 m_0} \cos(k_r \zeta), \qquad (8.86)$$

which takes the familiar pendulum form. The bucket parameters (height and synchrotron frequency) derived from Eq. (8.86) are functions of the radiation vector potential A_r, and thus become larger as the radiation field is amplified.

For a low-gain system, typified by an **FEL oscillator** placed inside an optical resonator and fed by a pulse train of electron bunches synchronized to the round-trip frequency of the circulating photon pulse, the relative growth of the radiation field is small over the length of the interaction. In this case, the electron beam is injected slightly above the resonant value (for a specified injected light wavelength, $\gamma_0 \cong \sqrt{\lambda_u (1 + a_u^2)/2\lambda_r}$), and the beam undergoes roughly one-half of a synchrotron oscillation, with most of the electrons ending up below the resonant energy, as is shown in Fig. 8.13. Even though the bunching induced is large, the growth in the radiated field due to the bunching is small.

However, in a very high gain system, the amplification of the field itself, with its associated longitudinal bunching of the electrons, takes place over a short time, roughly a quarter of a beam synchrotron oscillation, $\int \omega_s(t)\, dt \cong \pi/2$. The field growth associated with this process has similarities to adiabatic capture—the field turns on exponentially—but much faster, and therefore the longitudinal phase space dynamics are non-adiabatic. In addition, the exponential growth of the field arises from the beam's radiation, and this causes a non-negligible phase shift in the total radiation. This phase shift allows the particles to lose net energy to the wave more effectively than in the model shown by Fig. 8.13. In a system starting from noise in the spontaneous undulator radiation spectrum (**self-amplified spontaneous emission**, or SASE), there is no external signal that defines the amplified wavelength—it is simply that of the resonant undulator radiation. Thus, one would naively expect as many particles to absorb energy (accelerate) as radiate energy, and the gain to be therefore zero. The phase shift associated with the radiation process ensures that this does not happen, by shifting the phase of the wave continually backwards in the ζ frame. At the point where small amplitude electrons have undergone approximately one-quarter of a small-amplitude synchrotron oscillation, the beam is as tightly bunched as it can be, and the high-gain FEL amplifier is saturated.

Free-electron lasers, in particular, those based on the high-gain SASE process, are of potentially enormous importance in future light source production. The fact that radiation can be tuned to arbitrary wavelengths by simply changing the beam energy means that existing electron linacs can be used to create FELs operating from the millimeter to Angstrom wavelength regions of the spectrum. A particularly ambitious program is now underway at Stanford to build a hard X-ray FEL called the Linear Coherent Light Source (LCLS) using the SLAC linac. X-ray lasers are naturally based on the SASE FEL concept, because there are no good sources of external signals or mirrors in the X-ray region, nor are there many useful quantum systems with such large (>kV) energy transitions.

The X-ray SASE FEL is a perfect note upon which to end this volume, as it emphasizes the interplay between charged particles, photon beams, magnets, and accelerators. The FEL begins with electrons liberated by ultraviolet photons and violently accelerated in an rf photocathode gun, which is a resonant rf structure containing centimeter wavelength photons. The low-emittance, high-current beam is then accelerated to ultra-relativistic energies, where some of its energy is converted back to X-ray photons through the interaction of the electron beam, radiation beam, and undulator magnet. This interaction, furthermore, can be described using concepts developed initially for understanding gentle acceleration processes in heavy ion linear accelerators. With a rich application such as the X-ray SASE FEL to understand, it is perhaps easier to appreciate the motivation to write this book.

8.10   Summary and suggested reading

This chapter has been an exploration of the physics of photons, both as beams in their own right, and as the phenomena of photon emission and absorption affect charged particle beams. Our discussion began with the introduction of the photon, a categorization of its wave and particle-like properties, and a comparison of these properties with those of charged particles. The particle-like properties of the photon lead naturally to the description of light transport in terms of ray optics. Our treatment of ray optics has been based in this chapter on as many analogies as possible with the methods of charged particle optics introduced in Chapter 3, including a matrix description of linear ray transformations. These transformations were examined for basic optical elements, such as mirrors, lenses, and gradient index media. The additional complications encountered in describing photons due to the need to keep track of polarization have been investigated.

The matrix description naturally allows analysis of the optical resonator, which is a periodic system whose optics are reminiscent of those found in circular accelerators. We have reviewed here the basic results of this analysis, including resonator stability, and classes of commonly found resonator design. We have discussed longitudinal modes in such open devices, and generalized the notion of the cavity Q to apply to this type of resonator.

It has been seen that it is surprisingly easy to take the results of our ray-based analysis, and apply it as a tool for understanding the propagation of coherent photon beams, the simplest case of which is the Gaussian beam.

The ray optics of photons have been connected to the Gaussian beam descrip tion through introduction of a complex source point. An even greater surprise perhaps, was found in noting the analogies between classical descriptions of charged particle beam distributions, and intensity profiles of coherent photon beams. These analogies notably include the inherent emittance one may assign a photon, which we have shown is simply result of the Heisenberg uncertainty principle.

The Gaussian beam formalism has been generalized to higher-order Hermite–Gaussian modes. These modes have enabled us to discuss diffraction in coherent beams, and note the difference between aperturing of a coher ent beam and a classical particle beam. The Hermite–Gaussian modes also provided the context for introduction of manipulations such as relay imaging (which mitigates diffraction from apertures), and spatial filtering of higher-order modes.

The introduction of coherent beams raises the question of how to create them. At optical frequencies, the answer is almost assuredly the laser. Some basics of the role that quantum mechanical states play in the lasing process have been reviewed in this chapter. A detailed treatment of the lasing process has been omitted here in favor of a heuristic model based on concepts in waves and resonant systems introduced in Chapter 7.

Our discussion of photon beams, and their similarities and differences from charged particle beams then gave way to a short introduction to photon emis sion from relativistic charged particle beams, synchrotron radiation. The main effects of synchrotron radiation on storage rings—energy loss, and damping of transverse and longitudinal phase spaces—have been reviewed at a rudimentary level.

In this chapter, our primary emphasis was on photon beams, and so the short introduction to synchrotron radiation served also to introduce the creation of photon beams from electron beams. The use of synchrotron radiation to create coherent photon beams is now being pursued in the context of the free-electron laser, which is based on an intense electron beam traversing an undulator mag net. We have reviewed here the basic spectral characteristics of the FEL, and discussed in some detail the longitudinal dynamics of electrons taking part in the undulator–radiation-beam interaction. The similarities between FEL lon gitudinal dynamics, and particles in accelerator buckets has been noted, and exploited for analysis purposes.

As this chapter is primarily concerned with what is traditionally a separate field from charged particle beams—laser and photon beam physics—our brief introductory discussion is deficient in numerous ways. It can be supplemented by beginning with the references listed below. Likewise, our discussion of synchrotron radiation effects and free-electron laser physics serves only as an orientation on basic physical processes. More detailed treatments are recommended in the following list:

1. M. Born and E. Wolf, *Principles of Optics; Electromagnetic Theory of Propagation, Interference, and Diffraction of Light*, 7th edn (Cambridge Univiversity Press, 1999).
2. A. Yariv, *Quantum Electronics*, 3rd edn (Wiley, 1989). A good deriva tion of the paraxial wave equation solutions is found here. An advanced undergraduate text.

3. A.E. Siegman, *Lasers* (University Science Books, 1986). An encyclopedic survey of lasers, with many technical details discussed.

4. M. Sands, *The Physics of Electron Storage Rings: an Introduction* (Stanford, 1970). The definitive treatment of synchrotron radiation in storage rings and damping based on this radiation.

5. P.J. Duke, *Synchrotron Radiation: Production and Properties* (Oxford University Press, 2000).

6. C. Brau, *Free-electron Lasers* (Academic Press, 1990). A clear introduction to the principle underlying the free-electron laser. Written at the graduate level.

7. E.L. Saldin, E.A. Schneidmiller, and M.V. Yurkov, *The Physics of Free-electron Lasers* (Springer, 2000). Detailed survey of free-electron laser physics, with particularly up-to-date treatment of high-gain amplifiers and SASE. Written at the graduate-to-professional level.

Exercises

(8.1) In the planar mirror array shown in Fig. 8.14, one must apply the reflection transformation on the *p*-directed ray three times, and so the coordinate system is reversed every pass through the system. Explain, in terms of the wave's electric field vector why this is so.

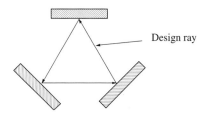

Fig. 8.14 Three-mirror array for reference in Exericise 8.1.

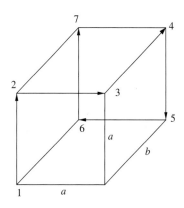

Fig. 8.15 Three-dimensional mirror array for reference in Exercise 8.2.

(8.2) In the optical array shown in Fig. 8.15, planar mirrors are used to move from one corner of the "box" (of dimensions $a \times a \times b$) to another in 90° turns (45° mirror tilt).

(a) Write the 4×4 transformation matrix (see the discussion in Section 3.8 for clarification of matrices with higher dimension than 2×2 that mix phase planes) that switches the *s*- and *p*-directed ray designations that is needed when the propagation axis is turned out of the *a* plane (e.g. after the mirror at point 3).

(b) Write out the full 4×4 matrix that describes the ray transformation from point 1 to point 7.

(8.3) Show that the focal length in the *s*-plane due to a tilted mirror is as indicated by Eq. (8.16), $f = R/2 \cos(\theta_M)$.

(8.4) For spherical dielectric interfaces (lens surfaces) that are rotated (tilted) there are changes in focal properties that are analogous to those discussed in Eqs (8.15) and (8.16). Derive, as in Eq. (8.13), the transformation matrices associated with a rotated lens. How does the focusing differ when compared with the unrotated lens?

(8.5) For an optical fiber with 100-μm diameter, with $n_0 = 1.5$ and $n_2 = 20$ m^{-2} at the wavelength of interest, what is the maximum angle of ray that propagates without encountering the boundary of the fiber?

(8.6) Consider a linear optical resonator consisting of one focusing mirror of fixed radius of curvature R_1, and another mirror, which may be focusing or defocusing with radius $(\pm)R_2$, placed a distance L from the first mirror. Plot the boundary where the stable region is found on a two-dimensional graph with axes L/R_1 and R_2/R_1.

(8.7) The plane-parallel resonator shown in Fig. 8.6(a) is referred to as a **Fabry–Perot** system. Assume each loss-less mirror has an amplitude transmission coefficient t, with a reflection coefficient r, such that energy is conserved at the mirror, $r^2 + t^2 = 1$. Consider the transmission of a plane wave (spatial dependence $A \exp(ikz)$, where A is an overall amplitude) through the Fabry–Perot resonator. Note that at a mirror interface, one must multiply the amplitude by r for the reflection and t or the transmission. Thus, for the transport of radiation entering one side of the cavity, we have the amplitude $A\sqrt{1 - r^2}$, and after one-round trip around the cavity it becomes $Ar^2\sqrt{1 - r^2} \exp(i2kL)$, after n round trips, $Ar^{2n}\sqrt{1 - r^2} \exp(i2nkL)$, etc.

(a) Taking the limit of an infinite number of round trips, write the sum of these waves, and evaluate the resulting infinite series to find the amplitude of the light as a function of k. Show that there is a narrow resonance in the response for values of k such that kL is an integer multiple of π.

(b) What is the total reflected amplitude of the wave incident on the Fabry–Perot resonator? Why is the reflected wave small at resonance? What is the total transmitted amplitude?

(c) From the square of stored wave amplitude, evaluate the frequency dependence of the circulating power in the resonator. Using the half-power definition, find the quality factor Q. If you use the definition given in Eq. (8.22) (and Eq. (7.61)) do you get the same answer?

(8.8) Show that Eq. (8.27) is, in fact, a solution to Eq. (8.24).

(8.9) The electron beam at the Stanford Linear Collider has a **normalized** rms emittance $\varepsilon_n = 3 \times 10^{-5}$ m rad. At the full energy of 50 GeV, find the wavelength of photon that has the same effective emittance as the Stanford electron beam.

(8.10)

(a) Using the **ABCD** transformation, find an **equilibrium** beam size $w = w_{eq}$ for light of 800 nm wavelength (in vacuum), which propagates without changing inside of an optical fiber with $n_0 = 1.5$ and $n_2 = 20$ m^{-2}.

(b) Compare this result to that obtained by using the equivalent envelope equation,

$$w'' + \kappa_f^2 w = \frac{\lambda^2}{n_0^2 \pi^2 w^3}.$$

(c) Discuss why this is the correct form of the Gaussian paraxial beam envelope equation.

(8.11) For the case discussed in Exercise 8.7, if the beam is launched from a **mismatched** waist $w \neq w_{eq}$ at $z = z_0$:

(a) Write the evolution of q as a function of z.

(b) Find the evolution of w as a function of z for this mismatched case.

(8.12) Consider a symmetric optical resonator made from two focusing mirrors with radii of curvature $(\pm)R$, separated by a length L.

(a) What is the ratio of the beam size w at the mirrors to that at the waist as a function of L?

(b) In some situations, for instance, when the optical power is to be minimized at the mirrors due to material damage considerations, one desires the beam to be very large at the mirror. For 2-in. diameter mirrors with $R = 50$ cm, and a photon wavelength of 1 μm, what distance L gives a beam size w that is equal to one-fourth of the mirror diameter? What is the waist size w_0 in this case?

(8.13) For the case of the optical resonator in Exercise 8.8, assuming a Gaussian beam profile:

(a) Find the fraction of the optical power outside of the mirror edge (you must integrate the intensity from the mirror edge to infinity in this case, cf. Eq. (8.33)). This power loss may, in fact, be desirable, as it couples power out of the cavity.

(b) What is the Q of a resonator with the fractional round-trip power loss given by twice the fractional loss (two mirrors) given in part (a)?

(8.14) A coherent paraxial light beam with wavelength λ and a round Gaussian profile is launched with flat phase fronts, and width $w_{x0} = w_{y0} = w_0$ at $z = 0$.

(a) Now consider the insertion of a slit aperture, which cuts off the intensity profile at $x = w_0$. Find the Hermite–Gaussian representation of the wave (the c_{nm}) just after its encounter with the slits. It is best to keep $w_x = w_0$ to specify the Hermite–Gaussian representation after the slits.

(b) A long way (the asymptotic limit, $z \gg Z_R$) from the slits, the wave profile expands without significantly changing form. In other words, the wave's **angular** profile $f(\theta_x, \theta_y, z) \cong u(x/R, y/R, z)$ does not change in this region. Find this angular profile for the case in part (a), and compare the result to the profile in the case where the beam encounters no slits.

(c) Now imagine that the light beam is composed of classical particles with the same effective emittance as the paraxial wave, and a bi-Gaussian trace space distribution. Apply the same slit aperture to this classical beam, and write the form of the spatial profile far from the aperture. How does this form compare to the profiles found in part (b), and the classical distribution without the aperture?

(8.15) For the light beam introduced in Exercise 8.14, evaluate the effective rms emittances in both transverse dimensions of the light beam after collimation by the slits. To do this, use the asymptotic angular spectrum $f(\theta_x, \theta_y)$ to calculate the effective second moments $\langle x'^2 \rangle$ and $\langle y'^2 \rangle$ present at the slits (where $\langle xx' \rangle = \langle yy' \rangle = 0$), and calculate $\langle x^2 \rangle$ and $\langle y^2 \rangle$ in the usual fashion. How do these values compare to $\lambda/4\pi$?

(8.16) Show that for the optics configurations given in Fig. 8.16 (with representative rays shown), that one can obtain the relay imaging condition, and find the magnification of the image with respect to the object size.

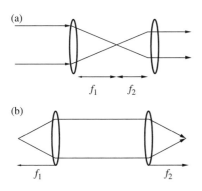

(a)

(b)

Fig. 8.16 Relay imaging optics configurations for Exercise 8.16.

(8.17) The Fourier transform of the transverse wave profile functions given by Eq. (8.46) is $v_{x,n}(k_x, z) = (2\pi)^{-1/2} \int_{-\infty}^{\infty} u_{x,n}(x, z) \exp(ik_x x)\, dx$, with an inverse transformation $u_{x,n}(x, z) = (2\pi)^{-1/2} \int_{-\infty}^{\infty} v_{x,n}(k_x, z) \times \exp(-ik_x x)\, dk_x$. Show, using the Fourier transform characteristics of the Hermite–Gaussian functions, that the phase-shifts coming into or leaving a waist condition, $\Delta\varphi_{g,n}(z) \to \pi/2 + n\pi$ over a distance $\Delta z \gg Z_R$, and change the wave profile into its Fourier transform.

(8.18) Explain why a two-level quantum system cannot achieve population inversion. Why do you think a three-level system (where state j is the ground state) is generally inferior to a four-level system in maintaining population inversion?

(8.19) Consider an optical resonator of the type shown in Fig. 8.9, where the cavity length is $L = 50$ cm and the gain material length is $L_m = 1$ cm long. The photon wavelength associated with the energy transition in the inverted population is $1.053\,\mu$m.

(a) Assume the inverted population density is $n_{inv} = 10^{20}$ cm^{-3}, and the interaction cross-section is $\Sigma_c = 10^{-25}$ cm^{-2}, find the cavity loss factor δ_c and the cavity Q needed for cw operation.

(b) If 40 percent of the cavity losses are due to power absorption in the mirrors and the gain medium, and the output power is 1 W, what is the circulating power in the cavity?

(8.20) Ignoring spatial dependences, solve the system of Eqs (8.65)–(8.66) with the initial conditions $n_i = n_0$, $n_j = 0$, and $I(\omega_{ij}) = I_0$.

(a) Show that for short time scales, the wave grows exponentially.

(b) Find the final saturated intensity.

(8.21) The LEP electron–positron collider is a synchrotron storage ring located at the European high-energy physics laboratory, CERN, which is sited just outside of Geneva, Switzerland. It has a circumference of 27 km.

(a) In order to create $Z^{\pm}$ particle pairs, the collider must operate with 91 GeV energy per beam. What is the average energy loss per turn, per particle?

(b) To replace this energy loss, a massive rf system (similar to that of a good sized linac) must be put in place. Because a storage ring is run in continuous mode, this system is superconducting to avoid power dissipation issues in the cavities. Assuming $\sin(\phi_0) = 0.4$, find the rf voltage needed for LEP.

(8.22) Using the results of Exercise 8.21, estimate the damping time of the LEP collider.

(8.23) The spectral characteristics of undulator radiation can be deduced simply by use of Lorentz transformations of the $(\omega, \vec{k})$ four-vector. The physical model adopted here is that the magnetostatic undulator field, from the point of view of the electron in its "rest frame," is seen as an oncoming electromagnetic wave. This wave is back scattered (Compton scattering) by the electron. When observed in the laboratory frame, this back scattered photon has the correct form predicted by Eq. (8.79). To verify this model,

(a) Transform the $(\omega, \vec{k})$ four-vector associated with the magnetostatic field in the lab frame ($\omega = 0, k_z = k_u$) to the rest frame of the electron beam.

(b) Check to see, by use of Eq. (1.58), the degree to which a planar magnetic field ($\vec{B} \cong B_0 \sin(k_u z)\hat{y}$) appears to be a traveling electromagnetic wave (virtual photons) with the correct relationship of the magnetic and electric fields, $\omega\vec{B} = \vec{k} \times \vec{E}$.

(c) Assuming the change in wavelength of the virtual photons (Compton shift) is negligible upon scattering, and assuming further that they are then real photons, find the frequency of these photons in the laboratory frame, for arbitrary angle θ.

(8.24) Show that Eq. (8.85) is equivalent to the condition given by Eq. (8.79) for $\theta = 0$.

(8.25) The LCLS uses a beam energy of 15 GeV, and an undulator with wavelength $\lambda_u = 3$ cm, and peak field of 1.32 T.

  (a) What is the fundamental wavelength of the undulator radiation? What is the photon energy associated with this wavelength?

  (b) The electron beam inside of the 100-m-long(!) undulator has an average Twiss parameter $\beta_x \cong 18$ m, and a normalized rms emittance of $\varepsilon_{n,x} \cong 2 \times 10^{-6}$ m rad. What is the rms transverse electron beam size?

  (c) Assuming the FEL radiation beam size is roughly equal to that of the electron beam, what is the Rayleigh range of the radiation beam? The exponential gain length (Γ^{-1}) in this system is 5.8 m. What is the growth in the radiation beam due to diffraction over this length?

Appendix A: selected problem solutions

The following solutions to selected problems have been primarily generated by the author's assistants, who include a number of students who have taken the UCLA Department of Physics and Astronomy beam physics course, Physics 150. The varying styles of solution reflect the diverse approaches to the material taken by different contributors.

Chapter 1

Exercise 1.1

In a three-dimensional simple harmonic oscillator where $T = \frac{1}{2}m\dot{\vec{x}}^2$ and $V = \frac{1}{2}k\vec{x}^2$:

(a) The Lagrangian is

$$L(\vec{x}, \dot{\vec{x}}) = T - V = \frac{1}{2}m\dot{\vec{x}}^2 - \frac{1}{2}k\vec{x}^2.$$

In Cartesian coordinates, $\vec{x} = x\hat{x} + y\hat{y} + z\hat{z}$. Therefore,

$$L(\vec{x}, \dot{\vec{x}}) = \frac{1}{2}m(\dot{x}^2 + \dot{y}^2 + \dot{z}^2) - \frac{1}{2}k(x^2 + y^2 + z^2).$$

(b) Canonical momenta are given by $p_{c,i} = \partial L/\partial \dot{x}_{c,i}$ (subscript c denotes canonical variable). For this system

$$p_{c,x} = \frac{\partial L}{\partial \dot{x}} = m\dot{x}, \quad p_{c,y} = \frac{\partial L}{\partial \dot{y}} = m\dot{y}, \quad p_{c,z} = \frac{\partial L}{\partial \dot{z}} = m\dot{z}.$$

The Hamiltonian is

$$H(\vec{x}_c, \vec{p}_c) = \vec{p}_c \cdot \dot{\vec{x}} - L$$

$$= (m\dot{x})\dot{x} + (m\dot{y})\dot{y} + (m\dot{z})\dot{z} - \frac{1}{2}m(\dot{x}^2 + \dot{y}^2 + \dot{z}^2) + \frac{1}{2}k(x^2 + y^2 + z^2)$$

$$= \frac{1}{2}m(\dot{x}^2 + \dot{y}^2 + \dot{z}^2) + \frac{1}{2}k(x^2 + y^2 + z^2),$$

$$H(\vec{x}_c, \vec{p}_c) = \frac{1}{2m}(p_{c,x}^2 + p_{c,y}^2 + p_{c,z}^2) + \frac{1}{2}k(x^2 + y^2 + z^2).$$

The total energy of the system is conserved if $\partial H/\partial t = 0$. Since this condition is true for the Hamiltonian given above, the Hamiltonian is a constant of the motion. Notice also that the portion of the Hamiltonian governing horizontal motion alone is constant, $\partial H_x/\partial t = 0$ (and similarly for H_y and H_z), and the energy in each phase plane (e.g. $(x_c, p_{c,x})$) is conserved. Therefore, we can write,

$$H = H_x + H_y + H_z,$$

where $H_x = (p_{c,x}^2/2m) + \frac{1}{2}kx^2$, etc.

(c) In cylindrical coordinates

$$x = \rho \cos \phi, \quad y = \rho \sin \phi, \quad z = z,$$

where $\vec{x}^2 = x^2 + y^2 + z^2 = \rho^2 \cos \phi + \rho^2 \sin^2 \phi + z^2 = \rho^2 + z^2$ and

$$\dot{\vec{x}} = \dot{\rho}(\cos \phi \hat{x} + \sin \phi \hat{y}) + \rho \dot{\phi}(-\sin \phi \hat{x} + \cos \phi \hat{y}) + \dot{z}\hat{z}$$
$$= \dot{\rho}\hat{\rho} + \rho \dot{\phi}\hat{\phi} + \dot{z}\hat{z},$$

so the Lagrangian is

$$L(\vec{x}, \dot{\vec{x}}) = T - V = \frac{1}{2}m\dot{\vec{x}}^2 - \frac{1}{2}k\vec{x}^2$$

or

$$L(\vec{x}, \dot{\vec{x}}) + \frac{1}{2}m(\dot{\rho}^2 + \rho^2 \dot{\phi}^2 + \dot{z}^2) - \frac{1}{2}k(\rho^2 + z^2).$$

(d) The canonical momenta, $p_{c,i} = \partial L/\partial \dot{x}_{c,i}$, in cylindrical coordinates are

$$p_{c,\rho} = \frac{\partial L}{\partial \dot{\rho}} = m\dot{\rho}, \quad p_{c,\phi} = \frac{\partial L}{\partial \dot{\phi}} = m\dot{\phi}, \quad p_{c,z} = \frac{\partial L}{\partial \dot{z}} = m\dot{z},$$

so the Hamiltonian, $H(\vec{x}_c, \vec{p}_c)$, where $\vec{x}_c = \rho \hat{\rho} + \phi \hat{\phi} + z\hat{z}$ and $\vec{p}_c = p_{c,\rho}\hat{\rho} + p_{c,\phi}\hat{\phi} + p_{c,z}\hat{z}$ is

$$H(\vec{x}_c, \vec{p}_c) = \vec{p}_c \cdot \dot{\vec{x}}_c - L$$
$$= (m\dot{\rho})\dot{\rho} + (m\rho^2 \dot{\phi})\dot{\phi} + (m\dot{z})\dot{z} - \frac{1}{2}(\dot{\rho}^2 + \rho^2 \dot{\phi}^2 + \dot{z}^2) + \frac{1}{2}k(\rho^2 + z^2)$$
$$= \frac{1}{2}m(\dot{\rho}^2 + \rho^2 \dot{\phi}^2 + \dot{z}^2) + \frac{1}{2}k(\rho^2 + z^2),$$

or in terms of canonical variables,

$$H(\vec{x}_c, \vec{p}_c) = \frac{1}{2m}\left(p_{c,\rho}^2 + \frac{p_{c,\phi}^2}{\rho^2} + p_{c,z}^2\right) + \frac{1}{2}k(\rho^2 + z^2).$$

From inspection of the Hamiltonian,

$$\frac{\partial H}{\partial \phi} = 0 \Rightarrow \dot{p}_{c,\phi} = -\frac{\partial H}{\partial \phi} = 0 \Rightarrow p_{c,\phi}$$

is a constant of motion.

(e) In spherical coordinates,

$$x = r \sin \theta \cos \phi, \quad y = r \sin \theta \sin \phi, \quad z = r \cos \theta,$$
$$\vec{x}^2 = r^2$$

where

$$\dot{\vec{x}} = \dot{r}\hat{r} + r\dot{\theta}\hat{\theta} + r \sin \theta \dot{\phi}\hat{\phi}$$

and the Lagrangian is

$$L(\vec{x}, \dot{\vec{x}}) = T - V = \tfrac{1}{2}m\dot{\vec{x}}^2 - \tfrac{1}{2}k\vec{x}^2$$

or

$$L(\vec{x}, \dot{\vec{x}}) = \tfrac{1}{2}m(\dot{r}^2 + r^2\dot{\theta}^2 + r^2 \sin^2 \theta \dot{\phi}^2) - \tfrac{1}{2}k(r^2).$$

(f) The canonical momenta, $p_{c,i} = \partial L / \partial \dot{x}_{c,i}$, in spherical coordinates are

$$p_{c,r} = \frac{\partial L}{\partial \dot{r}} = m\dot{r}, \quad p_{c,\theta} = \frac{\partial L}{\partial \dot{\theta}} = mr^2\dot{\theta}, \quad p_{c,\phi} = \frac{\partial L}{\partial \dot{\phi}} = mr^2 \sin^2 \theta \dot{\phi},$$

so the Hamiltonian, $H(\vec{x}_c, \vec{p}_c)$, where $\dot{\vec{x}}_c = \dot{r}\hat{r} + \dot{\theta}\hat{\theta} + \dot{\phi}\hat{\phi}$ $\vec{p}_c = p_{c,r}\hat{r} + p_{c,\theta}\hat{\theta} + p_{c,\phi}\hat{\phi}$ is

$$\begin{aligned}
H(\vec{x}_c, \vec{p}_c) &= \vec{p}_c \cdot \dot{\vec{x}}_c - L \\
&= ((m\dot{r})\dot{r}) + ((mr^2\dot{\theta})\dot{\theta}) + ((mr^2 \sin \theta \dot{\phi})\dot{\phi}) - L \\
&= \tfrac{1}{2}m(\dot{r}^2 + r^2\dot{\theta}^2 + r^2 \sin^2 \theta \dot{\phi}^2) + \tfrac{1}{2}kr^2, \\
H(\vec{x}_c, \vec{p}_c) &= \frac{1}{2m}\left(p_{c,r}^2 + \frac{p_{c,\theta}^2}{r^2} + \frac{p_{c,\phi}^2}{r^2 \sin^2 \theta}\right) + \frac{1}{2}k(r^2)
\end{aligned}$$

From inspection of the Hamiltonian,

$$\frac{\partial H}{\partial \phi} = 0 \implies \dot{p}_{c,\phi} = -\frac{\partial H}{\partial \phi} = 0 \implies p_{c,\phi}$$

is a constant of motion.

Exercise 1.3

In a sextupole magnet with a force, $F_x = -ax^2$ (where a is constant);
    (a) the Hamiltonian associated with one-dimensional motion is obtained by first solving for the Lagrangian—the difference between the kinetic and

potential energy of the system, $L = T - V$, where

$$T = \frac{1}{2}m\dot{x}^2, \quad V = -\int F_x \, dx = \frac{ax^3}{3} \quad \Rightarrow \quad L = \frac{1}{2}m\dot{x}^2 - \frac{1}{3}ax^3$$

The Hamiltonian is

$$H = p_x \cdot \dot{x} - L = \left(\frac{\partial L}{\partial \dot{x}}\right)\dot{x} - L \quad \text{or} \quad H = \frac{p_x^2}{2m} + \frac{1}{3}ax^3.$$

(b) In order to plot representative constant H curves, first solve for the parameter p_x, and set H, m, and a to arbitrary constants,

$$p(x) = \sqrt{(2mH - \tfrac{2}{3}amx^3)}.$$

Below, $H = 1$, 4, and 8, respectively, while $m = 1$ and $a = 1$. The curves are not closed, indicating that the particle motion is unbounded.

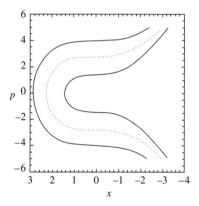

In an octupole field, the force is $F_x = -ax^3$. Following the above method, $L = T - V$, where

$$T = \frac{1}{2}m\dot{x}^2, \quad V = -\int F_x \, dx = \frac{ax^4}{4} \quad \Rightarrow \quad L = \frac{1}{2}m\dot{x}^2 - \frac{1}{4}ax^4.$$

The Hamiltonian is

$$H = p_x \cdot \dot{x} - L = \left(\frac{\partial L}{\partial \dot{x}}\right)\dot{x} - L \quad \text{or} \quad H = \frac{p_x^2}{2m} + \frac{1}{4}ax^4.$$

In an octupole-like field, the equation for plotting the phase space becomes

$$p(x) = \sqrt{(2mH - \tfrac{1}{2}amx^4)}.$$

We consider the sign of the constant a, to find if the particle motion is affected. For $a = -1$ (i.e. $a < 0$), $m = 1$ and values of $H = 0.25$, 1, and 4, respectively, the motion is unbounded.

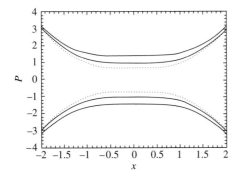

For $a = 1$ (i.e. $a > 0$) and the same values assumed for m and H, the motion is bounded. The closed curves are not ellipses, but have a "race-track" appearance.

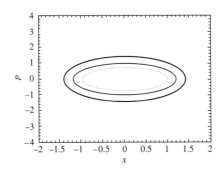

Exercise 1.5

The Taylor series expansion for a small x is

$$f(x) = \sum_{n=0}^{\infty} \frac{x^n}{n!} f^{(x)}(0).$$

The equation $U = \gamma m_0 c^2 = m_0 c^2 (1 - (v^2/c^2))^{-1/2}$ can be expanded using v/c, since $v/c \ll 1$. Therefore

$$U = m_0 c^2 \left[1 + \frac{1}{2} \left(\frac{v^2}{c^2} \right) + \cdots \right],$$

and

$$U \approx m_0 c^2 \left[1 + \frac{1}{2} \left(\frac{v^2}{c^2} \right) \right]$$

$$= m_0 c^2 + \tfrac{1}{2} m_0 v^2$$

The first term in the mechanical energy represents the rest energy and the second term represents the familiar non-relativistic kinetic energy.

Chapter 2

Exercise 2.1

Given

$$L(x, \dot{x}) = \frac{-m_0 c^2}{\gamma} + q\vec{A} \cdot \dot{\vec{x}}.$$

(a) Show $\nabla \times \vec{A} = \vec{B}$ given $\vec{A} = (B_0/2)\rho\hat{\phi}$.
In cylindrical coordinates,

$$\vec{\nabla} \times \vec{A} = \left[\frac{1}{\rho} \frac{\partial A_z}{\partial \phi} - \frac{\partial A_\phi}{\partial z} \right] \hat{\rho} + \left[\frac{\partial A_\rho}{\partial z} - \frac{\partial A_z}{\partial \rho} \right] \hat{\phi} + \frac{1}{\rho} \left[\frac{\partial}{\partial \rho}(\rho A_\phi) - \frac{\partial A_\rho}{\partial \phi} \right] \hat{z}$$

$$= \frac{1}{\rho} \left[\frac{\partial}{\partial \rho}(\rho A_\phi) \right] \hat{z} = \frac{1}{\rho} \left[\frac{\partial}{\partial \rho} \left(\frac{B_0}{2} \rho^2 \right) \right] \hat{z} = \frac{1}{\rho} \frac{B_0}{2}(2\rho) = B_0 \hat{z}.$$

(b) Write the Lagrangian in cylindrical coordinates:

$$x^2 = \rho^2 + z^2,$$

$$\dot{x} = \dot{\rho}\hat{\rho} + \dot{z}\hat{z} + \rho\dot{\phi}\hat{\phi},$$

$$L = \frac{-m_0 c^2}{\gamma} + q\vec{A} \cdot \dot{\vec{x}}$$

$$= \frac{-m_0 c^2}{\gamma} z + q \left(\frac{B_0 \rho}{2} \right) \hat{\phi} \cdot (\dot{\rho}\hat{\rho} + \dot{z}\hat{z} + \rho\dot{\phi}\hat{\phi}),$$

$$L = \frac{-m_0 c^2}{\gamma} + \frac{q B_0 \rho^2 \dot{\phi}}{2},$$

where

$$\gamma = \frac{1}{\sqrt{(1 - (1/c^2))(\dot{\rho}^2 + \rho^2 \dot{\phi}^2 + \dot{z}^2)}}.$$

(c) Lagrange–Euler equations,

$$\frac{\partial L}{\partial x_{i,c}} - \frac{d}{dt} \frac{\partial L}{\partial \dot{x}_{i,c}} = 0$$

$$\frac{\partial L}{\partial \rho} - \frac{d}{dt} \frac{\partial L}{\partial \dot{\rho}} = q B_0 \rho \dot{\phi} + \gamma m_0 \rho \dot{\phi}^2 - \frac{d}{dt}(\gamma m_0 \dot{\rho}) = 0,$$

$$\frac{\partial L}{\partial \phi} - \frac{d}{dt} \frac{\partial L}{\partial \dot{\phi}} = 0 - \frac{d}{dt} \left[\left(\gamma m_0 \rho^2 \dot{\phi} + \frac{1}{2} q B_0 \rho^2 \right) \right] = 0,$$

$$\frac{\partial L}{\partial z} - \frac{d}{dt} \frac{\partial L}{\partial \dot{z}} = 0 - \frac{d}{dt}(\gamma m_0 \dot{z}) = 0.$$

Since we are considering a pure magnetic field, the total energy of this problem is conserved, and $\gamma = $ constant, or $d\gamma/dt = 0$.

The Lagrange–Euler equations become, after use of the constants of the motion:

$$\gamma m_0 \ddot{\rho} = qB_0\rho\dot{\phi} + \gamma m_0\rho\dot{\phi}^2,$$

$$\gamma m_0\rho^2\dot{\phi} = \tfrac{1}{2}qB_0\rho^2 = \text{constant},$$

and

$$\gamma m_0\dot{z} = \text{constant}.$$

For circular motion, the equilibrium solution requires that $\ddot{\rho} = \dot{\rho} = 0$; therefore, ρ must be a constant.

Thus, we have $\dot{\phi} = -qB_0/\gamma m_0 = -\omega_c$.

Now, using the fact that the total energy is conserved, $U = \gamma m_0 c^2 = $ constant, and $1/\gamma^2 = 1 - (\rho^2\dot{\phi}^2/c^2)$ is also constant.

Using $\dot{\phi} = -qB_0/\gamma m_0$ and $\rho = R$ (radius of curvature), we have

$$\frac{\gamma^2 - 1}{\gamma^2} = \frac{R^2q^2B_0^2}{c^2\gamma^2m_0^2},$$

$$qB_0R = m_0c\sqrt{\gamma^2 - 1} = m_0c\gamma\sqrt{1 - \gamma^2} = m_0c\gamma\left(\frac{v_\perp}{c}\right) = m_0v_\perp\gamma = p_\perp,$$

so $p_\perp = qB_0R$, which is the same as given in Eq. (2.7).

Exercise 2.2

Starting with $p_0(t) = eB_0(t)R \ldots$

(a) Differentiate the above equation with respect to time

$$\frac{dp_0(t)}{dt} = \frac{d}{dt}(eB_0(t)R) = eR\frac{d}{dt}B_0(t) = eE_\phi,$$

$$E_\phi = -R\frac{d}{dt}B_0(t).$$

(a) Using Maxwell's equations we have that $\nabla \times E = -\partial B/\partial t$. Let us apply Stokes Theorem

$$\int_s \nabla \times E \cdot dA = \oint E \cdot dl = -\int_s \frac{\partial B}{\partial t} \cdot dA,$$

$$\oint E \cdot dl = -\frac{\partial \bar{B}}{\partial t}(\pi R^2),$$

$$\int_0^{2\pi} E_\phi R\, d\phi = -(2\pi R^2)\frac{\partial}{\partial t}\bar{B}_0(t),$$

$$\frac{d}{dt}\bar{B}_0(t) = \frac{\int d\vec{R}/dt \cdot d\vec{A}}{2\pi R^2}.$$

Using the definition of the average magnetic field given in the problem

$$\dot{\bar{B}} \equiv \frac{\int \dot{\vec{B}} \cdot d\vec{A}}{\pi R^2},$$

we obtain that $B_0(t) = (\bar{B}(t)/2) + $ constant. The constant of integration can be chosen to be non-zero. It, in practice, often is made non-zero to avoid saturation effects in the iron. This is called a reverse-biased betatron. For a description of saturation see discussion in Chapter 6.

Exercise 2.9

(a) Starting with Eq. (2.29) (for motion in a static, constant electric field)

$$\gamma(z) = \gamma_0 + \gamma' z = \gamma_0 + \frac{qE_0}{m_0 c^2} z.$$

Take the time derivative

$$\frac{d\gamma}{dt} = \gamma' \frac{dz}{dt} = \gamma' v_z.$$

Using the paraxial approximation:

$$v_z \cong v = \beta c = c\sqrt{1 - \frac{1}{\gamma^2}} = c\sqrt{\frac{\gamma^2 - 1}{\gamma^2}} = \frac{c}{\gamma}\sqrt{\gamma^2 - 1}.$$

With this expression we obtain

$$\frac{d\gamma}{dt} = \frac{qE_0}{m_0 c} \frac{\sqrt{\gamma^2 - 1}}{\gamma}.$$

(a) Integrate this equation:

$$\int \frac{qE_0}{m_0 c} dt = \int \frac{\gamma}{\sqrt{\gamma^2 - 1}} d\gamma,$$

to find

$$\frac{qE_0}{m_0 c} t + C_1 = \sqrt{\gamma^2 - 1}.$$

Here, C_1 is a constant of integration, but since $\gamma(0) = 1$, we have $C_1 = 0$. Solving for γ, we obtain

$$\gamma = \sqrt{\left(\frac{qE_0}{m_0 c} t\right)^2 + 1}.$$

(b) Now, let us directly integrate Eq. (2.25)

$$\frac{dp_z}{dt} = qE_0, \quad p_z = qE_0 t + C_2.$$

Here, C_2 is a constant of integration but since $\gamma(0) = 1$ and $v_z(0) = 0$, then $C_2 = 0$, and

$$\gamma m_0 v_z = qE_0 t \quad \text{or} \quad \gamma = \frac{qE_0 t}{m_0 v_z}.$$

Using the paraxial approximation,

$$v_z \cong v = \beta c = c\sqrt{1 - \frac{1}{\gamma^2}} = c\sqrt{\frac{\gamma^2 - 1}{\gamma^2}} = \frac{c}{\gamma}\sqrt{\gamma^2 - 1},$$

we have

$$\gamma = \frac{qE_0 t}{m_0 c} \frac{\gamma}{\sqrt{\gamma^2 - 1}}.$$

Solving for γ we obtain

$$\gamma = \sqrt{\left(\frac{qE_0 t}{m_0 c}\right)^2 + 1}.$$

This matches the previous result.

Exercise 2.11

First let us compare the equations of motion for a particle traveling in $\hat{z}$ and its antiparticle traveling in the opposite direction $-\hat{z}$.

The particle has charge q, mass m_0, and momentum $p_z = \gamma m_0 v_z \hat{z}$. The antiparticle has charge $-q$, mass m_0, and momentum $-p_z = \gamma m_0 v_z (-\hat{z})$.

For magnetic focusing channel:

$$\vec{B} = 2a_2(-x\hat{x} + y\hat{y}) - 2b_2(y\hat{x} + x\hat{y}).$$

In this case we need to omit the skew term ($a_2 = 0$) and keep the normal term ($b_2 > 0$)

$$\vec{B} = -2b_2(y\hat{x} + x\hat{y}).$$

The force is given by $\vec{F}_\perp = q(\vec{v}_z \times \vec{B})$.

For the particle
$$\begin{cases} \vec{F}_\perp = -qv_z 2b_2(y(\hat{z} \times \hat{x}) + (x(\hat{z} \times \hat{y})), \\ \vec{F}_\perp = 2qv_z b_2(x\hat{x} - y\hat{y}). \end{cases}$$

For the anti-particle
$$\begin{cases} \vec{F}_\perp = -(-qv_z 2b_2)(y(-\hat{z} \times \hat{x}) + (x(-\hat{z} \times \hat{y})), \\ \vec{F}_\perp = 2qv_z b_2(x\hat{x} - y\hat{y}). \end{cases}$$

and the equations of motion are the same.

Now let us do the same thing for the electric quadrupole fields:

$$\vec{E} = 2a_2(-x\hat{x} + y\hat{y}) + 2b_2(y\hat{x} + x\hat{y}).$$

In this case we set the skew term $b_2 = 0$, and keep the normal term ($a_2 > 0$)

$$\vec{E} = 2a_2(-x\hat{x} + y\hat{y}).$$

The force is given by $\vec{F}_\perp = q\vec{E}$.

For the particle $\vec{F}_\perp = 2qa_2(-x\hat{x} + y\hat{y})$

For the antiparticle $\vec{F}_\perp = -2qa_2(-x\hat{x} + y\hat{y}) = 2qa_2(x\hat{x} - y\hat{y})$

In this case the dimensions of focusing and defocusing are exchanged for the particle and antiparticle.

Exercise 2.14

(a) Consider a cylindrically symmetric beam in a solenoid. Using Gauss' Law:

$$\int \vec{E}_\rho \cdot \mathrm{d}s = \frac{Q_{enc}}{\varepsilon_0},$$

$$\vec{E}_\rho (2\pi\rho l) = \frac{1}{\varepsilon_0} \lambda(\rho)l,$$

where $\lambda(\rho)$ is the enclosed charge per unit length, related to the beam density n_b by $\lambda(\rho) = q n_b \pi \rho^2$. Thus,

$$\vec{E}_\rho = \frac{q n_b}{2\varepsilon_0}\rho.$$

The radial space charge electric force is

$$F = q E_\rho = \frac{q^2 n_b}{2\varepsilon_0}\rho = \frac{m_0}{2}\omega_{\rho_0}^2 \rho,$$

where

$$\omega_{\rho_0}^2 = \frac{q^2 n_b}{m_0 \varepsilon_0}.$$

(a) Due to length contraction of the moving cylinder ($l = l'/\gamma$) the density in the rest frame is given by

$$n_b' = \frac{N}{\pi r^2 l'} = \frac{N}{\pi r^2 l \gamma} = \frac{n_b}{\gamma}.$$

In the rest frame

$$E_\rho' = \frac{q}{2\varepsilon_0}\frac{n_b}{\gamma}\rho,$$

$$B' = 0,$$

$$\varphi' = -\int E' \cdot \mathrm{d}\rho = -\frac{q n_b}{4\varepsilon_0}\rho^2.$$

Transform to the lab frame:

$$\begin{pmatrix} \varphi \\ cA_x \\ cA_y \\ cA_z \end{pmatrix} = \begin{pmatrix} \gamma & 0 & 0 & -\beta\gamma \\ 0 & 0 & 0 & 0 \\ 0 & 0 & 0 & 0 \\ -\beta\gamma & 0 & 0 & \gamma \end{pmatrix} \begin{pmatrix} \phi' \\ 0 \\ 0 \\ 0 \end{pmatrix}.$$

In the lab frame

$$\phi = \gamma\phi' = -\frac{q n_b}{4\varepsilon_0}\rho^2, \qquad cA_z = -\beta\gamma\phi' = \beta\frac{q n_b}{4\varepsilon_0}\rho^2.$$

Then

$$\vec{E} = -\vec{\nabla}\phi = \frac{q n_b}{2\varepsilon_0}\rho\hat{\rho}, \qquad \vec{B} = \vec{\nabla}\times\vec{A} = -\frac{\beta}{c}\frac{q n_b}{2\varepsilon_0}\rho\hat{\phi}.$$

Now let us look at the forces

$$\vec{F}_{\rm E} = q\vec{E} = \frac{q^2 n_{\rm b}}{2\varepsilon_0}\rho\hat{\rho},$$

$$\vec{F}_{\rm B} = q\vec{v} \times \vec{B} = qv_zB_\phi = -q^2(\beta c)\frac{\beta}{c}\frac{n_{\rm b}}{2\varepsilon_0}\rho\hat{\rho},$$

where $v_z \cong \beta c$.

We have the compact result that the total electromagnetic force is given by

$$\vec{F}_{\rm EM} = \vec{F}_{\rm E} + \vec{F}_{\rm B} = \frac{q^2 n_{\rm b}}{2\varepsilon_0}(1 - \beta^2)\rho\hat{\rho} = \frac{q\vec{E}}{\gamma^2}.$$

This result could also have been obtained by direct calculation of the radial electric and azimuthal magnetic field in the lab frame.

(b) We also have two other forces: the solenoid force

$$\vec{F}_{\rm sol} = q\vec{v} \times \vec{B}_{\rm sol} = -qv_\phi B_0 = -q(\omega_{\rm r}\rho)B_0\hat{\rho} \quad \text{where } v_\phi = \omega_{\rm r}\rho,$$

and the centripetal force

$$\vec{F}_{\rm cent} = \frac{\gamma m_0 v_\phi^2}{\rho}\hat{\rho} = \frac{\gamma m_0 (\omega_{\rm r}\rho)^2}{\rho}\hat{\rho} = \gamma m_0\omega_{\rm r}^2\rho\hat{\rho}.$$

The total force is

$$\vec{F}_{\rm net} = \vec{F}_{\rm EM} + \vec{F}_{\rm sol} + \vec{F}_{\rm cent},$$

$$F_{{\rm net},\rho} = q^2\frac{n_{\rm b}}{2\varepsilon_0}\frac{1}{\gamma^2}\rho - q\omega_{\rm r}B_0\rho + \gamma m_0\omega_{\rm r}^2\rho.$$

At equilibrium:

$$F_{{\rm net},\rho} = q^2\frac{n_{\rm b}}{2\varepsilon_0}\frac{1}{\gamma^2}\rho - q\omega_{\rm r}B_0\rho + \gamma m_0\omega_{\rm r}^2\rho = 0,$$

or

$$\frac{\omega_\rho^2}{2} - \omega_{\rm c}\omega_{\rm r} + \omega_{\rm r}^2 = 0.$$

Here we have used

$$\omega_{\rm c} = \frac{qB_0}{\gamma m_0} \quad \text{and} \quad \omega_\rho^2 = \frac{\omega_{\rho 0}^2}{\gamma^3} = \frac{q^2 n_{\rm b}}{m\varepsilon_0}\frac{1}{\gamma^3}.$$

In more standard form, we have

$$\omega_{\rm r}^2 = -\frac{\omega_{\rm p}^2}{2} + \omega_{\rm r}\omega_{\rm c}.$$

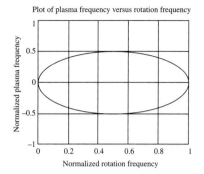

(c) Plot ω_p/ω_c versus ω_r/ω_c.

The rotation frequency that maximizes the equilibrium density of the beam is $\omega_r = \omega_c/2$.

Exercise 2.16

The helical undulator has on on-axis field given by

$$\vec{B}(z) = \frac{B_0}{\sqrt{2}}[\sin(k_u z)\hat{x} + \cos(k_u z)\hat{y}].$$

The vector potential in cylindrical coordinates is

$$A_\phi = -\frac{B_0}{k_u}[I_0(k_u\rho) + I_2(k_u\rho)]\cos(\phi - k_u z),$$

$$A_\rho = -\frac{B_0}{k_u}[I_0(k_u\rho) - I_2(k_u\rho)]\sin(\phi - k_u z),$$

where $I_\nu(x)$ are the modified Bessel functions which obey he following relations:

$$I_0(x) - I_2(x) = \frac{2}{x}I_1(x), \quad I_1'(x) = I_0(x) - \frac{1}{x}I_1(x),$$

$$I_0(x) + I_2(x) = 2I_1'(x), \quad I_1'(x) = I_2(x) + \frac{1}{x}I_1(x), \quad I_0'(x) = I_1(x).$$

Let us use some abbreviations to simplify notation

$$B_0 \rightarrow B, \quad k_u \rightarrow k$$

$$I_\nu(k_u\rho) \rightarrow I_\nu, \quad \frac{\partial}{\partial\rho}I_\nu(k_u\rho) \rightarrow kI_\nu'.$$

(a) Show that the $\vec{\nabla}^2\vec{A} = 0$. Since there are no magnetic source terms, $\vec{\nabla} \times \vec{B} = \vec{\nabla} \times (\vec{\nabla} \times \vec{A}) = \vec{\nabla}(\vec{\nabla} \cdot \vec{A}) - \vec{\nabla}^2\vec{A} = 0$. Therefore, $\vec{\nabla}(\vec{\nabla} \cdot \vec{A}) = \vec{\nabla}^2\vec{A}$.

First let us calculate $\vec{\nabla} \cdot \vec{A}$ (in cylindrical coordinates):

$$
\begin{aligned}
\vec{\nabla} \cdot \vec{A} &= \frac{1}{\rho} \frac{\partial}{\partial \rho}(\rho A_\rho) + \frac{1}{\rho} \frac{\partial A_\phi}{\partial \phi} + \frac{\partial A_z}{\partial z} \\
&= \frac{1}{\rho} \frac{\partial}{\partial \rho} \left[-\frac{B}{k} \sin(\phi - kz) \rho (I_0 + I_2) \right] \\
&\quad + \frac{1}{\rho} \frac{\partial}{\partial \phi} \left[-\frac{B}{k} \cos(\phi - kz) \rho (I_0 + I_2) \right] \\
&= -\frac{B}{k} \sin(\phi - kz) \frac{1}{\rho} \frac{\partial}{\partial \rho} \left[\rho \frac{2}{k\rho} I_1 \right] + \frac{1}{\rho} \frac{B}{k} \sin(\phi - kz)[I_0 + I_2] \\
&= \frac{B}{k\rho} \sin(\phi - kz) \left\{ -\frac{\partial}{\partial \rho} \left[\frac{2}{k} I_1 \right] + [I_0 + I_2] \right\} \\
&= \frac{2B}{k\rho} \sin(\phi - kz) \{ -I_1' + I_1' \} = 0.
\end{aligned}
$$

So from $\vec{\nabla}(\vec{\nabla} \cdot \vec{A}) = \vec{\nabla}^2 \vec{A}$ we can conclude that $\vec{\nabla}^2 \vec{A} = 0$.

(b) Next we calculate the field components in cylindrical coordinates:

$$
\vec{B} = \vec{\nabla} \times \vec{A} = \left[\frac{1}{\rho} \frac{\partial A_z}{\partial \phi} - \frac{\partial A_\phi}{\partial z} \right] \hat{\rho} + \left[\frac{\partial A_\rho}{\partial z} - \frac{\partial A_z}{\partial \rho} \right] \hat{\phi} + \frac{1}{\rho} \left[\frac{\partial}{\partial \rho}(\rho A_\phi) - \frac{\partial A_\rho}{\partial \phi} \right] \hat{z}.
$$

Let us calculate each term individually. The ρ-component is:

$$
B_\rho = \left[\frac{1}{\rho} \frac{\partial A_z}{\partial \phi} - \frac{\partial A_\phi}{\partial z} \right] B_\rho = 2B \left[I_0(k_u \rho) - \frac{1}{k_u \rho} I_1(k_u \rho) \right] \sin(\phi - kz).
$$

The ϕ-component is:

$$
B_\phi = \frac{\partial A_\rho}{\partial z} - \frac{\partial A_z}{\partial \rho} = \frac{2B_0}{k_u \rho} I_1(k_u \rho) \cos(\phi - kz).
$$

The z-component is:

$$
B_z = \frac{1}{\rho} \left[\frac{\partial}{\partial \rho}(\rho A_\phi) - \frac{\partial A_\rho}{\partial \phi} \right] = -2B_0 I_1(k_u \rho) \cos(\phi - k_u z).
$$

(c) First we can use small angle approximations for the modified Bessel functions since $k_u \rho \ll 1$.

$$
I_0 \rightarrow 1 + \left(\frac{k\rho}{2} \right)^2, \qquad I_2 \rightarrow \frac{1}{2} \left(\frac{k\rho}{2} \right)^2.
$$

In the first-order approximation these become

$$
I_0 \rightarrow 1, \qquad I_2 \rightarrow 0.
$$

We need to transform the vector potential from cylindrical coordinates to Cartesian coordinates:

$$A_x = A_\rho \cos\phi - A_\phi \sin\phi$$

$$= -\frac{B}{k}\{[I_0 - I_2]\cos\phi\sin(\phi - k_u z) - [I_0 + I_2]\sin\phi\cos(\phi - k_u z)\}$$

$$\cong -\frac{B}{k}\{\cos\phi(\sin\phi\cos kz - \sin kz\cos\phi)$$

$$- \sin\phi(\cos\phi\cos kz + \sin\phi\sin kz)\} = \frac{B}{k}\sin kz,$$

$$A_y = A_\phi \cos\phi - A_\rho \sin\phi$$

$$= -\frac{B}{k}\{[I_0 - I_2]\sin\phi\sin(\phi - k_u z) + [I_0 + I_2]\cos\phi\cos(\phi - k_u z)\}$$

$$\cong -\frac{B}{k}\{\sin\phi(\sin\phi\cos kz - \sin kz\cos\phi)$$

$$- \cos\phi(\cos\phi\cos kz + \sin\phi\sin kz)\} = -\frac{B}{k}\cos kz,$$

$$A_z = 0.$$

The Hamiltonian for this system is given by

$$H = \sqrt{(\vec{p} - q\vec{A})^2 c^2 + m^2 c^4} = \sqrt{(p_x - qA_x)^2 c^2 + (p_y - qA_y)^2 c^2 + p_z^2 c^2 + m^2 c^4}$$

The equations of motion can be derived from:

$$\dot{x} = \frac{\partial H}{\partial p_x} = \frac{c^2}{H}(p_x - qA_x), \quad \dot{y} = \frac{\partial H}{\partial p_y} = \frac{c^2}{H}(p_y - qA_y), \quad \dot{z} = \frac{\partial H}{\partial p_y} = \frac{c^2}{H}p_z.$$

The canonical momenta are:

$$\dot{p}_x = -\frac{\partial H}{\partial x} = \frac{c^2}{H}(p_x - qA_x)\left(-q\frac{\partial A_x}{\partial x}\right) = 0,$$

$$\dot{p}_y = -\frac{\partial H}{\partial y} = \frac{c^2}{H}(p_y - qA_y)\left(-q\frac{\partial A_y}{\partial y}\right) = 0, \quad \dot{p}_z = -\frac{\partial H}{\partial z} = 0.$$

The Hamiltonian is constant and $H = \gamma mc^2$. Since $\dot{p}_z = 0$, we have $p_z = \gamma m\beta_z c$ and formally

$$\dot{z} = \frac{c^2}{\gamma mc^2}p_z = \beta_z c, \quad z(t) = \beta_z ct.$$

A particle entering a helical undulator's magnetic field is assumed to have no transverse velocity components. This means that $p_x = p_y = 0$.

Introducing the undulator parameter $K = qB/mkc$, we now have:

$$\dot{x} = -\frac{qB}{\gamma mk}\sin(kz) = -\frac{cK}{\gamma}\sin(kz), \quad x(t) = \frac{cK}{\gamma k}\cos(kz),$$

$$\dot{y} = \frac{qB}{\gamma mk}\cos(kz) = \frac{cK}{\gamma}\cos(kz), \quad y(t) = \frac{cK}{\gamma k}\sin(kz).$$

The equations for $x(t)$ and $y(t)$ describe the motion of a helix with radius $\rho = cK/k\gamma = $ constant.

From the equations of motion we can see that the mechanical angular momentum $p_{\phi,\text{mech}} = \gamma m \rho \dot{\phi}$ is constant. Let us return to our coordinate transformations and differentiate with respect to time realizing that ρ is a constant.

$$x = \rho \cos \phi, \qquad y = \rho \sin \phi,$$

$$\dot{x} = -\rho \dot{\phi} \sin \phi, \qquad \dot{y} = \rho \dot{\phi} \cos \phi.$$

Combining this and the previous expressions for $\dot{x}$ and $\dot{y}$, we obtain two expressions for $\dot{\phi}$.

$$\dot{\phi} \sin \phi = \frac{cK}{\lambda \rho} \sin kz, \qquad \dot{\phi} \cos \phi = \frac{cK}{\lambda \rho} \cos kz.$$

The left-hand side of these equations can be written as total derivatives of sine and cosine functions:

$$\frac{\mathrm{d}}{\mathrm{d}t}(\cos \phi) = \frac{cK}{\lambda \rho} \sin kz, \qquad \frac{\mathrm{d}}{\mathrm{d}t}(-\sin \phi) = \frac{cK}{\lambda \rho} \cos kz.$$

Recall that in the paraxial approximation, we have

$$\frac{\mathrm{d}}{\mathrm{d}t} = \frac{\mathrm{d}z}{\mathrm{d}t} \frac{\mathrm{d}}{\mathrm{d}z} = v_z \frac{\mathrm{d}}{\mathrm{d}z} = \beta_z c \frac{\mathrm{d}}{\mathrm{d}z}.$$

Using this we obtain,

$$\frac{\mathrm{d}}{\mathrm{d}z}(\cos \phi) = \frac{cK}{\lambda \rho} \sin kz,$$

which can be integrated to find

$$\cos \phi = \frac{K}{\gamma \rho \beta_z k}(-\cos kz).$$

Combining these expressions we have,

$$\dot{\phi} \cos \phi = \frac{cK}{\lambda \rho} \cos kz,$$

$$\dot{\phi} \cdot \frac{K}{\gamma \rho \beta_z k}(-\cos kz) = \frac{cK}{\lambda \rho} \cos kz,$$

or $\dot{\phi} = -ck\beta_z = \text{constant}$. Recall that $p_{\phi,\text{mech}} = \text{constant}$. Throughout this exercise we have kept terms only up to first order in $k_u \rho$.

Chapter 3

Exercise 3.2

Using the equipartition theorem,

$$\frac{p_x^2}{2m} = \frac{p_y^2}{2m} = \frac{1}{2}kT,$$

and assuming that

$$x'_{rms} = \sqrt{\frac{kT}{m_0c^2}} \cdot \frac{1}{\beta\gamma} = y'_{rms}.$$

To calculate the field index, n, required to guarantee that the betatron beam obeys $x_{rms} = 2y_{rms}$, start with:

$$x_{rms} = \frac{R \cdot x'_{rms}}{v_x}, \qquad y_{rms} = \frac{R \cdot y'_{rms}}{v_y}.$$

The tunes are $v_x = \sqrt{1-n}$ and $v_y = \sqrt{n}$.
  Following the given requirements with the above relations:

$$x_{rms} = 2y_{rms}$$

and

$$\frac{R \cdot x'_{rms}}{v_x} = \frac{2R \cdot y'_{rms}}{v_y}, \quad \text{with } x'_{rms} = y'_{rms}$$

gives $2v_x = v_y$, which can be written as $2\sqrt{1-n} = \sqrt{n}$, or $1-n = n/4$, and so the required field index is $n = 4/5$.

Exercise 3.4

(a) $\vec{x} = \mathbf{M_T} \cdot \vec{x}_0 = \mathbf{M_O} \cdot \mathbf{M_F} \cdot \vec{x}_0$
The total transformation matrix for one period is $\mathbf{M}_T = \mathbf{M_O} \cdot \mathbf{M_F}$

$$\mathbf{M_T} = \begin{pmatrix} 1 & L \\ 0 & 1 \end{pmatrix} \cdot \begin{pmatrix} 1 & 0 \\ -\frac{1}{f} & 1 \end{pmatrix} = \begin{pmatrix} 1 - \frac{L}{f} & L \\ -\frac{1}{f} & 1 \end{pmatrix}.$$

(b) To ensure linear stability, $|\text{Tr}\,\mathbf{M_T}| \leq 2$ giving in this case

$$\left| 1 - \frac{L}{f} + 1 \right| \leq 2 \quad \text{or} \quad \left| 2 - \frac{L}{f} \right| \leq 2,$$

which is simply

$$\frac{L}{f} \leq 4.$$

Exercise 3.12

Consider the FOFO focusing system:

(a) The transformation matrix of the FOFO system is given by

$$\mathbf{M_T} = \mathbf{M_O} \cdot \mathbf{M_F} = \begin{pmatrix} 1 & l \\ 0 & 1 \end{pmatrix} \cdot \begin{pmatrix} \cos(\kappa_0 l) & \frac{1}{k_0}\sin(\kappa_0 l) \\ -\kappa_0 \sin(k_0 l) & \cos(k_0 l) \end{pmatrix}$$

$$= \begin{pmatrix} \cos(\kappa_0 l) - \kappa_0 l \sin(\kappa_0 l) & \frac{1}{k_0}\sin(\kappa_0 l) + l\cos(\kappa_0 l) \\ -\kappa_0 l \sin(\kappa_0 l) & \cos(\kappa_0 l) \end{pmatrix}.$$

The phase advance, μ, is a function of $\kappa_0 l$:

$$\cos\mu = \frac{1}{2}\mathrm{Tr}\,\mathbf{M_T} \quad \text{or} \quad \cos\mu = \cos(\kappa_0 l) - \frac{\kappa_0 l}{2}\sin(\kappa_0 l).$$

(b) In the limit of $\kappa_0 l \ll$, use the following expansions up to fourth order:

$$\cos(\kappa_0 l) \approx 1 - \frac{(\kappa_0 l)^2}{2} + \frac{(\kappa_0 l)^4}{24}, \qquad \sin(\kappa_0 l) \approx \kappa_0 l - \frac{(\kappa_0 l)^3}{6},$$

and

$$\cos\mu \approx 1 - \frac{\mu^2}{2} + \frac{\mu^4}{24}.$$

Substituting the above relations into $\cos\mu = \cos(\kappa_0 l) - (\kappa_0 l/2)\sin(\kappa_0 l)$ yields

$$1 - \frac{(\kappa_0 l)^2}{2} + \frac{(\kappa_0 l)^4}{24} - \frac{(\kappa_0 l)^2}{2} + \frac{(\kappa_0 l)^4}{24} = 1 - \frac{\mu^2}{2} + \frac{\mu^4}{24}.$$

Simplifying, we have

$$\mu^2 - \frac{\mu^4}{12} = 2(\kappa_0 l)^2 - \frac{1}{4}(\kappa_0 l)^4,$$

which yields

$$\mu^2 = 2(\kappa_0 l)^2 + \frac{(\kappa_0 l)^4}{12}.$$

Now, evaluate this expression for k_{sec}^2:

$$k_{\mathrm{sec}}^2 = \frac{\mu^2}{L_{\mathrm{p}}^2} = \frac{\mu^2}{(2l)^2} \quad \text{or} \quad k_{\mathrm{sec}}^2 = \frac{\kappa_0^2}{2} + \frac{1}{48}\kappa_0^4 l^2.$$

(c) Now define κ_0^2 as a harmonic series

$$\kappa^2(z) = \begin{cases} \kappa_0^2, & 0 \le z \le l, \\ 0, & l \le z \le 1. \end{cases}$$

The Fourier series is useful:

$$\kappa^2(z) = \kappa_0^2 \left[\frac{a_0}{2} + \sum_{n=1}^{\infty} a_n \cos\left(\frac{n\pi z}{l}\right) + b_n \sin\left(\frac{n\pi z}{l}\right) \right].$$

The coefficients are:

$$a_0 = \frac{1}{(\kappa_0^2 l)} \int_0^1 \kappa_0^2 \, dz = 1,$$

$$a_n = \frac{1}{(\kappa_0^2 l)} \int_0^1 \kappa_0^2 \cos\left(\frac{n\pi z}{l}\right) dz = \frac{1}{n\pi}[\sin(n\pi) - \sin(0)] = 0,$$

$$b_n = \frac{1}{\kappa_0^2} \sin\left(\frac{n\pi z}{l}\right) dz = \frac{1}{n\pi}[\cos(0) - \cos(n\pi)]$$

$$= \begin{cases} 0, & n \text{ even}, \\ \dfrac{2}{n\pi}, & n \text{ odd} \end{cases}$$

and so

$$\kappa^2(z) = \kappa_0^2 \left[\frac{1}{2} + \sum_{n=1,\text{odd}}^{\infty} \frac{2}{n\pi} \sin\left(\frac{n\pi z}{l}\right) \right].$$

(d) The Fourier series solution is a superposition of a uniform focusing system of strength $\kappa_0^2/2$ (first term on the RHS) and an FD system of the same strength (the second term on the LHS). The averaged focusing strength is

$$\langle \kappa^2 \rangle = \frac{\kappa_0^2}{2} + \frac{\kappa_0^4}{2k_p^2} \sum_{n=1,\text{odd}}^{\infty} \frac{b_n^2}{n^2}.$$

Using $b_n = 2/n\pi$ and $k_p = 2\pi/L_p = \pi/l$,

$$\langle \kappa^2 \rangle = \frac{\kappa_0^2}{2} + \frac{2\kappa_0^4 l^2}{\pi^4} \sum_{n=1,\text{odd}}^{\infty} \frac{1}{n^4}$$

or

$$\langle \kappa^2 \rangle = \frac{\kappa_0^2}{2} + \frac{2\kappa_0^4 l^2}{48}.$$

(e) Solutions to (b) and (d) are identical.

Exercise 3.15

In the smooth approximation,

$$k_{\text{sec},x} = \frac{v_x}{R_0}.$$

Referring to the evolution of horizontal dispersion, Eq. (3.65),

$$\eta'' + k_x^2 \eta_x = \frac{1}{R_0}.$$

Solutions are given by Eq. (3.66),

$$\eta_x = A\cos(\kappa_b s) + B\sin(\kappa_b s) + \frac{1}{\kappa_b^2 R_0}.$$

The solution is given in two parts. The homogeneous solution is given by the first two terms with coefficients A and B. The final term on the RHS is the particular solution,

$$\eta_x = \frac{1}{k_x^2 R_0}.$$

The given Tevatron values are $v_x = 19.4$ and $R_0 = 1\,\text{km}$. The approximate constant value of the dispersion can be estimated for the Tevatron using only the particular solution

$$\eta_x = \frac{10^3\,\text{m}}{19.4^2} = 2.66\,\text{m}.$$

Exercise 3.18

In Fig. 3.13, the transport line translates the beam to the side. The dispersion and its derivative vanish at the second bend exit. The dispersion also vanishes at the mid-point between the magnets.

(a) The given conditions are

$$\eta_x(0) = \eta_x'(0) = 0,$$

$$\eta_x(\text{end}) = \eta_x'(\text{end}) = 0,$$

$$\eta_x(\text{mid}) = 0.$$

The focal length is derived using

$$\begin{pmatrix} \eta_x(s) \\ \eta_x'(s) \\ 1 \end{pmatrix} = \mathbf{M_T} \cdot \begin{pmatrix} \eta_x(0) \\ \eta_x'(0) \\ 1. \end{pmatrix}.$$

Since the system is completely symmetrical about the midpoint, this expression can be simplified by rewriting the expression for the focal length,

$$\begin{pmatrix} \eta_x(\text{mid}) \\ \eta_x'(\text{mid}) \\ 1 \end{pmatrix} = \mathbf{M_{T/2}} \cdot \begin{pmatrix} \eta_x(0) \\ \eta_x'(0) \\ 1 \end{pmatrix},$$

with

$$\mathbf{M_{T/2}} = \mathbf{M_{o,b}} \cdot \mathbf{M_F} \cdot \mathbf{M_{o,a}} \cdot \mathbf{M_{bend}}$$

$$= \begin{pmatrix} 1 & b & 0 \\ 0 & 1 & 0 \\ 0 & 0 & 0 \end{pmatrix} \cdot \begin{pmatrix} 1 & 0 & 0 \\ -1/f & 1 & 0 \\ 0 & 0 & 1 \end{pmatrix} \cdot \begin{pmatrix} 1 & a & 0 \\ 0 & 1 & 0 \\ 0 & 0 & 1 \end{pmatrix}$$

$$\cdot \begin{pmatrix} \cos\theta_b & R\sin\theta_b & R(1-\cos\theta_b) \\ -(1/R)\sin\theta_b & \cos\theta_b & \sin\theta_b \\ 0 & 0 & 1 \end{pmatrix}.$$

In calculating the midpoint expression, it is only necessary to solve $M_{13} = 0$,

$$\begin{pmatrix} 0 \\ \eta_x'(\text{mid}) \\ 1 \end{pmatrix} = \mathbf{M_{T/2}} \cdot \begin{pmatrix} 0 \\ 0 \\ 1 \end{pmatrix} = \begin{pmatrix} M_{13} \\ M_{23} \\ M_{33} \end{pmatrix}$$

or $\quad M_{13} = \left(1 - \dfrac{b}{f}\right)(R(1-\cos\theta_b) + a\sin\theta_b) + b\sin\theta_b = 0.$

To simplify notation, normalize all terms to R:

$$f' = \frac{f}{R}, \quad a' = \frac{a}{R}, \quad b' = \frac{b}{R}.$$

Solving for the focal length f' in terms of (normalized) a and b,

$$f' = \frac{b'(1 - \cos\theta_b + a'\sin\theta_b)}{1 - \cos\theta_b + (a' + b')\sin\theta_b}.$$

(b) The expression for the momentum compaction parameter is

$$\alpha_c = 2 \cdot \left[\frac{1}{R\theta_b} \int_0^{R\theta_b} \frac{\eta_x(s)}{R} ds \right] = 2 \cdot \left[\frac{1}{R\theta_b} \int_0^{R\theta_b} \frac{R(1 - \cos(s/R))}{R} ds \right],$$

or

$$\alpha_c = 2 \cdot \left(1 - \frac{\sin\theta_b}{\theta_b} \right).$$

Chapter 4

Exercise 4.1

The drift tube linac has an electric field profile given by

$$E_z(z) = \begin{cases} E_0, & 0 \leq z < d/2 \\ 0, & d/2 \leq z < d \end{cases}.$$

(a) Since the system is periodic,

$$E_z(z) = \text{Im} \sum_{n=-\infty}^{\infty} a_n e^{i(2n\pi/d)x}$$

The coefficients a_n are given by

$$a_n = \frac{1}{d}\text{Im} \int_{-d/2}^{d/2} E_i(z) e^{i(2n\pi/d)z} dz$$

$$= \frac{1}{d}\text{Im} \int_0^{d/2} E_0 e^{i(2n\pi/d)z} dz = \frac{E_0}{d}\text{Im} \int_0^{d/2} e^{i(2n\pi/d)z} dz$$

$$= \frac{E_0}{d} \left[\frac{id}{2\pi n} e^{i(2n\pi/d)z} \right]_0^{d/2} = \text{Im}(i(e^{in\pi} - 1))$$

$$= \frac{E_0}{2n\pi}(\cos(n\pi) - 1) = \frac{E_0}{2n\pi} \begin{cases} -2, & n = \text{odd} \\ 0, & n = \text{even} \end{cases}.$$

Thus,

$$E_z(z) = -E_0 \cdot \text{Im} \sum_{n=-\infty,\text{odd}}^{\infty} \frac{1}{n\pi} e^{i(2n\pi/d)z}.$$

(b) Assuming the particle is synchronous with the first harmonic ($n = 1$), the particle velocity can be written as

$$v_z \approx v_{\phi,1} = \frac{\omega}{k_{z,1}} = \frac{d\omega}{2\pi},$$

since $k_{z,n} = 2n\pi/d$. The relationship between the period and the frequency is,

$$d = \frac{2\pi v_z}{\omega} = \frac{v_z}{f}.$$

(c) For a partially relativistic ion beam, the periodicity must change to keep particles in resonance. The relationship between kinetic energy and velocity is derived from the energy expression $u = T + m_0c^2 = \gamma m_0 c^2$.
Solving for γ, we have $\gamma = (T/m_0 c^2) + 1$.
We were given $T_0 = 5\,\text{MeV}$ and $T_f = 250\,\text{MeV}$. Recall that the rest mass of a proton is $m_p c^2 = 939\,\text{MeV}$, and thus

$$\gamma_0 = \frac{5}{939} + 1 \quad \text{and} \quad \gamma_f = \frac{250}{939} + 1.$$

Then the velocity can be found from

$$\frac{v_z}{c} = \beta = \sqrt{1 - \frac{1}{\gamma^2}} \quad \text{and} \quad \beta_0 = 0.1, \ \beta_f = 0.61.$$

Using the results from part (b) we obtain:

$$d_0 = \frac{\beta_0 c}{f} = 6\,\text{cm} \quad \text{and} \quad d_f = \frac{\beta_f c}{f} = 36.6\,\text{cm}.$$

Exercise 4.2

Evaluate α_{rf}:

$$\alpha_{rf} = \frac{qE_0}{k_z m_0 c^2}.$$

(a) For a high gradient electron linac: $f = \omega/2\pi = 2856\,\text{MHz}$, $E_0 = 50\,\text{MV/m}$, $qE_0 = 50\,\text{MeV/m}$, $U = 1\,\text{GeV}$, $m_e c^2 = 0.511\,\text{MeV}$.
Determine the value of k_z: Begin with

$$v_z = \frac{\omega}{k_z}, \qquad \gamma = \frac{1}{\sqrt{1 - (v^2/c^2)}},$$

and using the paraxial approximation $v \approx v_z$:

$$v_z = c\sqrt{1 - \frac{1}{\gamma^2}}.$$

We can express γ in terms of the energy:

$$U = \gamma m_0 c^2 \text{ or } \gamma = \frac{U}{\gamma m_0 c^2} \quad \text{and} \quad v_z = c\sqrt{1 - \frac{(m_0 c^2)^2}{U^2}} = c\sqrt{1 - \left(\frac{.511}{10^3}\right)^2} \cong c.$$

Now evaluate α_{rf}:

$$\alpha_{rf} = \frac{cqE_0}{\omega m_0 c^2} = 1.63.$$

(b) For a moderate gradient proton linac: $f = \omega/2\pi = 805\,\text{MHz}$, $qE_0 = 8\,\text{MeV/m}$, $T = 200\,\text{MeV}$ (here T is the kinetic energy), and $m_p c^2 = 939\,\text{MeV}$. Recall that the total energy $U = \gamma m_0 c^2 = T + m_0 c^2$ or $\gamma = 1 + (T/m_0 c^2)$.

Plugging this value into

$$v_z = c\sqrt{1 - \frac{1}{\gamma^2}}$$

we find

$$v_z = c\sqrt{1 - \left(1 + \frac{T}{m_0 c^2}\right)^{-2}} = 0.57c.$$

For a proton linac α_{rf}:

$$\alpha_{rf} = \frac{v_z q E_0}{\omega m_0 c^2} = 3 \times 10^{-4}.$$

Exercise 4.3

The Hamiltonian for this problem is given by

$$\tilde{H}(\chi, \phi) = m_0 c^2 [\chi^{-1} + \alpha_{rf} \cos \phi],$$

where

$$\chi = \sqrt{\frac{1 + \beta_z}{1 - \beta_z}}.$$

(a) For optimum capture, in the final state $\phi_f = \pi/2$, and $v_z \approx c$, so $\chi^{-1} \Longrightarrow 0$, and

$$\tilde{H}\left(\chi \to \infty, \phi_f = \frac{\pi}{2}\right) = 0.$$

The Hamiltonian is a constant of motion, so in the initial state we have

$$\tilde{H}(\chi, \phi) = m_0 c^2 [1 + \alpha_{rf} \cos \phi_0] = 0.$$

The minimum value of α_{rf} which allows solution of this expression is clearly 1.

(b) From the above final state Hamiltonian, we have

$$\phi_0 = \arccos\left(-\frac{1}{\alpha_{rf}}\right).$$

(c) The initial phase for the electron linac: $f = \omega/2\pi = 2856\,\text{MHz}$, with $E_0 = 50\,\text{MV/m}$, or $qE_0 = 50\,\text{MeV/m}$, and $m_e c^2 = 0.511\,\text{MeV}$, we have $\alpha_{rf} = 1.63$. So $\phi_0 = 127.57°$.

(d) **Hamiltonian method**: Since the Hamiltonian is a constant, the initial and final values are the same,

$$1 + \alpha_{rf} \cos \phi_i = \alpha_{rf} \cos \phi_f$$

Using $\phi_i = 127.57° - 1° = 126.57°$ (one degree early):

$$\phi_f = \arccos\left(\frac{1}{\alpha_{rf}} + \cos \phi_i\right),$$

$$\phi_f = 89.2°.$$

First-order analysis

$$\frac{\Delta\phi_f}{\Delta\phi_i} = \frac{\partial\phi_f}{\partial\phi_i} = \frac{(\partial H/\partial\phi_i)_i}{(\partial H/\partial\phi_f)_f},$$

$$\frac{\Delta\phi_f}{\Delta\phi_i} = \frac{-m_0 c^2 \alpha_{rf} \sin \phi_i}{-m_0 c^2 \alpha_{rf} \sin \phi_f},$$

$$\Delta\phi_f = \frac{\sin \phi_i}{\sin \phi_f} \cdot \Delta\phi_i.$$

The parameters given are

$$\phi_i = 127.57°, \quad \Delta\phi_i = 1°, \quad \phi_f = \frac{\pi}{2} - \Delta\phi_f$$

and

$$\Delta\phi_f = \frac{\sin \phi_i}{\sin \phi_f} \cdot \Delta\phi_i = \frac{\sin(127.57)}{\sin(\pi/2)} \cdot 1° = 0.8°,$$

$$\phi_f = \frac{\pi}{2} - \Delta\phi_f = 90° - 0.8° = 89.2°.$$

Both of these methods yield the same result.

Exercise 4.4

The parameters given in Exercise 4.2(b) are:

$$\alpha_{rf} = 3 \times 10^{-4}, \quad \gamma_0 = 1.21, \quad \beta_0 = 0.57, \quad k_z = 29.7 \, \text{m}^{-1}$$

The momentum is $p_0 = \gamma_0 m_0 \beta_0 c$.
(a) The bucket height is given by Eq. (4.32):

$$\frac{\delta p_{max}}{p_0} = \pm \sqrt{\frac{4\alpha_{rf}\gamma_0}{\beta_0^2}},$$

with our parameters this becomes:

$$\frac{\delta p_{max}}{p_0} = \pm 6.7\%.$$

(b) The area of the bucket is given by Eq. (4.33):

$$A_b = \frac{16 p_0}{k_z} \sqrt{\frac{\alpha_{rf} \gamma_0}{\beta_0^2}} = 0.018 m \cdot p_0.$$

We want to calculate the momentum in units of MeV/c.

$$p_0 \left(\frac{MeV}{c} \right) = \gamma_0 \frac{m_0 c^2}{c} \beta_0 = 1.21 \left(939 \frac{MeV}{c} \right) 0.57 = 648 \frac{MeV}{c}.$$

Combining the above equations we obtain:

$$A_b = 3.9 \times 10^{-8} \text{ MeV} \cdot \text{s}.$$

(c) The synchrotron frequency is given by Eq. (4.36):

$$\omega_s = \sqrt{\frac{\alpha_{rf}}{\gamma_0^3} \frac{\omega}{\beta_0}}.$$

The normalized synchrotron frequency is:

$$\frac{\omega_s}{\omega} = 0.023 = 2.3\%.$$

Exercise 4.5

$$H = m_0 c^2 \left[\frac{\beta_0^2}{2 \gamma_0 p_0^2} (\delta p)^2 + \alpha_{rf} [\cos(k_z \zeta) + 1] \right].$$

The equation of motion is

$$\dot{\zeta} = \frac{\partial H}{\partial (\delta p)} = \frac{m_0 c^2 \beta_0^2}{\gamma_0 p_0^2} (\delta p) = \frac{1}{\gamma_0^3 m_0} (\delta p),$$

so the Hamiltionian can be written as

$$H = m_0 c^2 \left[\frac{\gamma_0^3 m_0^2}{2 c^2} (\dot{\zeta})^2 + \alpha_{rf} [\cos(k_z \zeta) + 1] \right].$$

Solving for $\dot{\zeta}$, we have

$$\dot{\zeta} = \sqrt{\frac{2 c^2}{\gamma_0^3 m_0^2}} \sqrt{\frac{H}{m_0 c^2} - \alpha_{rf} [\cos(k_z \zeta) + 1]} = \frac{d\zeta}{dt}.$$

To obtain a dependence on amplitude, we need to evaluate H at a turning point $H(\zeta_{max}, 0)$

$$\frac{d\zeta}{dt} = \sqrt{\frac{2 c^2 \alpha_{rf}}{\gamma_0^3 m_0^2}} \sqrt{\cos(k_z \zeta_{max}) - \cos(k_z \zeta)},$$

or

$$\frac{d\zeta}{dt} = \sqrt{\frac{2 c^2 \alpha_{rf}}{\gamma_0^3 m_0^2}} \sqrt{\cos(k_z \zeta_{max}) - 1 + 2 \sin^2 \left(\frac{k \zeta}{2} \right)},$$

which can be written

$$\frac{d\zeta}{dt} = \sqrt{\frac{2c^2\alpha_{rf}(\cos(k_z\zeta_{max}) - 1)}{\gamma_0^3 m_0^2}} \sqrt{1 + \frac{2\sin^2(k\zeta/2)}{\cos(k_z\zeta_{max}) - 1}}.$$

Let us now set

$$\sqrt{\frac{2}{\cos(k_z\zeta_{max}) - 1}} = b,$$

$$\frac{d\zeta}{dt} = \sqrt{\frac{2c^2\alpha_{rf}(\cos(k_z\zeta_{max}) - 1)}{\gamma_0^3 m_0^2}} \sqrt{1 + b^2 \sin^2\left(\frac{k\zeta}{2}\right)}.$$

Integrating, we obtain

$$\int_0^{T/4} dt = \sqrt{\frac{\gamma_0^3 m_0^2}{2c^2\alpha_{rf}(\cos(k_z\zeta_{max}) - 1)}} \int_0^{\zeta_{max}} \frac{1}{\sqrt{1 + b^2 \sin^2(k\zeta/2)}} d\zeta,$$

or

$$\frac{T}{4} = \sqrt{\frac{\gamma_0^3 m_0^2}{2c^2\alpha_{rf}(\cos(k_z\zeta_{max}) - 1)}} \int_0^{\zeta_{max}} \frac{1}{\sqrt{1 + b^2 \sin^2\left(\frac{k\zeta}{2}\right)}} d\zeta$$

Thus,

$$\int_0^{\zeta_{max}} \frac{1}{\sqrt{1 + b^2 \sin^2(k\zeta/2)}} d\zeta = E(\zeta_{max}, b)$$

is an elliptical integral of the first order.

You can find the solution to this in a table. See Gradshteyn, I.S., *Table of Integrals, Series, and Products*, pp. 904–909.

The frequency is then given by

$$\omega = \frac{2\pi}{T} = \frac{\pi}{4} \sqrt{\frac{2c^2\alpha_{rf}(\cos(k_z\zeta_{max}) - 1)}{\gamma_0^3 m_0^2 E(\zeta_{max}, b)}}.$$

Chapter 5

Exercise 5.1

Equation (5.9) is

$$P(p_x) = C \exp\left(-\frac{p_x^2}{2\gamma_0 m_0 k_B T_x}\right) \equiv C_p \exp\left(-\frac{p_x^2}{2\sigma_p^2}\right).$$

Equation (5.11) is

$$X(x) = C_x \exp\left(-\frac{\gamma_0 m_0 v_0^2 \kappa_0^2 x^2}{2k_B T_x}\right) \equiv C_x \exp\left(-\frac{x^2}{2\sigma_x^2}\right).$$

Normalization constant for Eq. (5.9): In terms of σ_{p_x} we have

$$\int_{-\infty 0}^{\infty} C_p \exp\left(\frac{-p_x^2}{2\sigma_{p_x}^2}\right) dp_x = 1,$$

or

$$\rightarrow \sqrt{2\pi}\sigma_{p_x} C_p = 1 \quad \text{and} \quad \Rightarrow C_p = \frac{1}{\sqrt{2\pi}\sigma_{p_x}}.$$

In terms of T_x

$$\sigma_{p_x} = \gamma_0 m_0 k_B T_x \quad \Rightarrow \quad C_p = \frac{1}{\sqrt{2\pi}\gamma_0 m_0 k_B T_x}.$$

Normalization constant for Eq. (5.11): in terms of σ_x (same form, same solution)

$$\int_{-\infty}^{\infty} C_x \exp\left(\frac{-x^2}{2\sigma_x^2}\right) dx \quad \Rightarrow \quad C_x = \frac{1}{\sqrt{2\pi}\sigma_x}.$$

In terms of T_x

$$\sigma_x = \frac{\gamma_0 m_0 v_0^2 \kappa_0^2}{k_B T_x} \quad \Rightarrow \quad C_x = \frac{1}{\sqrt{2\pi}}\left(\frac{k_B T_x}{\gamma_0 m_0 v_0^2 \kappa_0^2}\right).$$

Exercise 5.6

An initial correlated bi-Gaussian distribution function is written in standard form as

$$f_x(x_0, x_0') = \frac{1}{2\pi\sigma_x\sigma_{x'}} \exp\left[-\frac{\gamma_x(z)x^2 + 2\alpha_x(z)xx' + \beta_x(z)x'^2}{2\varepsilon_x}\right].$$

Given the transformation

$$\vec{x}_f = M \cdot \vec{x}_0, \qquad \vec{x} = (x, x'),$$

we can also write the inverse transformation

$$\vec{x}_0 = M^{-1} \cdot \vec{x}_f, \quad \text{with } M^{-1} = \begin{bmatrix} M_{22} & -M_{12} \\ -M_{21} & M_{11} \end{bmatrix}.$$

The final distribution function can be written as

$$f_x(x_f, x_f')$$

$$= \frac{1}{2\pi\sigma_x\sigma_{x'}} \exp\left[\begin{array}{l} -\frac{\gamma_{x0}(M_{22}x_f - M_{12}x_f')^2 + 2\alpha_{x0}(M_{22}x_f - M_{12}x_f')(-M_{21}x_f + M_{11}x_f')}{2\varepsilon_x} \cdots \\ -\frac{\beta_{x0}(-M_{21}x_f + M_{11}x_f')^2}{2\varepsilon_x} \end{array} \right]$$

$$= \frac{1}{2\pi\sigma_x\sigma_{x'}} \exp\left[-\left[\begin{array}{l} \frac{(\gamma_{x0}M_{22}^2 - 2\alpha_{x0}M_{21}M_{22} + \beta_{x0}M_{21}^2)x_f^2}{2\varepsilon_x} \\ +\frac{(-\gamma_{x0}M_{22}M_{12} + 2\alpha_{x0}(M_{22}M_{11} + M_{12}M_{21}) - \beta_{x0}M_{21}M_{11})x_f x_f'}{2\varepsilon_x} \\ +\frac{(\gamma_{x0}M_{12}^2 - 2\alpha_{x0}M_{12}M_{11} + \beta_{x0}M_{11}^2)x_f'^2}{2\varepsilon_x}, \end{array} \right] \cdots \right]$$

which is a correlated bi-Gaussian.

The projection integrals in such correlated bi-Gaussians can be done by simply completing the squares in the exponential arguments:

$$n(x_f) = \int_{-\infty}^{\infty} f_x(x_f, x_f')\, dx' = \frac{1}{2\pi \sqrt{\beta_{xf}\varepsilon_x}} \exp\left[-\frac{x_f^2}{2\beta_{xf}\varepsilon_x}\right],$$

with

$$\beta_{xf} = M_{11}^2 \beta_{x0} - 2M_{11}M_{12}\alpha_{x0} + M_{12}^2 \gamma_{x0};$$

$$\Psi(x_f') = \int_{-\infty}^{\infty} f_x(x_f, x_f')\, dx = \frac{1}{2\pi \sqrt{\gamma_{xf}\varepsilon_x}} \exp\left[-\frac{x_f'^2}{2\gamma_{xf}\varepsilon_x}\right],$$

with

$$\gamma_{xf} = M_{21}^2 \beta_{x0} - 2M_{21}M_{22}\alpha_{x0} + M_{22}^2 \gamma_{x0}.$$

For more information on such Twiss parameter transformations, see the solution to Exercise 5.9.

Exercise 5.9

Equation (5.32) gives a relationship between the Twiss parameters,

$$\gamma_x(z)x^2 + 2\alpha_x(z)xx' + \beta_x(z)x'2 = \varepsilon_x.$$

Given the trace space vector transformation

$$\vec{x}_f = M \cdot \vec{x}_0, \vec{x} = (x, x'),$$

we can also write the inverse transformation

$$\vec{x}_0 = M^{-1} \cdot \vec{x}_f, \quad \text{with } M^{-1} = \begin{bmatrix} M_{22} & -M_{12} \\ -M_{21} & M_{11} \end{bmatrix}.$$

We can thus write out our first equation in terms of initial Twiss parameters and final phase space variables as

$$\gamma_{x0}(M_{22}x_f - M_{12}x_f')^2 + 2\alpha_{x0}(M_{22}x_f - M_{12}x_f')(-M_{21}x_f + M_{11}x_f')$$
$$+ \beta_{x0}(-M_{21}x_f + M_{11}x_f')^2 = \varepsilon_x.$$

We next need to collect terms proportional to x_f^2, $x_f x_f'$, and $x_f'^2$, to give

$$(\gamma_{x0}M_{22}^2 - 2\alpha_{x0}M_{21}M_{22} + \beta_{x0}M_{21}^2)x_f^2$$
$$+ (-\gamma_{x0}M_{22}M_{12} + 2\alpha_{x0}(M_{22}M_{11} + M_{12}M_{21}) - \beta_{x0}M_{21}M_{11})x_f x_f' + \cdots$$
$$+ (\gamma_{x0}M_{12}^2 - 2\alpha_{x0}M_{12}M_{11} + \beta_{x0}M_{11}^2)x_f'^2 = \varepsilon_x.$$

Identifying the coefficients of the terms with the appropriate Twiss parameters, we can write

$$\begin{pmatrix} \beta_x \\ \alpha_x \\ \gamma_x \end{pmatrix}_f = \begin{bmatrix} M_{11}^2 & -2M_{11}M_{12} & M_{12}^2 \\ -M_{11}M_{21} & M_{11}M_{22} + M_{12}M_{21} & -M_{12}M_{22} \\ M_{21}^2 & -2M_{21}M_{22} & M_{22}^2 \end{bmatrix} \begin{pmatrix} \beta_x \\ \alpha_x \\ \gamma_x \end{pmatrix}_0$$

as desired.

Exercise 5.12

Equation (5.48) is

$$\sigma_{xf}^2 = \sigma_{x0}^2 \left[1 + (z - z_0) \frac{\sigma_{x'0}^2}{\sigma_{x0}^2} \right],$$

$$\sigma_{xf}^2 = \sigma_{x0}^2,$$

$$\sigma_{xx'f} = 2(z - z_0)\sigma_{x0}^2.$$

The envelope equation is

$$\sigma_x'' + \kappa_x^2 \sigma_x = \frac{\varepsilon_x^2}{\sigma_x^3}.$$

The following is a general method for the solution for the force-free case:

$$\sigma_x'' = \frac{\varepsilon_x^2}{\sigma_x^3} \quad \text{or} \quad \frac{d}{dz}\left(\frac{1}{2}\sigma_x'^2\right) = \frac{\varepsilon_x^2}{\sigma_x^3}$$

and the first integral is

$$\sigma'^2 = C - \frac{\varepsilon_x^2}{\sigma_x^2} = \varepsilon_x^2 \left(\frac{1}{\sigma_{x,0}^2} - \frac{1}{\sigma_x^2} \right),$$

where we have substituted the waist condition. The final integration is

$$\int_{\sigma_0}^{\sigma_x} \frac{d\tilde{\sigma}_x}{\varepsilon_x \sqrt{(1/\sigma_{x,0}^2) - (1/\tilde{\sigma}_x^2)}} = z - z_0 \quad \text{or} \quad \frac{\sigma_{x,0}}{\varepsilon_x} \int_{\sigma_0}^{\sigma_x} \frac{\tilde{\sigma}_x d\tilde{\sigma}_x}{\sqrt{\tilde{\sigma}_x^2 - \sigma_{x,0}^2}} = z - z_0,$$

which can be written as

$$\tilde{\sigma}_x^2 - \sigma_{x,0}^2 = (z - z_0)^2 \cdot \frac{\varepsilon_x^2}{\sigma_{x,0}^2} \quad \text{or} \quad \tilde{\sigma}_x^2 = \sigma_{x,0}^2 \left[1 + (z - z_0)^2 \frac{\sigma_{x',0}^2}{\sigma_{x,0}^2} \right]$$

as expected from the matrix treatment.

Exercise 5.14

Equation (5.67) is

$$\varepsilon_{n,x}^2 \equiv (\beta\gamma)^2 [\langle x^2\rangle \langle x'^2\rangle - \langle xx'\rangle^2] = (m_0 c)^{-2}[\langle x^2\rangle \langle p_x^2\rangle - \langle xp_x\rangle^2],$$

where $\varepsilon_n = $ const., if conserved

$$\frac{d}{dz}\langle x^2\rangle\langle x'^2\rangle - \langle xx'\rangle^2 = 2\langle xx'\rangle\langle x'^2\rangle + 2\langle x^2\rangle\langle x'\rangle\langle x''\rangle - 2\langle xx''\rangle\langle xx'\rangle = 0$$

For linear transformations, $x'' = -k_x^2 x$, and the right-hand side of the equation is

$$2k_x^2\langle x^2\rangle\langle xx'\rangle - 2\langle x^2\rangle\langle xx'\rangle k_x^2 = 0,$$

so

$$\frac{d}{dz}\langle x^2\rangle\langle x'^2\rangle - \langle xx'\rangle^2 = 0$$

Exercise 5.17

We have

$$\beta_x(z) = \beta_p + \Delta\beta_0 \cos(2\kappa_0 z) - \frac{\Delta\alpha_0}{\kappa_0} \sin(2\kappa_0 z),$$

$$\Delta\beta_0 = \beta_{x0} - \beta_p = \frac{\beta_{x0}}{2} - \frac{\gamma_0}{2\kappa_0^2},$$

where

$$\Delta\alpha_0 = \alpha_{x0} = 0,$$

$$\gamma_0 = \frac{1}{\beta_0}.$$

The following relation must be fulfilled

$$\beta_p < \Delta\beta_0 (= \beta_{x0} - \beta_p) \rightarrow 2\beta_p < \beta_{x0}$$

and so

$$\beta_{x0} + \frac{1}{\beta_{x0}^2 \kappa_0^2} > \beta_{x0} \quad \text{or} \quad \beta_{x0}\left(1 + \frac{1}{\beta_{x0}^2 \kappa_0^2}\right) > \beta_{x0},$$

and

$$\left(1 + \frac{1}{\beta_{x0}^2 \kappa_0^2}\right) > 1 \quad \Rightarrow \quad \frac{1}{\beta_{x0}^2 \kappa_0^2} > 0.$$

Exercise 5.17

(a) The path length through all four magnets in the chicane is

$$\Delta s = 4R\theta_b, \quad \text{where } \theta_b = L/R.$$

As an explicit function of the radius of curvature R, the path length is

$$\Delta s = 4R \sin^{-1}(L/R).$$

(b) The differential of the path length with respect to the radius of curvature is

$$\frac{\partial(\Delta s)}{\partial R} = 4\left[\sin^{-1}(L/R) - \frac{L}{R} \sec\left(\frac{L}{R}\right)\right],$$

$$= 4[\theta_b - \tan\theta_b].$$

Therefore, we have

$$\alpha_c = -\frac{\partial(\Delta s)}{\partial R}\frac{R}{\Delta s} = \frac{4R}{\Delta s}[\theta_b - \tan\theta_b]$$

and

$$R_{56} = \left(\alpha_c - \frac{1}{\gamma_0^2}\right)\Delta s$$

$$= 4R[\tan\theta_b - \theta_b] + \frac{4R\theta_b}{\gamma_0^2},$$

or

$$R_{56} = 4R[\tan\theta_b - \beta_0^2\theta_b].$$

Exercise 5.20

For a tightly bunched beam, the distribution is bi-Gaussian, and the emittance is given by

$$\varepsilon_{z,n} = \sigma_\zeta\sigma_{\delta p,b},$$

where

$$\sigma_{\delta p} = \sqrt{\frac{\alpha_{RF}\gamma_0^3 m_0^2 c^2}{\Gamma}}, \quad \sigma_{\delta\zeta} = k_z^{-1}\sqrt{\frac{2}{\Gamma_l}}, \quad \Gamma_1 = \frac{2\alpha_{RF}m_0 c^2}{k_B T_z}.$$

Thus, we have

$$\varepsilon_{z,n} = \sigma_\zeta\sigma_{\delta p,b} = \frac{\lambda_{rf}m_0 c}{2\pi\Gamma_1}\sqrt{2\alpha_{RF}\gamma_0^3}$$

$$= \frac{\lambda_{rf}k_B T_z}{4\pi\alpha_{RF}c}\sqrt{2\alpha_{RF}\gamma_0^3} = \frac{\lambda_{rf}k_B T_z}{4\pi c}\sqrt{\frac{2\gamma_0^3}{\alpha_{RF}}}.$$

Exercise 5.23

The given conditions are $\sigma_\zeta = 1$ mm and $\lambda_{rf} = 0.105$, so $\sigma_\varphi = k_z\sigma_\zeta = 2\pi\cdot 0.001/0.105 = 0.060$, and $\varphi_0 = 13\pi/36$.

(a) The longitudinal Twiss parameters are

$$\beta_\zeta = \frac{\sqrt{2}\sigma_\zeta}{(k_z\sigma_\zeta)^2} = 0.4\,\text{m}, \quad \alpha_\zeta = -\sqrt{2}\frac{\cot(\varphi_0)}{k_z\sigma_\zeta} = 11,$$

$$\gamma_\zeta = \frac{1+\alpha_\zeta^2}{\beta_\zeta} = 310\,\text{m}^{-1}.$$

(b) The needed value of the matrix element in question is calculated as

$$R_{56} = \frac{\alpha_\zeta}{\gamma_\zeta} = 3.6\,\text{cm}.$$

The radius of curvature of the 20 MeV beam in a 0.3 T magnetic field is $R = 22.2$ cm. From Exercise 3.19, we have derived that

$$R_{56} = 4R(\theta - \sin\theta).$$

Solving for the bend angle, we have $\theta = 0.65 = 37°$.

Exercise 5.27

(a) The square of the J matrix is given by

$$\mathbf{J}^2 = \begin{bmatrix} \alpha & \beta \\ -\gamma & -\alpha \end{bmatrix} \cdot \begin{bmatrix} \alpha & \beta \\ -\gamma & -\alpha \end{bmatrix} = \begin{bmatrix} \alpha^2 - \beta\gamma & 0 \\ 0 & \alpha^2 - \beta\gamma \end{bmatrix}$$

$$= \begin{bmatrix} -1 & 0 \\ 0 & -1 \end{bmatrix} = -\mathbf{I}.$$

(b) The quantity

$$[\mathbf{I}\cos(\mu_1) + \mathbf{J}\sin(\mu_1)][\mathbf{I}\cos(\mu_2) + \mathbf{J}\sin(\mu_2)]$$
$$= \mathbf{I}[\cos(\mu_1)\cos(\mu_2) - \sin(\mu_1)\sin(\mu_2)]$$
$$+ \mathbf{J}[\sin(\mu_1)\cos(\mu_2) + \sin(\mu_2)\cos(\mu_1)]$$
$$= \mathbf{I}\cos(\mu_1 + \mu_2) + \mathbf{J}\sin(\mu_1 + \mu_2).$$

(c) From part (b), the mapping of a matrix applied n times can be written compactly as

$$\mathbf{M}^n = [\mathbf{I}\cos(\mu) + \mathbf{J}\sin(\mu)]^n = \mathbf{I}\cos(n\mu) + \mathbf{J}\sin(n\mu).$$

This should not be surprising, given that μ is the phase advance per period.

Chapter 6

Exercise 6.2

(a) We know that the trajectory is

$$\vec{r} = \begin{pmatrix} a\cos(k_u z) \\ a\sin(k_u z) \\ z \end{pmatrix},$$

and the tangent unit vector is

$$\vec{t} = \frac{d\vec{r}}{dz} = \begin{bmatrix} -k_u a\sin(k_u z) \\ k_u a\cos(k_u z) \\ 1 \end{bmatrix}.$$

Thus, the trajectory tangent vector unit vector is

$$\hat{t} = \frac{1}{\sqrt{k_u^2 a^2 + 1}} \begin{bmatrix} -k_u a\sin(k_u z) \\ k_u a\cos(k_u z) \\ 1 \end{bmatrix}$$

Using Eq. (4.16)

$$\vec{B} = \frac{\mu_0}{4\pi} \iiint_V \frac{\vec{J} \times \vec{r}}{r^3} dV$$

and

$$\vec{J} = I\delta(x - a\cos(k_u z))\delta(y - a\sin(k_u z))\hat{t}$$
$$- I\delta(x + a\cos(k_u z))\delta(y + a\sin(k_u z))\hat{t}$$

$$\vec{J} \times \vec{r} = \frac{I\delta(x - a\cos(k_u z))\delta(y - a\sin(k_u z))}{\sqrt{k_u^2 a^2 + 1}}(\hat{t} \times \vec{r})$$
$$- \frac{I\delta(x + a\cos(k_u z))\delta(y + a\sin(k_u z))}{\sqrt{k_u^2 a^2 + 1}}(\hat{t} \times \vec{r})$$

$$\hat{t} \times \vec{r} = [\hat{x}(k_u az\cos(k_u z) - y) + \hat{y}(k_u az\sin(k_u z) - x)$$
$$+ \hat{z}(-k_u ay\sin(k_u z) - k_u ax\cos(k_u z))]$$

$$\iiint_V \frac{\vec{J} \times \vec{r}}{r^3} dV = -2\hat{x}\int \frac{a\sin(k_u z)}{(\sqrt{a^2 + z^2})^3}dz - 2\hat{y}\int \frac{a\cos(k_u z)}{(\sqrt{a^2 + z^2})^3}dz$$

$$- 2\hat{z}\int \frac{a^2 k_u}{(\sqrt{a^2 + z^2})^3}dz$$

$$-2\hat{x}\int \frac{a\sin(k_u z)}{(\sqrt{a^2 + z^2})^3}dz = 0$$

since sin is an odd function and $(\sqrt{a^2 + z^2})^3$ is an even function. The integral of an odd function over a symmetric integral is 0.

$$-2\hat{y}\int \frac{a\cos(k_u z)}{(\sqrt{a^2 + z^2})^3}dz = -\frac{8a^3}{k_u^2\sqrt{\pi}}\Gamma\left(\frac{3}{2}\right)K_{-2}(k_u a)\hat{y}$$

$$= \frac{8a^3}{k_u^2\sqrt{\pi}}(0.886227)K_{-2}(k_u a)\hat{y}$$

where $K_{-2}(k_u a)$ is the modified Bessel function.

$$-2\hat{z}\int \frac{a^2 k_u}{(\sqrt{a^2 + z^2})^3}dz = -\frac{2zk_u}{\sqrt{a^2 + z^2}}\hat{z}$$

Putting these expressions together, we obtain

$$\vec{B} = \frac{\mu_0}{4\pi}\iiint_V \frac{\vec{J} \times \vec{r}}{r^3}dV = \frac{\mu_0}{4\pi}\left[\frac{7.08916a^3}{k_u^2\sqrt{\pi}}K_{-2}(k_u a)\hat{y} - \frac{2zk_u}{\sqrt{a^2 + z^2}}\hat{z}\right].$$

$$B_z = \frac{\mu_0}{4\pi}\left[\frac{2zk_u}{\sqrt{a^2 + z^2}}\right]$$

is the longitudinal field.

(b) The return winding (of the opposite polarity) serves to cancel the net axial current in the device. If the helical undulator consisted of only one winding it would have, like a single wire, a field gradient that points away from the wire. This gradient serves either repel or attract the beam (like current will attract), and so the design trajectory is not stable.

Exercise 6.4

(a) In general the superposition of multipoles in the magnetic field is give by

$$B_\rho = \sum_j a_j \rho^{j-1} \cos(j\phi) + b_j \rho^{j-1} \sin(j\phi)$$

with the additional restriction $\vec{\Delta} \cdot \dot{B} = 0$. Therefore we have

$$\frac{1}{\rho} \frac{\partial(\rho B_\rho)}{\partial \rho} = -\frac{1}{\rho} \frac{\partial B_\phi}{\partial \phi},$$

and

$$B_\phi = \sum_j -a_j \rho^{j-1} \sin(j\phi) + b_j \rho^{j-1} \cos(j\phi)$$

The induced voltage is

$$V = \oint \vec{E} \cdot d\vec{s} = \int \dot{\vec{B}} \cdot dA$$

Using $\phi = \omega t$

$$\int \dot{\vec{B}} \cdot dA = 2Na\omega \int_0^{b/2} \sum_j \left(-ja_j \rho^{j-1} \cos(j\phi) - jb_j \rho^{j-1} \sin(j\phi) \right) d\rho$$

$$= Nab\omega \sum_j \left(ja_j \rho^{j-1} \cos(j\phi) - jb_j \rho^{j-1} \sin(j\phi) \right)$$

and the voltage

$$V = -Nab\omega \sum_j \left(-ja_j \rho^{j-1} \cos(j\phi) - jb_j \rho^{j-1} \sin(j\phi) \right)$$

The signal due to the j-th multipole contributes a component at j times the rotation frequency.

(b) If one was to place the coil at some position offset from the axis of a device one would measure the fields produced by that device and also the lower order fields. For example, if one placed the coil correctly inside of a pure quadrupole, one would only see the quadrupole field. If you didn't place the coil correctly inside the quadrupole you would also measure a dipole field.

Exercise 6.5

We know from Eq. (2.39)

$$\psi = \sum_{n=1}^{\infty} a_n \rho^n \cos(n\phi) + b_n \rho^n \sin(n\phi)$$

We are given that $b_3 = 10 \, \text{T/m}^2$ and the pole tip radius $a = 5 \, \text{cm}$, and so $\psi = b_3 a^3 \sin(3\phi)$.

For a sextupole $\phi = 30°$ and therefore $\psi = b_3 a^3$.

From Eq. (6.20) $\Delta \psi = I_{enc} \mu_0$ we get $2\psi = 2I_{pole} \mu_0$ or $\psi/\mu_0 = I_{pole}$, and

$$I_{pole} = \frac{b_3 a^3}{\mu_0} = 994.7 \, \text{A}$$

Exercise 6.6

(a) Since the H-magnet has a winding around each leg of the yolk and magnetic fields are vector fields that add linearly, the H-magnet gives twice as much field as the C-magnet.

(b) If one of the windings is turned off the magnet does not behave like a C-magnet. The flux produced by the remaining winding that is still on is free to travel through the low reluctance iron in the opposite yoke. Therefore one expects very little the magnetic field produced in the gap in this case.

Exercise 6.8

Given that B varies sinusoidally from 0.1 to 1 T

$$f = 60\,\text{Hz}, \quad \alpha_\text{m} = 0.1, \quad \mu/\mu_0 = 10^4.$$

(a)

$$\frac{dP}{dV}\bigg|_\text{hist} = \frac{f\alpha_\text{m}B_\text{max}^2}{2\mu} = 238.7\ N/m^2 s$$

(b) For ohmic losses,

$$\frac{dP}{dV}\bigg|_\text{ohmic} = \sigma_\text{c}\vec{E}_\text{induction}^2.$$

From the information given above we know $B_\text{gap} = 0.55\ \text{T} + 0.45\ \text{T}\sin(\omega t)$. We also know that $\mu/\mu_0 = 10^4$ in the iron and so

$$B_\text{iron} = \frac{0.55\ \text{T} + 0.45\ \text{T}\sin(\omega t)}{10^4}.$$

We want to find $\vec{E}_\text{induction}$:

$$\nabla \times \vec{E} = -\frac{\partial B}{\partial t}$$

$$\dot{B} = (4.5 \times 10^{-5}\text{T})\omega\cos(\omega t)$$

$$\iint \nabla \times E \cdot dA = \oint E \cdot dl = -\iint \dot{B} \cdot dA$$

$$E(2\pi r) = (4.5 \times 10^{-5}\ \text{T})\omega\cos(\omega t)(\pi r^2),\ \text{or}$$

$$E(r) = (2.25 \times 10^{-5}\ \text{T})\omega\cos(\omega t)(r),$$

and

$$\frac{dP}{dV}\bigg|_\text{ohmic} = \sigma_\text{c}(2.25 \times 10^{-5}\ \text{T})^2\omega^2\cos^2(\omega t)(r)^2.$$

Time averaging, $\langle\cos^2(\omega t)\rangle = 0.5$, and using $\sigma_\text{c} = 1.03 \times 10^3\ \text{m}/\Omega$ we have

$$\frac{dP}{dV}\bigg|_\text{ohmic} = \sigma_\text{c}(2.25 \times 10^{-5}\ \text{T})^2\omega^2(0.5)(r)^2 = 9.38 \times 10^{-4}r^2\ \text{W}/\text{m}^3.$$

Chapter 7

Exercise 7.2

Each scalar component of the electromagnetic wave (e.g. $u = E_x$), obeys the wave equation,

$$\left[\vec{\nabla}^2 - \frac{1}{c^2}\frac{\partial^2}{\partial t^2}\right]u = 0.$$

For the spherically symmetric case, the Laplacian operator takes a simple form that can be substituted to obtain,

$$\left[\frac{1}{r^2}\frac{\partial}{\partial r}r^2\frac{\partial}{\partial r} - \frac{1}{c^2}\frac{\partial^2}{\partial t^2}\right]u = 0.$$

The function $u = u_0 \exp[ik(r - ct)]/r$ has first derivative

$$\frac{\partial}{\partial r}\left[\frac{\exp[ik(r-ct)]}{r}\right] = ik\frac{\exp[ik(r-ct)]}{r} - \frac{\exp[ik(r-ct)]}{r^2},$$

and thus

$$\frac{1}{r^2}\frac{\partial}{\partial r}(ikr\exp[ik(r-ct)] - \exp[ik(r-ct)]) = -k^2\frac{\exp[ik(r-ct)]}{r},$$

and with $\omega^2 = k^2c^2$,

$$\left[\frac{1}{r^2}\frac{\partial}{\partial r}r^2\frac{\partial}{\partial r} - \frac{1}{c^2}\frac{\partial^2}{\partial t^2}\right]u = 0.$$

The condition $\hat{r} \times \vec{E} = 0$ can be obtained from Gauss' law by noting that

$$\vec{\nabla} \cdot \vec{E} = 0 \quad \text{or} \quad ikE_\mathrm{r} + \vec{\nabla}_\perp \cdot \vec{E}_\perp = 0.$$

But, since the transverse derivatives of the field components vanish, $\vec{\nabla}_\perp \cdot \vec{E}_\perp = 0$, and $E_\mathrm{r} = 0$, and $\hat{r} \times \vec{E} = 0$. Note that this is a general condition, applying to any case (such as plane waves) with uniform phase fronts.

Exercise 7.3

(a) In S-band, we have

$$R_\mathrm{s} = \frac{1}{2}\sqrt{\frac{\omega\mu_0}{2\sigma_\mathrm{c}}} = 6.87 \times 10^{-3}\,\Omega/\mathrm{m},$$

$$\omega\varepsilon/\sigma_\mathrm{c} = 2.66 \times 10^{-9} \quad \text{and} \quad \delta_\mathrm{s} = \sqrt{\frac{2}{\omega\mu_0\sigma_\mathrm{c}}} = 1.22\,\mu\mathrm{m}.$$

(b) For the $\lambda = 10.6\,\mu\mathrm{m}$ case (4 orders of magnitude smaller than S-band)

$$R_\mathrm{s} = 6.87 \times 10^{-1}\,\Omega/\mathrm{m}, \quad \omega\varepsilon/\sigma_\mathrm{c} = 2.66 \times 10^{-5}, \quad \delta_\mathrm{s} = 12.2\,\mathrm{nm}.$$

(c) $\omega\varepsilon/\sigma_\mathrm{c} = 1$ for $\lambda = 0.4\,\text{Å}$, which is in the X-ray regime. For this case, the Drude model is completely invalid, and the radiation frequency is in fact well above the metal's plasma frequency.

Exercise 7.4

If one begins with, $v_\phi v_g = c^2$, it helps to first write it explicitly as

$$\frac{\omega}{k}\frac{d\omega}{dk} = c^2 \quad \text{or} \quad \omega \, d\omega = k \, dk \, c^2.$$

Integrating once, we obtain

$$\omega^2 = k^2 c^2 + \text{constant},$$

which is the form we desire.

Exercise 7.7

(a) With a line charge charge density λ_e, the radial electric field is easily found by Gauss' law with cylindrical symmetry,

$$2\pi \rho D_\rho = \lambda_e \quad \text{or} \quad E_\rho = \frac{\lambda_e}{2\pi \varepsilon \rho}.$$

The total line charge density on the outside conductor's inner surface ($\rho = b$) must be equal to $-\lambda_e$ in order to cancel the field for $\rho > b$.

(b) The surface currents that are found on the conductors are

$$I = \pm \lambda_e v_\phi = \pm \frac{\lambda_e}{\sqrt{\mu_0 \varepsilon}}.$$

The magnetic field that is associated with these currents is contained between the conductors and is, from Ampere's law,

$$2\pi \rho H_\phi = \frac{\lambda_e}{\sqrt{\mu_0 \varepsilon}} \quad \text{or} \quad H_\phi = \frac{\lambda_e}{2\pi \rho \sqrt{\mu_0 \varepsilon}}.$$

The electric and magnetic fields have the same functional form, and their ratio is a constant.

(c) The wave impedance is defined to be this constant, which is

$$Z_0 \equiv \|\tilde{E}\| / \|\tilde{H}\| = E_\rho / H_\phi = \sqrt{\frac{\mu_0}{\varepsilon}},$$

as expected.

Exercise 7.8

For 500 MHz waves in copper, one has the attenuation constant of the coaxial line (2 mm, 5 mm inner/outer radius, $\kappa_e = 2.25$)

$$\eta_w = \frac{R_s}{Z_0} \cdot \frac{1 + (a/b)}{2a \ln(b/a)} = 0.0044 \text{ nepers/m}.$$

In a 100-m run of this coaxial line, the attenuation of the power is $\exp(-0.44) = 0.65$. The voltage on the line is only attenuated by the square root of this factor, or 0.80.

Exercise 7.11

We begin with the stored energy, in the rectangular cavity, for fixed L_z

$$U_{EM} = \varepsilon_0 E_0^2 \frac{L_x L_y L_z}{16} \left[1 - \left(\frac{L_x}{L_y} + \frac{L_y}{L_x} \right)^{-2} \right].$$

We have an additional constraint, which is that the frequency be held constant as the geometry is varied,

$$\omega_{lmn} = c \sqrt{ \left(\frac{l\pi}{L_x} \right)^2 + \left(\frac{m\pi}{L_y} \right)^2 + \left(\frac{n\pi}{L_z} \right)^2 },$$

$$l, m = 1, 2, 3, \ldots, \quad n = 0, 1, 2, \ldots, \text{ is constant.}$$

Without loss of generality, we may choose $n = 0$ (this is the obvious choice for an accelerating cavity), and $l = m = 1$,

$$\omega_{110} = \pi c \sqrt{ \left(\frac{1}{L_x} \right)^2 + \left(\frac{1}{L_y} \right)^2 } = \text{constant.}$$

We now introduce the aspect ratio $g = L_y/L_x$ and write the stored energy explicitly including the constraint of constant frequency

$$U_{EM} = \frac{1}{2} \varepsilon_0 E_0^2 \left(\frac{\pi c}{2\omega_{110}} \right)^2 \left(g + \frac{1}{g} \right) \left[1 - \left(g + \frac{1}{g} \right)^{-2} \right]$$

$$= \varepsilon_0 E_0^2 \left(\frac{\pi c}{2\omega_{110}} \right)^2 \left[\left(g + \frac{1}{g} \right) - \left(g + \frac{1}{g} \right)^{-1} \right].$$

We can vary the stored energy with respect to choice of g to find the minimum at

$$\frac{\partial U_{EM}}{\partial g} = \frac{1}{2} \varepsilon_0 E_0^2 \left(\frac{\pi c}{2\omega_{110}} \right)^2 \left[1 - \frac{1}{g^2} + \left(g + \frac{1}{g} \right)^{-2} \left(1 - \frac{1}{g^2} \right) \right] = 0.$$

This expression has as a solution $g = 1$. One should also check the second derivative to make sure that it is a minimum condition, not maximum.

Exercise 7.14

(a) Inclusion the azimuthal derivatives in the Laplacian operator in Eq. (7.1) (the wave equation in vacuum), gives

$$\left[\frac{1}{\rho} \frac{\partial}{\partial \rho} \left(\rho \frac{\partial}{\partial \rho} \right) + \frac{1}{\rho^2} \frac{\partial}{\partial \phi^2} + \frac{\partial^2}{\partial z^2} - \frac{1}{c^2} \frac{\partial^2}{\partial t^2} \right] E_z = 0.$$

(b) Separation of variables by the substitution $E_z = R(\rho)\Phi(\phi)Z(z)$ gives three equations:

$$\frac{d^2 Z}{dz^2} + k_{z,n}^2 Z = 0$$

$$\frac{d\Phi}{d\phi^2} + m^2 \Phi = 0,$$

and

$$\left[\rho^2 \frac{d^2}{d\rho^2} + \rho \frac{d}{d\rho} + m^2 + \rho^2 \left(k_{z,n}^2 - \frac{\omega^2}{c^2}\right)\right] R = 0.$$

The solutions to the first equation are simple harmonic $\exp(ik_{z,n}z)$, with $k_{z,n}$ determined from conducting boundaries in z.

The second equation has solutions of the form $\exp(im\phi)$, and are periodic in ϕ, which gives the eigenvalue requirement that m is an integer.

The final equation is a Bessel equation, which when m is not equal to zero admits solutions that do not necessarily have vanishing derivative (maximum or minimum) at $\rho = 0$. The value of m gives the form of the radial solution, as the Bessel function $J_m(k_r r)$. The order of the Bessel function is specified by m, but the argument is given by the factor multiplying the final term in the operator,

$$k_r^2 = \frac{\omega^2}{c^2} - k_{z,n}^2.$$

In practice, the radial wavenumber is determined by the transverse boundary conditions—metallic pipe radius in the simplest case, where the lth zero of the Bessel function must occur at the boundary R_w, and $k_{r,lm} = j_{0l}/R_w$. Thus, the frequency of a given mode is given by

$$\omega_{lmn}^2 = \left(k_{r,lm}^2 + k_{z,n}^2\right)c^2$$

(c) In Eq. (4.74), we are considering only speed-of-light modes, in which case $k_r^2 = (\omega^2/c^2) - k_{z,n}^2 \Rightarrow 0$. Thus, the Bessel functions all take on a limiting form as their arguments approach zero, i.e. $J_0 \Rightarrow 1, J_1 \propto \rho$, and the takes on the form of powers in ρ.

Exercise 7.16

The dispersion relation for such a cavity would be given by

$$\omega^2 = \left(k_r^2 + k_z^2\right)c^2,$$

where both k_r^2 and k_z^2 are non-vanishing, and positive (the radial wavenumber must always be positive to satisfy the boundary condition at cavity edge). Therefore, the phase velocity of the forward wave in such a cavity could not be made to equal the speed of light—it in fact must exceed the speed of light. Lining a simple cavity such as this with dielectric can slow the wave down (see Ex. 7.25).

Exercise 7.18

In a single cavity with are many resonances having distinct resonant frequencies ω_0, corresponding different values of lumped capacitance C and inductance L arise from different patterns of surface current and charge density associated with each mode. Thus each mode has a distinct "circuit" based on a different physical configuration of the sources of electric and magnetic fields.

Exercise 7.21

The VSWR is defined as

$$\text{VSWR} \equiv \frac{\|V_F\| + \|V_R\|}{\|V_F\| - \|V_R\|}.$$

The voltage at a given point in a waveguide is given by

$$V = \text{Re}\{V_R \exp[i(k_z z - \omega t)] + V_R \exp[i(k_z z - +\omega t)]\},$$

where V_F and V_R are complex voltage amplitudes.

The rms power voltage is obtained by taking the square average of this function, and averaging in time.

$$V_{\text{rms}}^2 = \tfrac{1}{2}\left[\|V_F\|^2 + \|V_R\|^2 + 2\text{Re}\{V_F V_R^* \exp[i2k_z z]\}\right]$$

This is a sinusoidal function of z. The time-averaged power as a function of z will be proportional to the rms voltage squared.

(b) From part (a), the rms voltage has minimum value of

$$\min V_{\text{rms}} = \tfrac{1}{\sqrt{2}}[\|V_F\| - \|V_R\|]$$

and a maximum value of

$$\max V_{\text{rms}} = \tfrac{1}{\sqrt{2}}[\|V_F\| + \|V_R\|].$$

Thus, the VSWR is given by the ratio of the maximum to minimum rms voltage

$$\text{VSWR} = \frac{\max V_{\text{rms}}}{\min V_{\text{rms}}} = \frac{\|V_F\| + \|V_R\|}{\|V_F\| - \|V_R\|}.$$

Determination of maximum and minimum voltage points is sufficient to determine the VSWR.

Exercise 7.24

The frequency of the nth mode is given by

$$\omega_n = \omega_0 \sqrt{1 + 4\kappa_c \sin^2\left(\frac{n\pi}{2(N_c - 1)}\right)} \cong \omega_0\left[1 + 2\kappa_c \sin^2\left(\frac{n\pi}{2(N_c - 1)}\right)\right]$$

Thus, the 12th, or π-mode, frequency is

$$\omega_\pi = \omega_0[1 + 2\kappa_c] = 2.998 \cdot 2\pi \times 10^9 \, \text{s}^{-1} \text{ (given)},$$

while the nearest neighbor mode (11th) is given by

$$\omega_{10\pi/11} \cong \omega_0\left[1 + 2\kappa_c \sin^2\left(\frac{5\pi}{11}\right)\right] \approx \omega_\pi\left(1 - \frac{\pi^2 \kappa_c}{242}\right).$$

We require a mode separation five times the resonance bandwidth, $5\omega_\pi/Q = \omega_\pi/1200$, so the minimum allowable coupling constant κ_c is given by the condition

$$\frac{\omega_\pi}{1200} \approx \omega_\pi\left(\frac{\pi^2 \kappa_c}{242}\right), \quad \text{or} \quad \kappa_{c,\text{min}} \approx \frac{242}{1200\pi^2} = 2.04 \times 10^{-2}.$$

Chapter 8

Exercise 8.2

Using the convention

$$\vec{p} = \begin{pmatrix} p \\ np' \end{pmatrix} \quad \text{and} \quad \vec{s} = \begin{pmatrix} s \\ ns' \end{pmatrix},$$

the transformation becomes:

$$\begin{pmatrix} \vec{p} \\ \vec{s} \end{pmatrix} = \begin{pmatrix} p \\ np' \\ s \\ ns' \end{pmatrix}.$$

At point z:

$$\begin{pmatrix} \vec{p} \\ \vec{s} \end{pmatrix}_f = \mathbf{M_T} \begin{pmatrix} \vec{p} \\ \vec{s} \end{pmatrix}_i.$$

The matrix for the in-plane bend is

$$\mathbf{M_{ipb}} = \begin{pmatrix} -\mathbf{I} & \mathbf{0} \\ \mathbf{0} & \mathbf{I} \end{pmatrix} \quad \text{where } \mathbf{I} = \begin{pmatrix} 1 & 0 \\ 0 & 1 \end{pmatrix} \text{ and } \mathbf{0} = \begin{pmatrix} 0 & 0 \\ 0 & 0 \end{pmatrix}.$$

Now lets look at the out-of-plane bend. We need to flip $\vec{s} \to \vec{p}$, by employing

$$\mathbf{M_{flip}} = \begin{pmatrix} 0 & I \\ I & 0 \end{pmatrix}.$$

The bend matrix ($\mathbf{M_{ipb}}$) is the same.

(a) Using this we obtain the matrix for the out of plane bend

$$\mathbf{M_{opb}} = \mathbf{M_{ipb}}\mathbf{M_{flip}} = \begin{pmatrix} 0 & -\mathbf{I} \\ \mathbf{I} & 0 \end{pmatrix}.$$

(b) All drifts are given by the matrix

$$\mathbf{M}_a = \begin{pmatrix} \mathbf{a} & 0 \\ 0 & \mathbf{a} \end{pmatrix} \quad \text{or} \quad \mathbf{M}_b = \begin{pmatrix} \mathbf{b} & 0 \\ 0 & \mathbf{b} \end{pmatrix},$$

where

$$\mathbf{a} = \begin{pmatrix} 1 & a \\ 0 & 1 \end{pmatrix} \quad \text{and} \quad \mathbf{b} = \begin{pmatrix} 1 & b \\ 0 & 1 \end{pmatrix}.$$

Using the above definitions the total transformation is

$$\mathbf{M_T} = \mathbf{M}_a\mathbf{M_{ipb}}\mathbf{M}_a\mathbf{M_{obp}}\mathbf{M}_a\mathbf{M_{obp}}\mathbf{M}_b\mathbf{M_{obp}}\mathbf{M}_a\mathbf{M_{ipb}}\mathbf{M}_a,$$

$$\mathbf{M_T} = \begin{pmatrix} 0 & 0 & -1 & -(5a+b) \\ 0 & 0 & 0 & -1 \\ 1 & 5a+b & 0 & 0 \\ 0 & 1 & 0 & 0 \end{pmatrix}.$$

Exercise 8.5

Given $R_0 = 50 \times 10^{-6}$ m, $n_2 = 20m^{-2}$, $n_0 = 1.5$.

Start with Eq. (8.20):

$$r'' + \kappa_f^2 r = 0$$

with

$$\kappa_f^2 = \frac{n_2}{n_0} \quad \text{or} \quad \kappa_f = \sqrt{\frac{20}{1.5}} m^{-1} = 3.65 m^{-1}.$$

This has the solution

$$r(z) = R_0 \cos(\kappa_f z).$$

In the paraxial approximation:

$$\psi \cong r' = \frac{dr}{dz},$$

$$r'(z) = -\kappa_f R_0 \sin(\kappa_f z).$$

r'_{max} is achieved when $\sin(\kappa_f z) = -1$; $r'_{max} = \kappa_f R_0 = 1.825 \times 10^{-4}$ rad, or $\psi \cong 0.02°$.

Exercise 8.6

The transformation matrix is $\mathbf{M}_T = \mathbf{M}_{f,2}\mathbf{M}_0\mathbf{M}_{f,1}$

$$
\mathbf{M}_T = \begin{pmatrix} 1 & 0 \\ \pm\frac{2}{R_2} & 1 \end{pmatrix} \begin{pmatrix} 1 & L \\ 0 & 1 \end{pmatrix} \begin{pmatrix} 1 & 0 \\ -\frac{2}{R_1} & 1 \end{pmatrix}
$$

$$
= \begin{pmatrix} 1 - \frac{2L}{R_1} & L \\ \mp\frac{2}{R_2}\left(1 - \frac{2L}{R_1}\right) - \frac{2}{R_1} & 1 \mp \frac{2L}{R_2} \end{pmatrix}.
$$

The stability condition is $|\mathrm{Tr}(\mathbf{M}_T)| \leq 2$

$$
\left| 1 - \frac{2L}{R_1} + 1 \mp \frac{2L}{R_2} \right| \leq 2 \quad \text{or} \quad \left| 1 - \frac{L}{R_1} \mp \frac{L}{R_2} \right| \leq 1.
$$

Now let $y = L/R_1$ and $x = R_2/R_1$. The stability condition becomes:

$$
\left| 1 - y \mp \frac{y}{x} \right| \leq 1 \quad \text{or} \quad -1 \leq 1 - y \mp \frac{y}{x} \leq 1, \quad \text{and} \quad -2 \leq -y \mp \frac{y}{x} \leq 0.
$$

Now let us plot for both cases $y(x)/f(x)$ versus x:

Case 1: defocusing

$$y(x) = \frac{2x}{x+1} \quad \text{and} \quad f(x) = 0$$

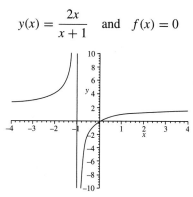

Case 2: focusing

$$y(x) = \frac{2x}{x-1} \quad \text{and} \quad f(x) = 0$$

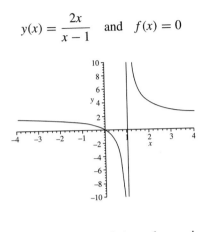

In both cases the stable region is found above the *x*-axis and below the curve.

Exercise 8.8

$$\left(\vec{\nabla}_\perp^2 + 2ik\frac{\partial}{\partial z}\right)u(x,y,z) = 0,$$

$$u(x,y,z) \cong \left[\frac{1}{R(z)} + i\frac{\lambda}{n_0\pi\omega^2(z)}\right]e^{ik((x^2+y^2)/2R(z))}\,e^{-((x^2+y^2)/\omega^2(z))}$$

First let us rewrite $u(x,y,z)$ in terms of $q(z)$

$$q(z) = \frac{1}{R(z)} + i\frac{\lambda}{n_0\pi\omega^2(z)},$$

$$\frac{ik}{2}\frac{1}{q(z)} = \frac{ik}{2R(z)} - \frac{k\lambda}{n_0\pi\omega^2(z)} = \frac{ik}{2R(z)} - \frac{1}{\omega^2(z)},$$

but

$$k = \frac{2\pi}{(\lambda/n_0)}, \quad \text{so } u(x,y,z) = \frac{1}{q(z)}e^{(ik/2q(z))(x^2+y^2)}$$

and with $\rho^2 = x^2 + y^2$

$$u(\rho,z) = \frac{1}{q(z)}e^{(ik/2q(z))(\rho^2)}.$$

Then, writing this in cylindrical coordinates

$$\nabla_\perp^2 = \frac{1}{\rho}\frac{\partial}{\partial\rho}\left(\rho\frac{\partial}{\partial\rho}\right), \quad \text{so } \nabla_\perp^2 u = \frac{1}{\rho}\frac{\partial}{\partial\rho}\left(\rho\frac{\partial}{\partial\rho}\frac{1}{q(z)}e^{(ik/2q(z))(\rho^2)}\right),$$

or

$$\nabla_\perp^2 u = \left(\frac{2ik}{q^2} - \frac{k^2\rho^2}{q^3}\right)e^{(ik/2q)\rho^2}.$$

With

$$\frac{\partial u}{\partial z} = \frac{\partial u}{\partial q}\frac{\partial q}{\partial z} = \frac{\partial u}{\partial q},$$

$$\frac{\partial u}{\partial z} = -\frac{1}{q^2}e^{(ik/2q)\rho^2} + \frac{1}{q^2}e^{(ik/2q)\rho^2}\left(\frac{ik}{2q}\rho^2\right)\left(-\frac{1}{q^2}\right),$$

or

$$\frac{\partial u}{\partial z} = e^{(ik/2q)\rho^2}\left(-\frac{1}{q^2} - \frac{ik}{2q^3}\rho^2\right),$$

we obtain

$$2ik\frac{\partial u}{\partial z} = e^{(ik/2q)\rho^2}\left(-\frac{2ik}{q^2} + \frac{k^2\rho^2}{2q^3}\right).$$

Combination of these expressions yields

$$\vec{\nabla}_\perp^2 u + 2ik\frac{\partial u}{\partial z} = e^{(ik/2q)\rho^2}\left(\frac{2ik}{q^2} - \frac{k^2\rho^2}{q^3} - \frac{2ik}{q^2} + \frac{k^2\rho^2}{q^3}\right) = 0.$$

Thus, $u(x, y, z)$ is a solution to the paraxial wave equation.

Exercise 8.12

(a) We want to find the ratio of the beam size at the mirrors to that at the waist.

$$R(z) = \frac{z_R^2}{z} + z,$$

$$R(L/2) = R = \frac{z_R^2}{L/2} + \frac{L}{2},$$

$$z_R^2 = \left(R - \frac{L}{2}\right)\frac{L}{2},$$

$$\omega^2(z) = \omega^2(0)\left[1 + \frac{z^2}{z_R^2}\right],$$

$$\frac{\omega^2(L/2)}{\omega^2(0)} = \left[1 + \frac{(L/2)^2}{(R - L/2)(L/2)}\right] = 1 + \frac{(L/2)}{(R - L/2)}$$

$$= \frac{R - L/2 + L/2}{R - L/2} = \frac{R}{R - L/2},$$

$$\frac{\omega(L/2)}{\omega_0} = \sqrt{\frac{R}{R - L/2}}.$$

(b) We are given: $R = 50\,\text{cm} = .5\,\text{m}$; $\lambda = 1 \times 10^{-6}\,\text{m}$; $R_M = 1\,\text{in.} = 2.54\,\text{cm} = 0.0254\,\text{m}$; $\omega(L/2) = (1/4)d_M = (1/2)R_M = 0.0127\,\text{m}$.

We know:

$$z_R = \frac{\pi n_0 \omega_0^2}{\lambda} \quad \text{and} \quad \omega^2(L/2) = \omega_0^2 \left(1 + \frac{(L/2)^2}{z_R^2}\right).$$

So we may deduce that

$$\omega^2(L/2) = z_R \frac{\lambda}{\pi n_0} \left(1 + \frac{(L/2)^2}{z_R^2}\right).$$

Rearranging this expression we obtain

$$z_R = \frac{\lambda}{\pi n_0 w^2(L/2)} \{z_R^2 + (L/2)^2\}$$

$$= \frac{\lambda}{\pi n_0 w^2(L/2)} \{(R - (L/2))(L/2) + (L/2)^2\},$$

$$\sqrt{(R - (L/2))(L/2)} = \left(\frac{\lambda R}{\pi n_0 w^2(L/2)}\right)(L/2).$$

Let us define

$$C = \left(\frac{\lambda R}{\pi n_0 w^2(L/2)}\right) \quad \text{and} \quad l = \frac{L}{2}$$

for simplicity.

$$(R - l)l = C^2 l^2,$$

$$l^2(1 + C^2) - Rl = 0 \quad \text{or} \quad l\{l(1 + C^2) - R\} = 0.$$

So either $l = 0$ (trivial case) or, more relevantly, $l = R/(1 + C^2)$.
   Finally, we may calculate

$$L = 2l = \frac{2R}{1 + (\lambda R/\pi n_0 w^2(L/2))^2} \cong 1 \text{ m},$$

$$\omega_0 = \omega(L/2)\sqrt{\frac{R - (L/2)}{R}} = \sqrt{1 - \frac{1}{\left(\frac{\lambda R}{\pi n_0 w^2(L/2)}\right)^2 + 1}} \cong 1.25 \times 10^{-5} \text{ m}.$$

Exercise 8.13

We know:

$$I = \frac{\varepsilon_0}{2} n_0 c u^2(x, y, z),$$

$$u = \left(\frac{1}{R(z)} + i\frac{\lambda}{\pi n_0 \omega^2(z)}\right) e^{ik(x^2+y^2)/2R(z)} e^{-(x^2+y^2)/\omega^2(z)},$$

$$u^2 = u^* u = \left(\frac{1}{R^2(z)} + \frac{\lambda^2}{(\pi n_0 \omega^2(z))^2}\right) e^{-2(x^2+y^2)/\omega^2(z)}.$$

(a) Power:

$$P = \int I \cdot dA = \frac{\varepsilon_0 n_0 c}{2} \left(\frac{1}{R^2(z)} + \frac{\lambda^2}{(\pi n_0 \omega^2(z))^2} \right) \iint e^{-2(x^2+y^2)/\omega^2(z)} \, dx \, dy$$

$$= \{\text{const}\} \int_0^{2\pi} \int_{\rho_0}^{\infty} e^{-2\rho^2/\omega^2} \rho \, d\rho \, d\phi,$$

$$\delta_c = \frac{P_{\text{loss}}}{P_{\text{total}}} = \frac{\{\text{const}\} \int_0^{2\pi} d\phi \int_{R_M}^{\infty} \rho e^{-2\rho^2/\omega^2} \, d\rho}{\{\text{const}\} \int_0^{2\pi} d\phi \int_0^{\infty} \rho e^{-2\rho^2/\omega^2} \, d\rho}.$$

Let $v = \rho^2$ and $dv = 2\rho \, d\rho$:

$$\int_{R_M}^{\infty} \rho e^{-2\rho^2/\omega^2} \, d\rho = \int_{R_M}^{\infty} \frac{1}{2} e^{-2v/\omega^2} \, dv = -\frac{\omega^2}{2} e^{-2\rho^2/\omega^2} \bigg|_{R_M}^{\infty}.$$

So, we have

$$\delta_c = \frac{-(\omega^2/4)\left(0 - e^{-2R_M^2/\omega^2}\right)}{-(\omega^2/4)(0 - 1)},$$

or simply

$$\delta_c = e^{-2R_M^2/\omega^2}.$$

(b)

$$Q = \frac{\omega U}{P} = \frac{\omega R_M}{c \delta_c} = \frac{\omega R_M}{c e^{-2R_M^2/\omega^2}} = \frac{\omega R_M}{c} \exp\left(\frac{2R_M^2}{\omega^2} \right)$$

.

Q grows very quickly with the mirror size.

Exercise 8.18

A two-level quantum system will not work as a laser. Since there are only two levels you cannot maintain population inversion. By the same considerations, a four-level quantum system works better than a three-level quantum system. The four-level quantum system has the extra level above ground state. Since the rate from the third level to the second level is very slow (this is the lasing transition) and the rate from the second level to ground is very fast the population of the second level is essentially zero. This helps when considering the single pass gain for the laser. Since the population of the level is approximately zero you do not have to consider it when calculating the rate equations.

Exercise 8.23

(a) The four-vector is

$$k = \begin{pmatrix} \omega \\ \vec{k}_u \end{pmatrix}.$$

We are given $\omega = 0, k_z = k_u$.

So the four-vector becomes

$$k = \begin{pmatrix} 0 \\ 0 \\ 0 \\ k_u \end{pmatrix}.$$

The Lorentz transformation is

$$L = \begin{pmatrix} \gamma_0 & 0 & 0 & -c\beta_0\gamma_0 \\ 0 & 1 & 0 & 0 \\ 0 & 0 & 1 & 0 \\ -\beta_0\gamma_0/c & 0 & 0 & \gamma_0 \end{pmatrix}.$$

In the rest frame of the electron

$$\beta_z = \beta_0 = 1 - \frac{1+k^2}{2\gamma} = 1 - \frac{1}{2\gamma}, \qquad \gamma_0 = \frac{\gamma}{1+k^2}.$$

Therefore,

$$k' = L \cdot k = \begin{pmatrix} -\beta_0\gamma_0 k_u c \\ 0 \\ 0 \\ \gamma_0 k_u \end{pmatrix}.$$

(b) The magnetic field is $\vec{B} = B_0 \sin(k_u z) \cdot \hat{y}$. The vector potential is

$$\vec{A} = \frac{B_0}{k_u} \begin{pmatrix} \cos(k_u z) \\ 0 \\ 0 \end{pmatrix} \quad \text{and} \quad \phi = 0.$$

The four-vector potential is

$$A = \begin{pmatrix} 0 \\ (B_0/k_u)\cos(k_u z) \\ 0 \\ 0 \end{pmatrix}.$$

A is a constant under Lorentz transformations.

But z transforms as $z = \gamma(z' + \beta_0 ct')$. This is a wave traveling in the $-\hat{z}$ direction. The field is then

$$B_y = (\nabla \times A)_x = B_0\gamma \sin(\gamma z' + \gamma\beta_0 ct'),$$

$$E_x = \frac{\partial A_x}{\partial t} = \gamma\beta_0 c B_0 \sin(\gamma z' + \gamma\beta_0 ct').$$

If $\beta_0 \approx 1$, these two equations define an EM wave.

(c) The four-vector is

$$\tilde{k} = \begin{pmatrix} \omega \\ k \sin \theta \\ 0 \\ k \cos \theta \end{pmatrix}.$$

As in (a) the transformation to the rest frame is $\tilde{k}' = L \cdot \tilde{k}$:

$$\tilde{k}' = \gamma_0 \omega - c\beta_0 \gamma_0 k \cos \theta.$$

The energy of the Compton scattering is conserved

$$\tilde{\omega} = c\beta_0 \gamma_0 k_u.$$

So $\beta_0 k_u = k(1 - \beta_0 \cos \theta)$. In this case $\beta_0 \approx 1$.
   Therefore, to first order,

$$\beta_0 k_u = k \left(1 - \left(1 - \frac{1 + a_u^2}{2\gamma^2} \right) \left(1 - \frac{\theta^2}{2} \right) \right).$$

Finally, we have

$$\lambda = \frac{\lambda_c}{2\gamma^2} (1 + a_u^2 + \gamma^2 \theta^2).$$

Exercise 8.24

We start with the Hamiltonian $H = \sqrt{(\vec{p} - e\vec{A})^2 c^2 + \omega^2 c^2}$, where

$$\vec{A} = \begin{pmatrix} (B_0/k_u) \sin(k_u z) \\ 0 \\ 0 \end{pmatrix},$$

$$\dot{x} = v_x = \frac{\partial H}{\partial p_x} = \frac{c^2 (p_x - eA_x)}{\gamma mc^2} \quad \text{and} \quad \beta_x = \frac{\dot{x}}{c} = \frac{\sqrt{2}a_u}{\gamma} \sin(k_u z).$$

As

$$\dot{p}_x = -\frac{\partial H}{\partial x} = 0$$

we can set $p_x = 0$, otherwise there would be off-axis motion and as $\dot{p}_y = 0$ and $\dot{y} = 0$, again we can set $p_y = 0, y = 0$.
   Energy is conserved, so we have

$$\beta^2 = 1 - \frac{1}{\gamma^2} \quad \text{and} \quad \beta_z = \sqrt{1 - \frac{1}{\gamma^2} - \beta_x^2} \cong 1 - \frac{1}{2\gamma^2} - \frac{1}{2}\beta_x^2,$$

or

$$v_z = c\langle \beta_z \rangle = c \left(1 - \frac{1 + a_u^2}{2\gamma^2} \right).$$

Resonance then requires

$$v_\phi = \frac{\omega}{k + k_u} = \frac{c}{1 + (k_u/k)} \cong c(1 + k_u/k) = c \left(1 - \frac{\lambda}{\lambda_u} \right)$$

be equal to v_z, which gives us

$$\lambda = \frac{\lambda_u}{2\gamma^2} (1 + a_u^2).$$

Exercise 8.25

(a)

$$a_u = \frac{1}{\sqrt{2}} \frac{eB_u}{mck_u} = 0.66 \cdot B_u[\text{T}] \cdot \lambda_u[\text{cm}] = 2.614,$$

$$\lambda = \frac{\lambda_u(1 + a_u^2)}{2\gamma^2} = 1.36\,\text{Å},$$

$$E = h\frac{c}{\nu} = 9.1\,\text{keV}.$$

(b)

$$\sigma_x = \sqrt{\frac{\beta_x \cdot \varepsilon_{x,u}}{\gamma}} = 35\,\mu\text{m},$$

$$\sigma_r = \sqrt{\sigma_x^2 + \sigma_y^2} = 49.5\,\mu\text{m}$$

since in this case $\sigma_x = \sigma_y$.

(c)

$$z_R = \frac{\Sigma}{\lambda} = \frac{\pi\sigma_r^2}{\lambda} = 56\,\text{m}.$$

The growth in the beam size is given by

$$\omega(z) = \omega_0\sqrt{1 + \frac{z^2}{z_R^2}},$$

$$\frac{\omega(5.8\,\text{m})}{\omega_0} = 1.0053 \text{ or } 0.53\%.$$

Appendix B: Advanced topics addressed in volume II

The second volume in this set of texts on beam physics is tentatively entitled "Collective Effects in Intense Beams." It will emphasize the basic physical processes that lay on the frontier in beam physics, which arise when the collective effects associated with the charge and current density of the beams themselves become equal in importance to effects due to the applied fields that were introduced in this volume. The subjects to be addressed in this second volume are more physically complex, and will be therefore treated at a higher level; the result will be appropriate for use as a graduate text, and as a professional reference.

Collective effects have historically been important in circular accelerators, as even relatively small perturbations due to the self-induced forces generated by circulating beams can, in an environment where the beam circulates for long times, eventually produce noticeable effects. Such effects have been studied in depth, with useful references and texts already in existence that summarize the present state of knowledge in these areas. Recent trends in the accelerator physics world, however, are moving toward very intense beams needed for advanced applications that primarily utilize linear accelerators—high-gain free-electron lasers, single-pass linear colliders, heavy ion fusion. Thus, the emphasis has shifted towards addressing extremely strong self-induced forces that can directly and immediately affect the dynamics of the beam. These types of systems are said to display collective beam dynamics.

In the volume following this text, we will examine the effects of such forces, whose common characteristic may be stated as **continuous action** on the beam itself, due to strong coupling of the beam to its immediate environment. In order to study such systems, one must first examine the fundamental forces and general dynamical responses involved in the processes. They can be due to either collective "static" or radiative fields, and depend on both the beam itself, and the response of the electromagnetic systems the beam comes in contact with. Examples include:

- space-charge fields leading to plasma behavior in beams;
- synchrotron radiation in bending and undulator magnets; energy exchange between a beam with its self-generated radiation field;
- the beam–plasma interaction; linear and nonlinear wake-field excitations;
- the beam–beam interaction in linear colliders, a relativistic "two-stream" system;

- beam–matter interaction: scattering, bremsstrahlung and Cerenkov radiation, transition radiation at metallic surfaces;
- electromagnetic radiation from nearby boundaries; electromagnetic wake-fields, diffraction radiation, resistive effects.

We note that all of the effects listed above are collective (with the exception of scattering and bremsstrahlung), in the sense that the beam-induced fields are macroscopic and coherent—they produce observable effects that depend on the macroscopic beam distribution, not simply on the *incoherent* fields due to individual particles. Such coherent effects include energy exchange (e.g. radiative energy loss) that depends not on the number of particles in the beam, but on the square of this number.

Once we have established a basic understanding of the relevant, fundamental collective processes, a number of phenomena specific to advanced laboratory topics can be investigated:

- space-charge-dominated beam physics: tune shifts, longitudinal and transverse plasma oscillations, emittance growth and compensation;
- ion trapping effects due to space-charge;
- the ultra-high-gain, single-pass free-electron laser, and other radiation-mediated instabilities;
- wake-field accelerators based on metallic structures and dielectrics;
- plasma wake-field acceleration; plasma focusing and self-consistent flows;
- beam–beam disruption, "beamstrahlung," and luminosity enhancement;
- fast longitudinal and transverse instabilities;
- coherent radiation production for diagnosis of ultra-fast beam phenomena.

These topics form a wide survey of coherent effects in state-of-the-art laboratory beams. They illuminate the importance of the general processes discussed, by providing clear examples of their impact on physical systems in actual application. It is, therefore, hoped that this text will provide a broad introduction, as well as a useful reference guide, to many of the fascinating high-intensity beam-based phenomena that make the field of beam physics so active at this time.

Index